AMERICAN MATHEMATICAL SOCIETY TRANSLATIONS

Series 2

Volume 82

SIXTEEN PAPERS ON NUMBER THEORY AND ALGEBRA

by

A. P. Birjukov
B. M. Bredihin
V. M. Busarkin
A. A. Karacuba
A. F. Lavrik
Ju. V. Linnik
G. A. Lomadze
A. I. Mal′cev
L. N. Ševrin
I. Š. Slavutskiĭ
A. I. Starostin
Ju. T. Tkačenko
V. I. Ušakov
È. B. Vinberg
A. I. Vinogradov
I. M. Vinogradov
D. A. Vladimirov

Published by the
AMERICAN MATHEMATICAL SOCIETY
Providence, Rhode Island
1969

Standard Book Number 821-81782-5

Library of Congress Catalog Number A51-5559

Printed in the United States of America

TABLE OF CONTENTS

ON EXTENSION TO THE LEFT HALFPLANE OF THE SCALAR PRODUCT OF HECKE L-SERIES WITH MAGNITUDE CHARACTERS

A. I. VINOGRADOV

UDC 511

(In deep respect to academician Ju. V. Linnik on his fiftieth birthday)

In this paper we prove that the scalar product of Hecke L-series can be extended to the line $\operatorname{Re} s = \frac{1}{2}$, and we use this fact to obtain some arithmetical consequences.

In 1920 Erich Hecke [1] introduced L-series with magnitude characters, proved them extendable to the whole plane, and obtained a functional equation for them. This result enabled him to create a multidimensional arithmetic of number fields.

The fundamental arithmetical problem for whose solution Hecke created this analytic theory is the following. We are given a field k of degree n. We shall consider a class $\mathbf{c}$ of ideal numbers of this field as a lattice in some n-dimensional non-Euclidean space t. The precise definition of this space is given in [1]. Let a finite region v be given in this space. We ask, how many prime points of our lattice will lie in the region v? In other words, we must find the number of solutions of the equation

$$p = |N\hat{\mu}| = |\hat{\mu}_1 \cdot \hat{\mu}_2 \cdot \cdot \cdot \hat{\mu}_n| = f(x_1, x_2, \ldots, x_n),$$

where p runs through all prime numbers of the rational field, N is the absolute norm, and $\hat{\mu}_1, \cdots, \hat{\mu}_n$ are all the conjugates of the ideal number $\hat{\mu} \in \mathbf{c}$, where each $\hat{\mu}_i$ is considered as the ith coordinate of the point $(\hat{\mu}_1, \cdots, \hat{\mu}_n)$ in the space t satisfying the condition $(\hat{\mu}_1, \cdots, \hat{\mu}_n) \in v$. f is a form of order n corresponding to the class $\mathbf{c}$.

We shall pose a general problem of that type.

Assume we are given r algebraic number fields $k_1, \cdots, k_r$ of degrees $n_1, \cdots, n_r$, respectively. In each of them we isolate a class $\mathbf{c}_i$ $(1 \leq i \leq r)$ of ideal numbers and examine it as a lattice in the corresponding n_1-dimensional non-Euclidean space t_i.

We give a region v_i in each space t_i. We ask how many prime numbers p of the rational field exist for which the system

$$p = |N_i\hat{\mu}_i| = |\hat{\mu}_{i,1} \cdot \hat{\mu}_{i,2} \cdots \hat{\mu}_{i,n_i}|$$
$$= f_i(x_{i,1}, x_{i,2}, \ldots, x_{i,n_i}), \quad 1 \leqslant i \leqslant r, \tag{A}$$

is soluble if it is compatible; here N_i is the absolute norm in the field k_i, the numbers $\hat{\mu}_{i,1}, \cdots, \hat{\mu}_{i,n_i}$ are the n_i conjugates of the number $\hat{\mu} \in \mathbf{c}_i$ of the field k_i, the point $(\hat{\mu}_{i,1}, \cdots, \hat{\mu}_{i,n_i}) \in v_i$ in the space t_i, and f_i is a form of degree n_i corresponding to the class $\mathbf{c}_i$ of the field k_i.

By well-known methods this problem can be reduced to investigating the properties of the scalar product of Hecke L-series with the product of two characters, one a Hecke magnitude character and one a character of the group of classes of ideal numbers in the corresponding fields:

$$\prod_{i=1}^{r} {}_c L(s, \lambda_i \cdot \chi_i, k_i) = \sum_{|N_1\hat{\mu}_1| = \ldots = |N_r\hat{\mu}_r|} \frac{\prod_{i=1}^{r} \lambda_i(\hat{\mu}_i) \cdot \chi_i(\hat{\mu}_i)}{|N_1\mu_1|^s}, \tag{1}$$

where $\hat{\mu}_i$ runs through all ideal numbers of the field k_i, λ_i is some Hecke magnitude character of the field k_i, and X_i is a character of the group of classes of ideal numbers of the field k_i.

The right side of (1) can be written in the form

$$\prod_{p=|N_1\hat{\pi}_1|=\ldots=|N_r\hat{\pi}_r|} \left[1 - \frac{1}{p^s} \prod_{i=1}^{r} \lambda_i(\hat{\pi}_i) \cdot \chi_i(\hat{\pi}_i)\right]^{-1} \varphi(2s), \tag{2}$$

where $\phi(2s)$ is an Euler product regular in the halfplane $\mathrm{Re}\, s > 1/2$ and $\hat{\pi}$ is a prime ideal number of the first degree of the field k_i $(1 \leq i \leq r)$.

Let K be the composite of the fields $(k_1, \cdots, k_r)$. We shall only examine the case when the degree n of the composite field K equals the product of the degrees of the subfields:

$$n = n_1 \cdot n_2 \cdots n_r. \tag{3}$$

The general case reduces to it.

We examine the following L-series for the field K:

$$L(s, \lambda \cdot \chi, K) = \sum_{\mu} \frac{\lambda(\hat{\mu}) \cdot \chi(\hat{\mu})}{|N\hat{\mu}|^s}, \tag{4}$$

where $\hat{\mu}$ runs through all ideal numbers of the field K and the characters λ and X are defined as follows:

$$\lambda(\hat{\mu}) = \prod_{i=1}^{r} \lambda_i(N_{K/k_i}\hat{\mu}), \tag{5}$$

$$\chi(\hat{\mu}) = \prod_{i=1}^{r} \chi_i (N_{K/k_i}\hat{\mu}); \tag{6}$$

here N_{K/k_i} is the relative norm, i.e. the norm of the ideal number $\hat{\mu}$ of the field K with respect to the subfield k_i.

We note a property of the L-series (4) which has fundamental significance in what follows. In the halfplane $\operatorname{Re} s > 1$

$$L(s, \lambda\cdot\chi, K) = \prod_{|N\hat{\pi}|=p} \left[1 - \frac{\lambda(\hat{\pi})\cdot\chi(\hat{\pi})}{p^s}\right]^{-1} \cdot \varphi_1(2s),$$

where $\phi_1(2s)$ is an Euler product regular in the halfplane $\operatorname{Re} s > \frac{1}{2}$ and $\hat{\pi}$ is a prime ideal number of the first degree of the field K. But by equation (3) and definitions (5) and (6) we obtain the fundamental relation of this paper:

$$\prod_{|N\hat{\pi}|=p} \left[1 - \frac{\lambda(\hat{\pi})\cdot\chi(\hat{\pi})}{p^s}\right]^{-1}$$

$$= \prod_{p=|N_1\hat{\pi}_1|=\ldots=|N_r\hat{\pi}_r|} \left[1 - \frac{1}{p^s}\prod_{i=1}^{r} \lambda_i(\hat{\pi}_i)\cdot\chi_i(\hat{\pi}_i)\right]^{-1}, \tag{7}$$

where the right part of (7) coincides with the product entering into (2). Consequently

$$\prod_{i=1}^{r} {}_cL(s, \lambda_i\cdot\chi_i, k_i) = L(s, \lambda\cdot\chi, K)\,\frac{\varphi\cdot(2s)}{\varphi_1(2s)}\,. \tag{8}$$

If we now show that (4) extends to the whole plane, then it immediately follows from (8) that (1) extends to the halfplane $\operatorname{Re} s > \frac{1}{2}$. This is entirely sufficient for arithmetic applications.

We first show that the character λ defined by equation (5) is a Hecke magnitude character of the field K. In fact, if η is any unit of K, then

$$\lambda(\eta\hat{\mu}) = \lambda(\hat{\mu}),$$
$$\lambda(\hat{\mu}\cdot\hat{\mu}') = \lambda(\hat{\mu})\cdot\lambda(\hat{\mu}'),$$
$$|\lambda(\hat{\mu})| = 1\,.$$

In distinction from the general Hecke magnitude character constructed in [1], character (5) is degenerate, since its actual construction does not involve the whole group of units of the field K, but only its subgroup composed of the units of the subfields k_i, $1 \le i \le r$. To finally reduce the problem to Hecke L-series with magnitude characters, we must still examine the properties of the character X defined by equation (6) in the field K.

Let H be the number of classes of ideal numbers of the field K and let $\hat{\gamma}_1, \cdots, \hat{\gamma}_m$ be their basis; then each ideal number $\hat{\mu}$ of the field K can be represented in the form

$$\hat{\mu} = \mu \cdot \hat{\gamma}_1^{a_1} \cdot \hat{\gamma}_2^{a_2} \ldots \hat{\gamma}_m^{a_m}, \tag{9}$$

where μ is an integer of the field K, the exponents a_ν run through the finite intervals $0 \leq a_\nu \leq H_\nu - 1$, $1 \leq \nu \leq m$, and $H = H_1 \cdots H_m$. Every class of ideal numbers is representable in the form (9) with some fixed system of exponents $(a_1, \cdots, a_m)$.

Let $\mathbf{C}_j$ $(1 \leq j \leq H)$ be a class of ideal numbers (9) with its unique system of exponents $(a_1, \cdots, a_m)$. We examine the relative norms of this class in the field k_i $(1 \leq i \leq r)$:

$$N_{K/k_i}(\mu \cdot \hat{\gamma}_1^{a_1} \cdot \hat{\gamma}_2^{a_2} \ldots \hat{\gamma}_m^{a_m}) = N_{K/k_i}(\mu) \cdot \prod_{\nu=1}^{m} (N_{K/k_i} \hat{\gamma}_\nu)^{a_\nu}. \tag{10}$$

Let h_i be the ideal class number of the field k_i and let $\hat{\beta}_1, \cdots, \hat{\beta}_{m_i}$ be a basis. Thus, each ideal $\hat{\mu}_i$ of the field k_i can be represented in the form

$$\hat{\mu}_i = \mu_i \cdot \hat{\beta}_1^{b_1} \cdot \hat{\beta}_2^{b_2} \ldots \hat{b}_{m_i}^{b_{m_i}}, \tag{11}$$

where μ_i runs through the integers of the field k_i and the exponents b_ν through the intervals $0 \leq b_\nu \leq l_\nu - 1$, $1 \leq \nu \leq m_i$, where $h_i = l_1 \cdots l_{m_i}$. All classes of ideal numbers of the field k_i for which the character X_i takes one and the same value can be represented in the form (11) with a fixed system of indices $(b_1, b_2, \cdots, b_{m_i})$.

Since in (10) the relative norms $N_{K/k_i} \hat{\gamma}_\nu$, $1 \leq \nu \leq m$, of the ideal basis numbers of the field K are ideal numbers of the field k_i or their associates, it follows that there exists a fully determined integer $\mu_{i,0}$ of the field k_i with its unique system of exponents $(b_1, \cdots, b_{m_i})$ such that

$$\prod_{\nu=1}^{m} (N_{K/k_i} \hat{\gamma}_\nu)^{a_\nu} = \mu_{i,0} \cdot \beta_1^{b_1} \cdot \beta_2^{b_2} \ldots \beta_{m_i}^{b_{m_i}}. \tag{12}$$

Consequently, for any ideal number $\hat{\mu}$ of the field K belonging to the class $\mathbf{C}_j$, we have, from (10) and (12),

$$\chi_i (N_{K/k_i} \hat{\mu}) = \chi_i (\beta_1^{b_1} \beta_2^{b_2} \ldots \beta_{m_i}^{b_{m_i}}) = \chi_i (\mathbf{c}_{ij}), \tag{13}$$

where $\mathbf{c}_{ij}$ is a class of ideal numbers (11) of the field k_i uniquely related to the class $\mathbf{C}_j$ of ideal numbers of the field K, by equation (12).

Making i run through its system of indices $1 \leq i \leq r$ in equation (13), we obtain the following assertion. The character χ, defined on ideal numbers of the field K by equation (6), takes one and the same value on all the numbers $\hat{\mu}$, of a class $\mathbf{C}_j$. This value $\chi(\hat{\mu})$ is defined by the equation

$$\chi(\hat{\mu}) = \chi(\mathbf{C}_j) = \prod_{i=1}^{r} \chi_i(\mathbf{c}_{ij}), \quad \hat{\mu} \in \mathbf{C}_j.$$

Using this equation, the L-series (4) can be transformed to

$$L(s, \lambda \cdot \chi, K) = \sum_{j=1}^{H} \chi(\mathbf{C}_j) \sum_{\hat{\mu} \in \mathbf{C}_j} \frac{\lambda(\hat{\mu})}{|N\hat{\mu}|^s}. \tag{14}$$

But in the right side of (14) we have obtained the Hecke L-series with degenerate magnitude character λ on the class $\mathbf{C}_j$ of ideal numbers of the field K, and this is an L-series which extends to the whole plane and has a functional equation. This is easily checked by repeating almost word for word Hecke's calculations in [1]. The most essential item is the verification ([1], §6) of the equation

$$E(u, \hat{\mu}, x_1, \ldots, x_l + 1, x_{l+1}, \ldots, x_r) = F(u, \eta\hat{\mu}, x_1, \ldots, x_r),$$

which is there designated (37). This equation is also fulfilled for the degenerate character λ, since for any unit η of the field K we have $\lambda(\eta) = 1$.

Thus we see that (1) can be extended to the left as far as the line Res = ½ and we have the modulus estimates necessary for arithmetic applications.

We note that the extension of (1) to the whole plane has in principle been realized. For this we must extend the relation $\phi(2s)/\phi_1(2s)$ to the left beyond the line $\mathrm{Re}\, s = 1/2$. This can be done if we unite the theory of Artin L-series having non-Abelian characters with the theory of Hecke L-series having magnitude characters. In this case infinitely many singular points (poles and branch points) can appear in the strip $0 < \mathrm{Re}\, s < 1/2$.

As already noted at the beginning of the article, the problem of finding asymptotically the number of solutions of system (A) only makes sense when this system is generally compatible; that is, when for given forms $f_1, \cdots, f_r$ induced by the classes $\mathbf{c}_1, \cdots, \mathbf{c}_r$ of the fields $k_1, \cdots, k_r$, the system

$$f_1 = f_2 = \cdots = f_r \tag{15}$$

has the asymptotics of the number of nontrivial solutions. The analytic apparatus obtained must distinguish the case of compatibility and the case of incompatibility of the system (15). We shall show that this occurs. We suppose that the equation $\phi(2s) = \phi_1(2s)$ is fulfilled (its denial causes technical trouble, but does not refute the following illustration). Using a well-known method we obtain

$$\sum_{f_1=f_2=\ldots=f_r} \frac{1}{f_1^s} = \left(\prod_{i=1}^{r} h_i\right)^{-1} \sum_{\chi_1} \ldots \sum_{\chi_r} \bar{\chi}_1(\mathfrak{c}_1) \ldots \bar{\chi}_r(\mathfrak{c}_r)$$

$$\times \sum_{j=1}^{H} \chi(\mathfrak{C}_j)\, \zeta_k(s, \mathfrak{C}_j) = \frac{\omega_k w_k}{s-1} + R(s), \tag{16}$$

where $R(s)$ is regular in the whole plane,

$$w_K = \sum_{j=1}^{H} \prod_{i=1}^{r} \left\{ \frac{1}{h_i} \sum_{\chi_i} \frac{\chi_i(\mathfrak{c}_{ij})}{\chi_i(\mathfrak{c}_i)} \right\},$$

$\omega_K > 0$ is the residue of $\zeta_K(s, \mathfrak{C}_j)$ at the point $s = 1$, and w_K differs from zero in the case when $\mathfrak{c}_i = \mathfrak{c}_{ij}$ for at least one j and for all i, $1 \leq i \leq r$.

It is of arithmetic interest to apply the concepts set forth above to finding an asymptotic formula for the representation of zero by an indefinite quadratic form composed of the difference of two binary quadratic forms. Let

$$F(x_1, x_2, x_3, x_4) = f_1(x_1, x_2) - f_2(x_3, x_4),$$

where

$$f_1(x_1, x_2) = a_1 x_1^2 + a_{12} x_1 x_2 + a_2 x_2^2,$$
$$f_2(x_3, x_4) = a_3 x_3^2 + a_{34} x_3 x_4 + a_4 x_4^2,$$

at least one of the two binary forms f_1 and f_2 being positive definite. Let the first form have this property:

$$d_1 = a_{12}^2 - 4a_1 a_2 < 0.$$

The discriminant of the second form can have any sign:

$$d_2 = a_{34}^2 - 4a_3 \cdot a_4 \gtrless 0.$$

We ask, how many solutions of the equation

$$F(x_1, x_2, x_3, x_4) = 0$$

exist under the condition $f_1(x_1, x_2) \leq N$?

Using (16), we obtain

$$\sum_{\substack{F(x_1,x_2,x_3,x_4)=0\\ f_1(x_1,x_2)\leqslant N}} 1 = \frac{1}{2\pi i}\int_{2-iN^2}^{2+iN^2} \frac{N^s}{s}\left(\sum_{f_1=f_2\neq 0} \frac{1}{f_1^s(x_1,x_2)}\right) ds + O(1)$$

$$= \frac{1}{h_1\cdot h_2}\sum_{\chi_1,\chi_2}\sum_{j=1}^{H} \frac{\chi_1(\mathfrak{c}_{1j})\cdot\chi_2(\mathfrak{c}_{2j})}{\chi_1(\mathfrak{c}_1)\cdot\chi_2(\mathfrak{c}_2)}\cdot\frac{1}{2\pi i}\int_{2-iN^2}^{2+iN^2}\frac{N^s}{s}\zeta_K(s, \mathfrak{C}_j)\, ds + O(1).$$

Here $\mathfrak{c}_1$ and $\mathfrak{c}_2$ are the classes of ideal numbers in the quadratic fields k_1 and k_2 which correspond to the forms f_1 and f_2, respectively. We move the contour of integration to the left onto the line at ½. In doing so, we pass through a pole at the point $s = 1$ with residue $\omega_K \cdot N$:

$$\sum_{\substack{F(x_1,x_2,x_3,x_4)=0\\ f_1(x_1,x_2)\leqslant N}} 1 = \omega_k\cdot w_k N + \frac{1}{h_1\cdot h_2}\sum_{\chi_1,\chi_2}\sum_{j=1}^{H}\frac{\chi_1(\mathfrak{c}_{1j})\cdot\chi_2(\mathfrak{c}_{2j})}{\chi_1(\mathfrak{c}_1)\cdot\chi_2(\mathfrak{c}_2)}\cdot$$

$$\times\frac{1}{2\pi i}\int_{1/2-iN^2}^{1/2+iN^2}\frac{N^s}{s}\zeta_K(s, \mathfrak{C}_j)\, ds + O(1).$$

The composite field K is a field of the fourth degree, so that it is possible to obtain an approximate functional equation for $\zeta_K(s, \mathfrak{C}_j)$ on the line at ½ with intervals of the corresponding series having length $|t|^2$, where $s = \sigma + it$. This is just the limiting length for obtaining the estimate

$$\frac{1}{2\pi i}\int_{1/2-iN^2}^{1/2+iN^2}\frac{N^s}{s}\zeta_K(s, \mathfrak{C}_j)\, ds \ll \sqrt{N}\cdot\ln^{C_4} N$$

on the line at ½.

Consequently,

$$\sum_{\substack{F(x_1,x_2,x_3,x_4)=0\\ f_1(x_1,x_2)\leqslant N}} 1 = \omega_K\cdot w_K\cdot N + O(\sqrt{N}\cdot\ln^{C_4} N).$$

Thus, a limiting remainder is obtained in the formula for the number of representations of zero for indefinite forms of this type. We have $w_k = 0$ in the case when the forms f_1 and f_2 are compatible.

Using the approximate functional equation for arbitrary K leads to the following general asymptotic formula for a system of forms:

$$\sum_{\substack{|f_1|=\ldots=|f_r|\\ |f_1|\leqslant N}} 1 = \omega_k\cdot w_k\cdot N + O(N^{1-\frac{r}{n}}\ln^{C_n} N),$$

where n is the degree of the composite field.

We note that the validity of the Riemann hypothesis for all zeta-functions and L-series of all fields would lead to a limiting remainder in the general case:

$$\sum_{\substack{|f_1|=\ldots=|f_r| \\ |f_1|\leqslant N}} 1 = \omega_k \cdot w_k \cdot N + O(N^{\frac{1}{2}+\varepsilon}).$$

Proceeding to calculate the number of solutions of a system of forms with the condition that the corresponding points $(\hat{\mu}_1, \ldots, \hat{\mu}_{n_i}) \in v_i$, as was formulated at the beginning of the paper, we naturally must worsen the remainder in the corresponding asymptotic formulas. Finally, in the most general case, when the requirement of primeness is also imposed, i..e.

$$p = |f_1| = \ldots = |f_r| \text{ and } (\hat{\mu}_1, \ldots, \hat{\mu}_{n_i}) \in v_i, \quad 1 \leqslant i \leqslant r,$$

the remainder in the asymptotic formulas has order

$$N \cdot e^{-a\sqrt{\ln N}},$$

since it can only be shown at the present time that the boundary of the zeros of Hecke L-series of the composite field K with magnitude character λ is subject to the condition

$$\sigma > 1 - \frac{c_n}{\ln|D| + \ln(|t|+1)}, \qquad s = \sigma + it,$$

where D is the discriminant of the field K.

BIBLIOGRAPHY

[1] E. Hecke, *Eine neue Art von Zetafunctionen und ihre Beziehungen zur Verteilung der Primzahlen. Zweite Mitteilung*, Math. Z. 6 (1920), 11–51.

Translated by:
N. Koblitz

ON THE DENSITY CONJECTURE FOR DIRICHLET L-SERIES

UDC 511

A. I. VINOGRADOV

(In deep respect to Academician Ju. V. Linnik on his fiftieth birthday)

In this paper we prove the density conjecture on zeros of Dirichlet L-series in the mean over all moduli.

§1

This paper is basically devoted to proving the following theorem.

Theorem 1. *Let $N_d(\sigma, t)$ be the number of zeros ρ of all Dirichlet L-series of a given modulus d which lie in the region* $\operatorname{Re} \rho \geq \sigma$, $|\operatorname{Im} \rho| \leq t$. *Then the following estimate is valid for d in the interval $D \leq d \leq 2D$ with the exception of no more than $D^{1-\epsilon/2}$ moduli:*

$$N_d(\sigma, t) < (t \cdot \ln D)^{C_0 \cdot \varepsilon^{-4}} \cdot D^{2(1+\varepsilon)\cdot(1-\sigma)}, \quad \frac{1}{2} \leqslant \sigma \leqslant 1, \quad t \geqslant 1,$$

where $\epsilon > 0$ is an arbitrarily small fixed number.

The result of this theorem was known as the density conjecture in the mean for Dirichlet L-series.

By a scheme worked out by M. B. Barban [1], Theorem 1 implies

Theorem 2. *For the distribution of prime numbers in a progression the following average asymptotic law is valid:*

$$\sum_{d \leqslant x^{\frac{1}{2}-\varepsilon}} \max_{(l, d)=1} \left| \pi(x, d, l) - \frac{1}{\varphi(d)} \operatorname{Li}(x) \right| < \frac{x}{(\ln x)^C},$$

where C is an arbitrarily large positive constant and $\epsilon > 0$ is an arbitrarily small fixed number.

Theorem 2 replaces the extended Riemann hypothesis in many problems of number theory. In particular, by [3] or [4], Theorem 2 leads to

Theorem 3. *For sufficiently large even m, the equation $m = p + P_3$ is soluble, where p is a prime and P_3 a near prime containing no more than three prime factors. The number of solutions of this equation is greater than $C_0 \cdot \mathfrak{S}(m) \cdot m/\ln^2 m$, where $C_0 > 0$ is an absolute constant and $\mathfrak{S}(m)$ is a special series.*

A similar result is true for the difference problem:

$$2k = p - P_3, \quad k = 1, 2, \ldots .$$

As an estimate from above for the binary problem we obtain

Theorem 4. *If m is even, then the number of solutions of the equation $m = p + q$ in prime numbers p and q does not exceed $(4 + \epsilon) \cdot \mathfrak{S}(m) \cdot m/\ln^2 m$, where $\epsilon > 0$ is an arbitrarily small number and $\mathfrak{S}(m)$ is a special series.*

There is a well-known connection between the behavior of the modulus of L-series and the boundary of the zeros. As far as the boundary of zeros it behaves as $O(\ln Dt)$. In this connection, using the scheme of [1], we can obtain an average law for a power of the divisor function $\tau_k^n(m)$ by analogy with Theorem 2:

$$\sum_{d \leqslant x^{\frac{1}{2}-\varepsilon}} \max_{(l,d)=1} \Big| \sum_{\substack{m \equiv l(d) \\ m \leqslant x}} \tau_k^n(m) - A_k^n(x, d) \Big| < \frac{x}{(\ln x)^C},$$

where $A_k^n(x, d)$ is the expected principal term in the sum for $\tau_k^n(m)$, and k and n are any fixed positive integers.

This form of the average law allows us to generalize Ju. V. Linnik's [6] asymptotic law as follows.

Theorem 5. *The following asymptotic equation holds:*

$$\sum_{m \leqslant x} \tau_k^n(m + l) \cdot \tau(m) \sim C_{k,n}(l) \cdot x \cdot \ln^{k^n} x.$$

We note that Theorem 2 leads to a very simple new proof (along lines followed by Titchmarsh [11] himself) of Linnik's theorem [7] solving Titchmarsh's problem of divisors:

$$\sum_{p \leqslant x} \tau(p - l) \sim E(l) \cdot x.$$

The plan of the argument we use to prove Theorem 1 is as follows. The whole difficulty of the first stage of the proof consists in establishing the following estimate: for any integer $n \geq 2$ and any Z in the interval $D^{1/n} \leq Z \leq D^{1/(n-1)}$ we have the inequality

$$\sum_{d=D}^{2D} \sum_{\chi_d \neq \chi_0} \Big| \sum_{m \leqslant Z} \chi_d(m) \Big|^2 \leqslant D^2 \cdot Z^n \exp\,[(\ln D)^{\varepsilon}]. \tag{1}$$

Estimate (1) means that in the average the sum of the values of a nonprincipal character is the square root of the length of the interval of summation.

In the second stage, we prove the validity of this estimate for any $n \geq 2$.

We note that estimate (1) was established by Linnik [5] for the case $n = 3$ and is used in an essential way in this paper, since the method proposed here is well able to copy with high moments for $n \geq 4$ but does not operate for $n = 3$ or 2.

This apparently strange phenomenen can be understood if the reasoning of §5 of this paper is analyzed.

The case $n = 2$ follows quite simply from the method of [5]. 1)

§2

We reduce Theorem 1 to estimate (1). We use the same plan of argument as in our note [2], but we essentially alter the length of the sum

$$\sum_{m \leqslant Q_0} \frac{\mu(m) \cdot \chi(m)}{m^\rho}.$$

Here we set $Q_0 = D^{(\epsilon/3)^2}$. Such a short length of this sum allows us to reduce the problem to "aliquot" sums. Ju. V. Linnik used this name for the selection from all sums

$$\sum_{m \leqslant x} a_m \cdot \chi(m)$$

of those for which $x = D^{1/n}$ with $n \geq 1$ integral because of their special importance.

The estimate in Theorem 1 is nontrivial only for the interval

$$\frac{1}{2} + \frac{\varepsilon}{2(1+\varepsilon)} \leqslant \sigma \leqslant 1.$$

Therefore, we shall henceforth assume that $\operatorname{Re} \rho \geq 1/2 + \epsilon/2(1 - \epsilon)$. With this condition, using the estimates of Vinogradov and Pólya, we obtain

$$\left| \sum_{m > D} \frac{\chi(m)}{m^\rho} \right| \leqslant |\rho| \cdot D^{-\frac{\varepsilon}{2(1+\varepsilon)}} \cdot \ln D. \qquad (2)$$

Henceforth, without stating it explicitly, we shall assume that the modulus d of the character χ lies in the interval $(D, 2D)$.

If ρ is a zero of the L-series, then, using estimate (2), we find

$$\sum_{m \leqslant D} \frac{\chi(m)}{m^\rho} = \theta \cdot |\rho| \cdot D^{-\frac{\varepsilon}{2(1+\varepsilon)}} \cdot \ln D, \quad |\theta| \leqslant 1. \qquad (3)$$

1) See the note at the end of the paper.

We set $Q_0 = D^{\epsilon_1^2}$, where $\epsilon_1 = \epsilon/3$. Multiplying both sides of equation (3) by the sum

$$\sum_{m \leqslant Q_0} \frac{\mu(m) \cdot \chi(m)}{m^\rho},$$

we obtain

$$1 + \sum_{m=D^{\varepsilon^2}}^{D^{1+\varepsilon^2}} \frac{a_m \cdot \chi(m)}{m^\rho} = \theta \cdot |\rho| \cdot D^{-\frac{\varepsilon - \varepsilon^2}{2(1+\varepsilon)}} \ln D, \qquad a_m = \sum_{\delta/m,\ \delta \leqslant Q_0} \mu(\delta). \tag{4}$$

We shall henceforth consider only those ρ which satisfy the condition

$$|\rho| \leqslant \frac{1}{2} D^{\frac{\varepsilon - \varepsilon_1^2}{2(1+\varepsilon)}} \cdot (\ln D)^{-1}, \tag{5}$$

since Theorem 1 is trivial otherwise.

But if estimate (5) is valid, then (4) leads to the inequality

$$\left| \sum_{m=D^{\varepsilon^2}}^{D^{1+\varepsilon_1^2}} \frac{a_m \cdot \chi(m)}{m^\rho} \right| \geqslant \frac{1}{2}, \quad \operatorname{Re} \rho \geqslant \sigma. \tag{6}$$

Applying partial summation, we find that

$$\frac{|S_\chi(D^{\varepsilon_1^2})|}{D^{\varepsilon_1^2 \cdot \sigma}} + \frac{|S_\chi(D^{1+\varepsilon_1^2})|}{D^{(1+\varepsilon_1^2)\sigma}} + \sum_{\nu=D^{\varepsilon_1^2}}^{D^{1+\varepsilon_1^2}} \frac{|S_\chi(\nu)|}{\nu^{1+\sigma}} \geqslant \frac{1}{2|\rho|}, \tag{7}$$

where

$$S_\chi(\nu) = \sum_{m \leqslant \nu} a_m \cdot \chi(m).$$

We use the inequality

$$\sum_{\nu=Q}^{2Q} \frac{S_\chi(\nu)}{\nu^{1+\sigma}} \leqslant \frac{1}{Q^{1+\sigma}} \sum_{\nu=Q}^{2Q} |S_\chi(\nu)|.$$

If we introduce the notation

$$T_\chi(Q) = \sum_{\nu=Q}^{2Q} |S_\chi(\nu)|$$

and set $Q_r = Q_0 2^r$, $Q_0 = D^{\epsilon_1^2}$, then from (7) we obtain

$$\frac{|S_\chi(D^{\varepsilon_1^2})|}{D^{\varepsilon_1^2\sigma}}+\frac{|S_\chi(D^{1+\varepsilon_1^2})|}{D^{(1+\varepsilon_1^2)\cdot\sigma}}+\sum_{r=0}^{r_0-1}\frac{T_\chi(Q_r)}{Q_r^{1+\sigma}}\geqslant\frac{1}{2|\rho|}, \tag{8}$$

where $r_0 \leq \ln D$ and $Q_{r_0} \leq 2 \cdot D^{1+\epsilon_1^2}$.

We shall consider zeros ρ with imaginary parts in the interval $t-1 \leq \operatorname{Im}\rho \leq t$ and with $\operatorname{Re}\rho \geq \sigma$. For each zero ρ, one of the following inequalities will always be fulfilled:

$$\frac{|S_\chi(D^{\varepsilon_1^2})|}{D^{\varepsilon_1^2\cdot\sigma}}\geqslant\frac{1}{6\sqrt{2t}},\qquad\frac{|S_\chi(D^{1+\varepsilon_1^2})|}{D^{(1+\varepsilon_1^2)\cdot\sigma}}\geqslant\frac{1}{6\sqrt{2t}}, \tag{9}$$

$$\sum_{r=0}^{r_0-1}\frac{T_\chi(Q_r)}{Q_r^{1+\sigma}}\geqslant\frac{1}{6\sqrt{2t}}. \tag{10}$$

In fact, if neither of these inequalities were fulfilled, then inequality (8) would be violated. We shall divide all zeros of all L-series of given modulus d, except for $L(s, \chi_0)$, which lie in the region $\operatorname{Re}\rho \geq \sigma$, and $t-1 \leq \operatorname{Im}\rho \leq t$, into three classes on the basis of inequalities (9) and (10). We put all zeros for which the first of inequalities (9) is valid into the first class, all for which the second of inequalities (9) is valid into the second class, and all for which inequality (10) is valid into the third class.

We set $\epsilon = n_0^{-1}$, where n_0 is a sufficiently large integer. For all zeros of the first class we then have

$$D^{-\frac{\sigma}{(3n_0)^2}}\left|S_\chi\left(D^{\frac{1}{(3n_0)^2}}\right)\right|\geqslant\frac{1}{2\sqrt{2t}}.$$

Raising both sides of this inequality to the power $2 \cdot (3n_0)^2$, we obtain

$$D^{-2\sigma}\Big|\sum_{m\leqslant D}b_m\cdot\chi(m)\Big|^2\geqslant(6\sqrt{2}\,t)^{-2\cdot(3n_0)^2}. \tag{11}$$

We let $N_d^{(1)}(\sigma, t)$ designate the number of zeros of the first class. We sum both sides of (11) over all zeros of the first class. Since each L-series has no more than $C_0 \cdot \ln Dt$ zeros in the unit interval, we obtain the estimate

$$N_d^{(1)}(\sigma, t)<C(\varepsilon)\cdot t^{C_0\cdot\varepsilon^{-2}}\cdot D^{-2\sigma}\sum_\chi\Big|\sum_{m\leqslant D}b_m\chi(m)\Big|^2\cdot\ln D.$$

Taking into account that $|b_m| \le \tau_{(3n_0)^2}(m) \cdot \tau(m)$, we find that

$$N_d^{(1)}(\sigma, t) < (t \cdot \ln D)^{C_0 \cdot \varepsilon^{-4}} \cdot D^{2(1-\sigma)}. \tag{12}$$

We obtain an estimate for the number of zeros of the second class completely analogously (the sum in the second inequality of (9) must be squared):

$$N_d^2(\sigma, t) \leqslant (t \cdot \ln D)^3 \cdot D^{2 \cdot (1+\varepsilon_1^2) \cdot (1-\sigma)}. \tag{13}$$

We consider zeros of the third class. Inequality (10) is valid for them. It follows from (10) that for every zero ρ of the third class there exists an r such that

$$Q_r^{-1-\sigma} T_\chi(Q_r) \geqslant (7\sqrt{2} \cdot t \cdot \ln D)^{-1}. \tag{14}$$

If such an r did not exist, then (10) would be violated. We divide all zeros of the third class into r_0 classes. We put in the class C_r all zeros for which inequality (14) is fullfilled with fixed r from the interval $0 \le r \le r_0 - 1$. We let $N_d(\sigma, t, C_r)$ designate the number of zeros in the class C_r. We examine separately two possibilities for Q_r: 1) Q_r lies in the interval

$$D^{\varepsilon_1^2} \leqslant Q_r \leqslant D^{\varepsilon}, \tag{15}$$

2) Q_r lies in the interval

$$D^{\varepsilon} < Q_r \leqslant D^{1+\varepsilon_1^2}. \tag{16}$$

Assuming Q_r lies in the interval (15), we let

$$Q_r = D^{\theta_r}, \quad \varepsilon_1^2 \leqslant \theta_r \leqslant \varepsilon.$$

Then for θ_r there exists an integer n_r in the interval $n_0 + 1 \le n_r \le (3n_0)^2$, where $\epsilon = n_0^{-1}$, such that

$$\frac{1}{n_r} \leqslant \theta_r \leqslant \frac{1}{n_r - 1}$$

and therefore

$$\theta_r = \frac{1}{n_r} + \frac{\theta}{n_r(n_r - 1)}, \quad 0 \leqslant \theta < 1. \tag{17}$$

Under these conditions we raise both sides of (14) to the power $2n_r$. We obtain

$$D^{-2n_r \theta_r \cdot (1+\sigma)} \cdot T_\chi^{2n_r}(Q_r) \geqslant (7\sqrt{2} \cdot t \cdot \ln D)^{-2n_r}. \tag{18}$$

We sum both sides of (18) over all zeros of the class C_r. Again taking into account that each L-series has no more than $C_0 \cdot \ln Dt$ zeros, we find

$$N_d(\sigma, t, C_r) < \frac{(7\sqrt{2}\cdot t \cdot \ln D)^{2n_r+1}}{D^{2n_r\theta_r\cdot(1+\sigma)}} \cdot \sum_\chi T_\chi^{2n_r}(D^{\theta_r}). \tag{19}$$

But

$$T_\chi^{2n_r}(D^{\theta_r}) = \Big(\sum_{\nu=D^{\theta_r}}^{2D^{\theta_r}} |S_\chi(\nu)|\Big)^{2n_r} \leqslant D^{\theta_r\cdot(2n_r-1)} \cdot \sum_{\nu=D^{\theta_r}}^{2D^{\theta_r}} |S_\chi(\nu)|^{2n_r}$$

$$= D^{\theta_r\cdot(2n_r-1)} \cdot \sum_{\nu=D^{\theta_r}}^{2D^{\theta_r}} \Big| \sum_{m\leqslant \nu^{n_r}} b_m \cdot \chi(m) \Big|^2,$$

from which it follows that

$$\sum_\chi T_\chi^{2n_r}(D^{\theta_r}) \leqslant D^{2\cdot\left(1+\frac{\theta}{n_r-1}\right)} \cdot D^{\theta_r\cdot 2n_r} \cdot \ln^{C_0 n_r^2} D.$$

Substituting this estimate in (19), we obtain

$$N_d(\sigma, t, C_r) \leqslant (7\sqrt{2}\cdot t \cdot \ln D)^{2n_r^2+1} \cdot D^{2\cdot\left(1+\frac{\theta}{n_r-1}\right)\cdot(1-\sigma)}.$$

But, by assumption, $n_r \geq n_0 + 1 = 1/\epsilon + 1$, so that

$$1 + \frac{\theta}{n_r - 1} \leqslant 1 + \theta\cdot\varepsilon < 1 + \varepsilon.$$

Consequently, for the classes whose θ_r lie in the interval (15), we obtain the estimate

$$N_d(\sigma, t, C_r) < (t\cdot \ln D)^{C_0\cdot\varepsilon^{-4}} \cdot D^{2(1+\varepsilon)\cdot(1-\sigma)}. \tag{20}$$

For the Q_r from the interval (16) whose exponents differ from "aliquot" points by no more than ϵ/n_r, i.e. for

$$\theta_r = \frac{1}{n_r} + \frac{\theta\varepsilon}{n_r}, \quad 1 \leqslant n_r \leqslant n_0, \quad 0 \leqslant \theta \leqslant 1,$$

concepts similar to those presented before lead to estimate (20).

Points θ_r far from "aliquot" points present special difficulty:

$$Q_r = D^{\theta_r}, \quad \theta_r = \frac{1}{n_r} + \frac{\theta}{n_r(n_r - 1)}, \tag{21}$$

where n_r lies in the interval $2 \leq n_r \leq n_0 = 1/\epsilon$ and

$$\varepsilon < \frac{\theta}{n_r - 1} < \frac{1}{n_r - 1}.$$

For such Q_r we transform $T_\chi(Q_r)$ as follows:

$$T_\chi(Q_r) = \sum_{\nu=Q_r}^{2Q_r} \Big| \sum_{m\leqslant\nu} a_m \cdot \chi(m) \Big|.$$

But

$$\sum_{m\leqslant\nu} a_m \cdot \chi(m) = \sum_{\delta\leqslant D^{\varepsilon_1^2}} \mu(\delta)\cdot\chi(\delta) \sum_{m\leqslant\nu/\delta} \chi(m) \leqslant \sum_{\delta\leqslant D^{\varepsilon_1^2}} \Big| \sum_{m\leqslant\nu/\delta} \chi(m) \Big|,$$

so that

$$T_\chi(Q_r) \leqslant \sum_{\nu=Q_r}^{2Q_r} \sum_{\delta\leqslant D^{\varepsilon_1^2}} \Big| \sum_{m\leqslant\nu/\delta} \chi(m) \Big|.$$

Raising both sides of this inequality to the power $2n_r$, we obtain

$$T_\chi^{2n_r}(Q_r) \leqslant D^{(2n_r-1)\cdot(\theta_r+\varepsilon_1^2)} \sum_{\nu=Q_r}^{2Q_r} \sum_{\delta\leqslant D^{\varepsilon_1^2}} \Big| \sum_{m\leqslant\nu/\delta} \chi(m) \Big|^{2n_r}. \tag{22}$$

Just as above, from inequalities (14) and (22) we deduce the estimate

$$N_d(\sigma, t, C_r) \leqslant C(\varepsilon)\cdot(t\cdot\ln D)^{2n_r+1} \frac{D^{(2n_r-1)\cdot(\theta_r+\varepsilon_1^2)}}{D^{2n_r\cdot\theta_r\cdot(1+\sigma)}}$$

$$\times \sum_{\nu=D^{\theta_r}}^{2D^{\theta_r}} \sum_{\delta\leqslant D^{\varepsilon_1^2}} \sum_{\chi\neq\chi_0} \Big| \sum_{m\leqslant\nu/\delta} \chi(m) \Big|^{2n_r}. \tag{23}$$

We now assume that we already have estimate (1), where

$$Z = D^{\frac{1}{n} + \frac{\theta}{n(n-1)}}, \quad n \geqslant 2, \quad 0 \leqslant \theta < 1.$$

In this case (1) implies the estimate

$$\sum_{\chi\neq\chi_0} \Big| \sum_{m\leqslant Z} \chi(m) \Big|^{2n} \leqslant D^{1+0.51\cdot\varepsilon}\cdot Z^n \tag{24}$$

for all moduli in the interval $(D, 2D)$ with the possible exception of no more than $D^{1-\epsilon/2}(\ln D)^{-1}$ of them. Substituting (24) in (23), we find

$$N_d(\sigma, t, C_r) \leqslant (t\cdot\ln D)^{C_0\varepsilon^{-2}}\cdot D^{1+0.51\cdot\varepsilon+n_r\theta_r-2n_r\cdot\theta_r\cdot\sigma+2n_2\cdot\varepsilon_1^2}.$$

But

$$1 + 0.51\varepsilon + n_r\cdot\theta_r - 2n_r\cdot\theta_r\cdot\sigma + 2n_r\cdot\varepsilon_1^2$$

$$= 2(1-\sigma) + 0.51\cdot\varepsilon + 2n_r\cdot\varepsilon_1^2 - \frac{\left(\sigma - \frac{1}{2}\right)\cdot 2\theta}{n_r - 1}$$

$$= 2(1+\varepsilon)\cdot(1-\sigma) + 0.51\cdot\varepsilon + 2n_r\frac{\varepsilon^2}{9}$$

$$-2\cdot\varepsilon\cdot(1-\sigma) - \frac{\left(\sigma - \frac{1}{2}\right)\cdot 2\theta}{n_r - 1}.$$

By assumption, our point is far away from the "aliquot" point, i.e. $\theta/(n_r - 1) > 8\epsilon/9$ and, in addition, $n_r \leq 1/\epsilon$. Therefore

$$0.51\cdot\varepsilon + 2n_r\cdot\frac{\varepsilon^2}{9} - 2\cdot\varepsilon\cdot(1-\sigma) - \frac{\left(\sigma - \frac{1}{2}\right)\cdot 2\theta}{n_r - 1} < 0. \tag{25}$$

Consequently estimate (20) is valid for our class C_r on all moduli d in the interval $(D, 2D)$ with the exception of "bad" ones numbering no more than $D^{1-\epsilon/2}\epsilon(\ln D)^{-1}$. We designate this set of "bad" moduli for the class C_r by Ω_r. Thus estimate (20) is valid for all classes C_r $(0 \leq r \leq r_0 - 1)$ on those moduli in the interval $(D, 2D)$ not contained in the set-theoretic union $\cup\Omega_r$. But there are no more than $\ln D$ classes in all and each Ω_r consists of no more than $D^{1-\epsilon/2}(\ln D)^{-1}$ elements. Therefore $\cup\Omega_r$ consists of no more than $D^{1-\epsilon/2}$ elements. Gathering estimate (20) over all classes (of which there are no more than $\ln D$), we obtain Theorem 1. Thus, confirming inequality (1) is actually the central part of the whole proof.

It is to this that we now proceed.

§3

From the point of view of method, inequality (1) is established in two cases: $n = 2$ or 3, and $n \geq 4$.

The case $n = 3$ was obtained by Ju. V. Linnik in his paper [5], as already noted at the beginning of the paper. The case $n = 2$ is a simplified variant of the case $n = 3$.

We examine the case $n \geq 4$. We first make the transformation

$$\sum_{d=D}^{2D}\sum_{\chi_d\neq\chi_0}\left|\sum_{m\leqslant Z}\chi_d(m)\right|^{2n} \leqslant 2D\sum_{d=D}^{2D}\frac{1}{\varphi(d)}\sum_{\chi_d\neq\chi_0}\left|\sum_{m\leqslant Z}\chi_d(m)\right|^{2n}. \tag{26}$$

We consider the sum in the right side of (26). We have

$$\sum_{d=D}^{2D} \frac{1}{\varphi(d)} \sum_{\chi_d \neq \chi_0} \Big| \sum_{m \leq Z} \chi_d(m) \Big|^{2n}$$

$$= \sum_{d=D}^{2D} \sum_{\substack{a_n - b_n = d \cdot u \\ (a_n \cdot b_n,\, d)=1}} 1 - \sum_{d=D}^{2D} \frac{1}{\varphi(d)} \Big(\sum_{\substack{m \leq Z \\ (m,\, d)=1}} 1 \Big)^{2n}, \qquad (27)$$

where $a_n, b_n = m_1 \cdot m_2 \cdots m_n$ with $m_i \leq Z$. The first double sum in the right side of (27) must be understood as a $(2n+2)$-fold sum with one constraint on the $(2n+2)$ variables, namely the equation $a_n - b_n = d \cdot u$. The intervals of summation for d and for the $2n$ variables entering into a_n and b_n are fixed, so that the interval of summation for u, because of the constraint $a_n - b_n = d \cdot u$, is also determined if u is considered as a free variable and the summation over u is made exterior (the first on the left in the multiple summation).

Henceforth, in equations (30), this sum must be understood precisely in the sense explained above. Moreover, beginning with sum (27), the multiple summation will systematically be replaced by a single symbol Σ to avoid excessively long formulas. The multiplicity of the summation symbol Σ will be given by the number of indices of summation below. Further, a single symbol Σ for all variables indicated below the symbol will often be replaced by two symbols Σ (see, for example, (33)), where the important index at the given stage, such as u in (33), has been especially isolated.

For the principal character we obtain

$$\sum_{\substack{m=Z \\ (m,\, d)=1}} 1 = \prod_{p/d} \Big(1 - \frac{1}{p}\Big) Z + O(\tau(d)).$$

Consequently

$$\sum_{d=D}^{2D} \frac{1}{\varphi(d)} \Big(\sum_{\substack{m=Z \\ (m,\, d)=1}} 1 \Big)^{2n} = Z^{2n} \sum_{d=D}^{2D} \frac{1}{\varphi(d)} \prod_{p/d} \Big(1 - \frac{1}{p}\Big)^{2n} + O(Z^{2n-1} \cdot \ln^{C_n} D). \qquad (28)$$

We compute the special series

$$\sum_{d=D}^{2D} \frac{1}{\varphi(d)} \prod_{p/d} \left(1 - \frac{1}{p}\right)^{2n}$$

$$= \frac{1}{2\pi i} \int_{1-iD^2}^{1+iD^2} \frac{(2D)^s - D^s}{s} \mathfrak{S}(1+s) \cdot \zeta(1+s)\, ds + O(D^{-1}), \tag{29}$$

where

$$\mathfrak{S}(1+s) = \prod_p \left[1 - \frac{1}{p^{1+s}} + \frac{1}{p^{1+s}} \left(1 - \frac{1}{p}\right)^{2n-1}\right].$$

In the halfplane $\operatorname{Re} s > \frac{1}{2} + \epsilon$ the product $\mathfrak{S}(1+s)$ is absolutely convergent and we have the bound $|\mathfrak{S}(1+s)| < C(\epsilon)$. We move the contour of integration (29) left to the line $\operatorname{Re} s = -\frac{1}{2} + \epsilon$. We thereby pass a pole at $s = 0$ with residue $\mathfrak{S}(1) \cdot \ln 2$. The integral obtained on the contour $\operatorname{Re} s = -\frac{1}{2} + \epsilon$ is estimated by the usual method, after which we have

$$\sum_{d=D}^{2D} \frac{1}{\varphi(d)} \prod_{p/d} \left(1 - \frac{1}{p}\right)^{2n} = \mathfrak{S}(1) \ln 2 + O(D^{-\frac{1}{2}+\varepsilon}).$$

Substituting the right side of this equation in (28), and then in (27), we find that

$$\sum_{d=D}^{2D} \frac{1}{\varphi(d)} \sum_{\chi_d \neq \chi_0} \Big| \sum_{m \leq Z} \chi_d(m) \Big|^{2n}$$

$$= \sum_{d=D}^{2D} \sum_{\substack{a_n - b_n = d \cdot u \\ (a_n \cdot b_n, d) = 1}} 1 - Z^{2n} \cdot \mathfrak{S}(1) \cdot \ln 2 + O(Z^{2n-1} \ln^{C_n} D). \tag{30}$$

Consequently, to prove estimate (1) we must find the asymptotics of the sum in the right side of (30), which must have the form

$$\sum_{d=D}^{2D} \sum_{\substack{a_n - b_n = d \cdot u \\ (a_n \cdot b_n, d) = 1}} 1 = Z^{2n} \mathfrak{S}(1) \cdot \ln 2 + O(D \cdot Z^n \ln^{C_n} D), \tag{31}$$

where

$$\mathfrak{S}(1) = \prod_p \left[1 - \frac{1}{p} + \frac{1}{p} \left(1 - \frac{1}{p}\right)^{2n-1}\right].$$

Such asymptotics will be established below.

§4

One of the fundamental technical difficulties on the path to attaining our goal lies in the proof that the numerical coefficient of Z^{2n} in the right side of (31) had precisely the value we need: $\mathfrak{S}(1)\cdot \ln 2$. Here, immense technical difficulties arise from the condition $(a_n\cdot b_n, d)=1$, which, in principle, is not very essential, in the sense of the growth of the principal term in (31). If this condition is discarded, then in the right side of (31), only the constant with Z^{2n} changes, while the order of the principal term remains the same: Z^{2n}. Thus, from a technical standpoint, calculating the sum in the right side of (31) is incomparably easier without the condition $(a_n\cdot b_n, d)=1$, since we can then proceed immediately to the congruences $a_n\equiv b_n(u)$. But we cannot completely discard the condition $(a_n\cdot b_n, d)=1$, since then the newly obtained sum would already have another constant attached to Z^{2n} in the right side of (31). It is therefore natural to try to replace the condition $(a_n\cdot b_n, d)=1$ by some new condition which would preserve the asymptotics of (31) and at the same time would facilitate obtaining equation (31) using a reasonable number of operations.

It turns out that such a replacement is possible, and this new condition is that $(a_n\cdot b_n, u)=1$. In other words, we now proceed to establish the equation

$$\sum_{d=D}^{2D}\sum_{\substack{a_n-b_n=d\cdot u\\(a_n\cdot b_n,d)=1}}1=\sum_{d=D}^{2D}\sum_{\substack{a_n-b_n=d\cdot u\\(a_n\cdot b_n,u)=1}}1+O(D\cdot Z^n\ln^{C_n}D), \tag{32}$$

which shows that replacing the condition $(a_n\cdot b_n, d)=1$ by the new condition $(a_n\cdot b_n, u)=1$ in our sum introduces an error which has the order of the remainder term in (31). Yet the condition $(a_n\cdot b_n, u)=1$ already allows a relatively simple calculation of the asymptotics of our sum, since the "inversion" principle allows us to go to the congruences $a_n\equiv b_n(u)$, $(a_n\cdot b_n, u)=1$, while the congruence $\operatorname{mod} u$, together with the condition $(a_n\cdot b_n, u)=1$, is "cut out" using characters $\operatorname{mod} u$:

$$\frac{1}{\varphi(u)}\sum_{\chi_u}\frac{\chi_u(a_n)}{\chi_u(b_n)}=\begin{cases}1, & \text{if } a_n\equiv b_n\,(u) \text{ and } (a_n\cdot b_n, u)=1,\\ 0 & \text{otherwise;}\end{cases}$$

the characters, in turn, make it possible to go to L-series, while the fourth moment of the L-series on the line at ½ already allows us to obtain our required asymptotics (31). The remainder of the paper is devoted to the detailed justification of what was said above.

Thus, we proceed to establish equation (32). It will be obtained by reducing both sums in (32) to one form without actually calculating them.

We examine the sum in the left side of (32):

$$\sum_{d=D}^{2D} \sum_{\substack{a_n - b_n = d\cdot u \\ (a_n\cdot b_n, d)=1}} 1 = \sum_{|u|=0}^{\frac{Z^n-1}{D}} \sum_{d=D}^{2D} \sum_{\substack{a_n - b_n = d\cdot u \\ (a_n\cdot b_n, d)=1}} 1. \tag{33}$$

The right side of (33) can be transformed to sums with only positive u: it will take the form

$$\sum_{d=D}^{2D} \sum_{\substack{a_n = b_n \\ (a_n, d)=1}} 1 + 2 \sum_{u=1}^{\frac{Z^n-1}{D}} \sum_{d=D}^{2D} \sum_{\substack{a_n - b_n = d\cdot u \\ (a_n\cdot b_n, d)=1}} 1. \tag{34}$$

The first sum in (34) has the estimate

$$\sum_{d=D}^{2D} \sum_{\substack{a_n = b_n \\ (a_n, d)=1}} 1 \leqslant D \sum_{m\leqslant Z^n} \tau_n^2(m) < D\cdot Z^n\cdot \ln^{C_4} D. \tag{35}$$

We examine the second term in (34). Let $(a_n, b_n) = \nu$. Since a_n and b_n are relatively prime with d, it follows from the equation $a_n - b_n = d\cdot u$ that ν must necessarily divide u. [1] Let

$$a_n = \nu\cdot a_\nu, \qquad b_n = \nu\cdot b_\nu, \qquad u = q\cdot\nu. \tag{36}$$

Since $a_n = m_1 \cdots m_n$ and $b_n = m_1' \cdots m_n'$, it follows that the equations $a_n = \nu\cdot a_\nu$ and $b_n = \nu\cdot b_\nu$, respectively, can be chosen in as many ways as there exist solutions of the two equations

$$\nu = \nu_1\nu_2 \ldots \nu_n,$$

$$\nu = \nu_1'\nu_2' \ldots \nu_n', \tag{36a}$$

where ν_i/m_i and ν_i'/m_i' respectively.

We note that for any two fixed sets of numbers $(\nu_1, \cdots, \nu_n)$ and $(\nu_1', \cdots, \nu_n')$ connected by relation (36a), the numbers a_ν and b_ν in equations (36) will generally be constructed in various ways out of a product of n factors, since

1) *Editor's note*: See the correction at the end of this article.

$$a_\nu = \frac{m_1}{\nu_1}\cdot\frac{m_2}{\nu_2}\cdots\frac{m_n}{\nu_n}, \quad b_\nu = \frac{m_1'}{\nu_1'}\cdot\frac{m_2'}{\nu_2'}\cdots\frac{m_n'}{\nu_n'}$$

and we would therefore have to introduce the system of indices $(\nu_1, \cdots, \nu_n)$ and $(\nu_1', \cdots, \nu_n')$ at the numbers a_ν and b_ν. But, henceforth, the concrete form of the numbers a_ν and b_ν will not be used, since the dependence of a_ν and b_ν on the individual collections of systems of indices $(\nu_1, \cdots, \nu_n)$ and $(\nu_1', \cdots, \nu_n')$ is necessary only to calculate the special series directly, i.e. starting from the condition $(a_n \cdot b_n, d) = 1$, and we are using a detour, i.e. the condition $(a_n \cdot b_n, u) = 1$. We shall therefore use the notation a_ν and b_ν and remember that the form of a_ν and b_ν depends on concrete solutions of (36a). What has been said should preclude any ambiguity.

Thus, the double sum from (34) can be transformed to the form

$$\sum_{\nu\leqslant\frac{Z^n-1}{D}} \sum_{\substack{\nu=\nu_1\cdot\nu_2,\ldots,\nu_n \\ \nu=\nu_1'\cdot\nu_2',\ldots,\nu_n'}} \sum_{q\leqslant\frac{Z^n-1}{D\cdot\nu}} \sum_{d=D}^{2D} \sum_{\substack{a_\nu-b_\nu=dq \\ (a_\nu\cdot b_\nu\cdot\nu,\, d)=1 \\ (a_\nu\cdot b_\nu,\, q)=1}} 1. \tag{37}$$

But the conditions $(a_\nu \cdot b_\nu \cdot \nu, d) = 1$ and $(a_\nu \cdot b \quad q) = 1$ can be replaced by the two new conditions: $(a_\nu, b_\nu) = 1$ and $(\nu, d) = 1$. In fact, the two new conditions obviously arise from the first two conditions with the equation $a_\nu - b_\nu = d \cdot q$ holding. The converse is also true–the two old conditions arise from the two new ones with the same equation holding.

To finally reduce the problem to the congruence $a_\nu \equiv b_\nu (q)$, we must still get rid of the condition $(\nu, d) = 1$ in the sum (37). To do this, we introduce there the discontinuous factor

$$\sum_{\delta/(\nu,d)} \mu(\delta) = \begin{cases} 1, & \text{if} \quad (\nu, d) = 1, \\ 0, & \text{if} \quad (\nu, d) > 1. \end{cases}$$

Then the sum (37) takes the form

$$\sum_{\delta\leqslant\frac{Z^n-1}{D}} \mu(\delta) \sum_{\nu\leqslant\frac{Z^n-1}{D\delta}} \sum_{\substack{\nu\delta=\nu_1\cdot\nu_2\ldots\nu_n \\ \nu\delta=\nu_1'\cdot\nu_2'\ldots\nu_n}} \sum_{q\leqslant\frac{Z^n-1}{D\delta\nu}} \sum_{d_1=D/\delta}^{2D/\delta} \sum_{\substack{a_{\nu\delta}-b_{\nu\delta}=d_1\cdot\delta q \\ (a_{\nu\delta},\, b_{\nu\delta})=1}} 1. \tag{38}$$

We examine the sum from the right side of (32):

$$\sum_{d=D}^{2D} \sum_{\substack{a_n-b_n=d\cdot u \\ (a_n\cdot b_n,\, u)=1}} 1 = 2 \sum_{1\leqslant u\leqslant\frac{Z^n-1}{D}} \sum_{d=D}^{2D} \sum_{\substack{a_n-b_n=d\cdot u \\ (a_n\cdot b_n,\, u)=1}} 1 + O(D\cdot Z^n \ln^{C_n} D). \tag{39}$$

Let $(a_n, b_n) = \nu$; then necessarily ν/d, and if

$$a_n = \nu \cdot a_\nu, \quad b_n = \nu \cdot b_\nu, \quad d = \nu \cdot d_1,$$

then the sum in the right side of (39) can be rewritten in the form

$$\sum_{\nu \leqslant \frac{Z^n - 1}{D}} \sum_{\substack{\nu = \nu_1 \cdot \nu_2 \ldots \nu_n \\ \nu = \nu_1' \cdot \nu_2' \ldots \nu_n}} \sum_{\substack{u \leqslant \frac{Z^n - 1}{D} \\ (u, \nu) = 1}} \sum_{d_1 = D/\nu}^{2D/\nu} \sum_{\substack{a_\nu - b_\nu = d_1 \cdot u \\ (a_\nu, b_\nu) = 1}} 1 + O(D \cdot Z^n \cdot \ln^{C_n} D).$$

Throwing away all equations with $\nu > (Z^n - 1)/D$ gives a quantity of order $D \cdot Z^n \cdot \ln^{C_n} D$.

We assume that we have somehow already established the equation

$$\sum_{d=Q}^{2Q} \sum_{\substack{a_\nu - b_\nu = d \cdot q \\ (a_\nu, b_\nu) = 1}} 1 = \frac{1}{\varphi(q)} \sum_{\substack{Q q \leqslant a_\nu - b_\nu \leqslant 2Qq \\ (a_\nu, b_\nu) = (a_\nu \cdot b_\nu, q) = 1}} 1 + O\left(\frac{Z^{2n-1}}{q \cdot \nu} \cdot \ln^{C_n} D\right) \tag{40}$$

(the actual proof of equation (40) will be given in §5 using the theory of Dirichlet L-series).

We introduce the notation $a_\nu - b_\nu \in \mathfrak{A}(M, q)$, meaning that the difference $a_\nu - b_\nu$ lies in the interval $M \leq a_\nu - b_\nu \leq 2M$ and, in addition, the following conditions of relative primeness are fulfilled:

$$(a_\nu, b_\nu) = (a_\nu \cdot b_\nu, q) = 1.$$

Using equation (40) we transform the sum (38) to the form

$$\sum_{\delta \leqslant \frac{Z^n - 1}{D}} \mu(\delta) \sum_{\nu \leqslant \frac{Z^n - 1}{D\delta}} \sum_{\substack{\nu = \nu_1 \cdot \nu_2 \ldots \nu_n \\ \nu = \nu_1' \cdot \nu_2' \ldots \nu_n'}} \sum_{q \leqslant \frac{Z^n - 1}{D\delta\nu}} \frac{1}{\varphi(q\delta)} \sum_{a_{\nu\delta} - b_{\nu\delta} \in \mathfrak{A}(Dq, \delta q)} 1$$
$$+ O(D \cdot Z^n \cdot \ln^{C_n} D), \tag{41}$$

and the sum (39) to the form

$$\sum_{\nu \leqslant \frac{Z^n - 1}{D}} \sum_{\substack{\nu = \nu_1 \cdot \nu_2 \ldots \nu_n \\ \nu = \nu_1' \cdot \nu_2' \ldots \nu_n'}} \sum_{\substack{\nu \leqslant \frac{Z^n - 1}{D} \\ (u, \nu) = 1}} \frac{1}{\varphi(u)} \sum_{a_\nu - b_\nu \in \mathfrak{A}\left(\frac{Du}{\nu}, u\right)} 1 + O(D \cdot Z^n \cdot \ln^{C_n} D).$$

In the last sum we introduce the discontinuous factor

$$\sum_{\delta \mid (u, \nu)} \mu(\delta) = \begin{cases} 1, & \text{if} \quad (u, \nu) = 1, \\ 0, & \text{if} \quad (u, \nu) > 1; \end{cases}$$

it then takes the form

$$\sum_{\delta\leqslant\frac{Z^n-1}{D}}\mu(\delta)\sum_{\nu\leqslant\frac{Z^n-1}{D\delta}}\sum_{\substack{\nu\delta=\nu_1\cdot\nu_2\cdots\nu_n\\ \nu\delta=\nu_1'\cdot\nu_2'\cdots\nu_n'}}\sum_{q_1\leqslant\frac{Z^n-1}{D\delta}}\frac{1}{\varphi(q_1\delta)}\sum_{a_{\nu\delta}-b_{\nu\delta}\in\mathfrak{A}\left(\frac{Dq_1}{\nu},q_1\delta\right)}1. \tag{42}$$

In (42) and (43) we use the equation

$$\frac{1}{\varphi(m)}=\frac{1}{m}\sum_{r|m}\frac{\mu^2(r)}{\varphi(r)}$$

and interchange the order of summing over r; (41) then takes the form [1]

$$\begin{aligned}&\sum_{r\leqslant\frac{Z^n-1}{D}}\frac{\mu^2(r)}{\varphi(r)}\sum_{r=r_1\cdot r_2}\sum_{\substack{\delta\leqslant\frac{Z^n-1}{D}\\ r_1|\delta}}\frac{\mu(\delta)}{\delta}\sum_{\nu\leqslant\frac{Z^n-1}{D\delta}}\sum_{\substack{\delta\nu=\nu_1\cdot\nu_2\cdots\nu_n\\ \delta\nu=\nu'_1\cdot\nu'_2\cdots\nu'_n}}\\ &\times\frac{1}{r_2}\sum_{q\leqslant\frac{Z^n-1}{D\delta\nu r_2}}\frac{1}{q}\sum_{a_{\nu\delta}-a_{\nu\delta}\in\mathfrak{A}(Dqr_2,\,q\delta r_2)}1.\end{aligned} \tag{43}$$

Analogously, sum (42) requires the form

$$\begin{aligned}&\sum_{r\leqslant\frac{Z^n-1}{D}}\frac{\mu^2(r)}{\varphi(r)}\sum_{r=r_1\cdot r_2}\sum_{\substack{\delta<\frac{Z^n-1}{D}\\ r_1|\delta}}\frac{\mu(\delta)}{\delta}\sum_{\nu\leqslant\frac{Z^n-1}{D\delta}}\sum_{\substack{\delta\nu=\nu_1\cdot\nu_2\cdots\nu_n\\ \delta\nu=\nu_1'\cdot\nu_2'\cdots\nu'_n}}\\ &\times\frac{1}{r_2}\sum_{q_1\leqslant\frac{Z^n-1}{D\delta r_2}}\frac{1}{q_1}\sum_{a_{\nu\delta}-b_{\nu\delta}\in\mathfrak{A}\left(\frac{Dq_1r_2}{\nu},\,q_1\cdot r_2\cdot\delta\right)}1.\end{aligned} \tag{44}$$

In sum (42) we introduce the discontinuous factor

$$\sum_{\substack{\pi\,|\,a_{\nu\delta}\cdot b_{\nu\delta}\\ \pi\,|\,q\cdot r_2\delta}}\mu(\pi)=\begin{cases}1, & \text{if}\quad (a_{\nu\delta}\cdot b_{\nu\delta},\,q\cdot r_2\cdot\delta)=1,\\ 0, & \text{otherwise.}\end{cases}$$

We introduce a similar discontinuous factor in sum (44) to avoid the condition $(a_{\nu\delta}\cdot b_{\nu\delta},q_1\cdot r_2\cdot\delta)=1$. Sum (43) then takes the form

$$\begin{aligned}&\sum_{\substack{\pi=\pi_1\cdot\pi_2\cdot\pi_3\\ \pi\leqslant\frac{Z^n-1}{D}}}\mu(\pi)\sum_{r\leqslant\frac{Z^n-1}{D}}\frac{\mu^2(r)}{\varphi(r)}\sum_{\substack{r=r_1\cdot r_2\\ \pi_2|r_2}}\sum_{\substack{r_1\cdot\pi_1|\delta\\ \delta\leqslant\frac{Z^n-1}{D}}}\frac{\mu(\delta)}{\delta}\sum_{\nu\leqslant\frac{Z^n-1}{D\delta}}\\ &\times\sum_{\substack{\delta\nu=\nu_1\cdot\nu_2\cdots\nu_n\\ \delta\nu=\nu_1'\cdot\nu_2'\cdots\nu'_n}}\frac{1}{r_2\pi_3}\sum_{q\leqslant\frac{Z^n-1}{D\cdot\delta\cdot\nu\cdot r_2\cdot\pi_3}}\frac{1}{q}\sum_{\substack{Dq\pi_3r_2\leqslant a_{\nu\delta}-b_{\nu\delta}\leqslant 2Dq\pi_3r_2\\ (a_{\nu\delta},\,b_{\nu\delta})=1,\ \pi\,|\,a_{\nu\delta}\cdot b_{\nu\delta}}}1.\end{aligned} \tag{43a}$$

---

1) *Editor's note*: See the correction at the end of the article.

Analogously, (44) takes the form

$$\sum_{\substack{\pi=\pi_1\cdot\pi_2\cdot\pi_3\\ \pi\leqslant\frac{Z^n-1}{D}}}\mu(\pi)\sum_{r\leqslant\frac{Z^n-1}{D}}\frac{\mu^2(r)}{\varphi(r)}\sum_{\substack{r=r_1\cdot r_2\\ \pi_2\mid r_2}}\sum_{\substack{r_1\cdot\pi_1\mid\delta\\ \delta\leqslant\frac{Z^n-1}{D}}}\frac{\mu(\delta)}{\delta}\sum_{\nu\leqslant\frac{Z^n-1}{D\delta}}$$
$$\times\sum_{\substack{\delta\nu=\nu_1\cdot\nu_2\ldots\nu_n\\ \delta\nu=\nu_1'\cdot\nu_2'\ldots\nu_n'}}\frac{1}{r_2\pi_3}\sum_{q_1\leqslant\frac{Z^n-1}{D\delta r_2\pi_3}}\frac{1}{q_1}\sum_{\substack{D\cdot\frac{q_1}{\nu}\cdot\pi_3\cdot r_2\leqslant a_{\nu\delta}-b_{\nu\delta}\leqslant 2\cdot D\cdot\frac{q_1}{\nu}\cdot\pi_3 r_2\\ (a_{\nu\delta},\, b_{\nu\delta})=1,\ \pi\mid a_{\nu\delta}\cdot b_{\nu\delta}}}1. \tag{44a}$$

We introduce the notation $a_\nu - b_\nu \in \mathfrak{A}_\pi(M)$, meaning that the difference $a_\nu - b_\nu$ lies in the interval $M \leq a_\nu - b_\nu \leq 2M$, the numbers a_ν and b_ν satisfy the condition of relative primeness $(a_\nu, b_\nu) = 1$, and their product is divisible by the number π.

We isolate from (44a) the inner double sum

$$\frac{1}{\pi_3 r_2}\sum_{q_1\leqslant\frac{Z^n-1}{D\cdot\delta\cdot r_2\cdot\pi_3}}\frac{1}{q_1}\sum_{a_{\nu\delta}-b_{\nu\delta}\in\mathfrak{A}_\pi\left(D\cdot\frac{q_1}{\nu}\cdot\pi_3 r_2\right)}1. \tag{45}$$

We divide all values of q_1 into progressions mod ν. Let $q_1 = \nu\cdot q + l$; then sum (45) takes the form

$$\frac{1}{r_2\pi_3}\sum_{l=1}^{\nu}\sum_{q\nu+l\leqslant\frac{Z^n-1}{D\cdot\delta\cdot r_2\cdot\pi_3}}\frac{1}{q\nu+l}\sum_{a_{\nu\delta}-b_{\nu\delta}\in\mathfrak{A}_\pi\left(Dq\pi_3 r_2+D\cdot\frac{l}{\nu}\cdot r_2\pi_3\right)}1. \tag{46}$$

We eliminate l in the inner sum. We have

$$\sum_{Dqr_2\pi_3\leqslant a_{\nu\delta}-b_{\nu\delta}\leqslant Dqr_2\pi_3+D\cdot\frac{l}{\nu}\cdot r_2\cdot\pi_3}1\ll\frac{D\cdot Z^n\cdot r_2\cdot\pi_3}{\nu\cdot\delta\cdot\pi}\cdot\tau_{n^2}(\pi)\cdot\ln^{C_n}D.$$

Similarly,

$$\sum_{2Dqr_2\pi_3\leqslant a_{\nu\delta}-b_{\nu\delta}\leqslant 2Dqr_2\pi_3+2D\cdot\frac{l}{\nu}\cdot r_2\cdot\pi_3}1\ll\frac{D\cdot Z^n\cdot r_2\pi_3}{\nu\cdot\delta\cdot\pi}\cdot\tau_{n^2}(\pi)\cdot\ln^{C_n}D.$$

Thus (46) can be rewritten in the form

$$\frac{1}{r_2\pi_3}\sum_{l=1}^{\nu}\sum_{q\nu+l\leqslant\frac{Z^n-1}{D\delta r_2\pi_3}}\frac{1}{q\nu+l}\sum_{a_{\nu\delta}-b_{\nu\delta}\in\mathfrak{A}_\pi(Dqr_2\pi_3)}1+O\left(\frac{D\cdot Z^n}{\nu\cdot\delta\cdot\pi}\cdot\tau_{n_2}(\pi)\cdot\ln^{C_n}D\right), \tag{47}$$

where the inner sum is already independent of l.

We divide the sum over q in (47) into two sums: for $q \geq 1$ and $q = 0$. Corresponding to this, we obtain

$$\frac{1}{r_2\pi_3}\sum_{l=0}^{\nu-1}\sum_{\substack{q\geqslant 1\\ q\nu+l\leqslant \frac{Z^n-1}{D\delta r_2\pi_3}}}\frac{1}{q\nu+l}\sum_{a_{\nu\delta}-b_{\nu\delta}\in\mathfrak{A}_\pi(Dqr_2\pi_3)}1+O\Big(\frac{1}{r_2\pi_3}\sum_{l=1}^{\nu}\frac{1}{l}\sum_{a_{\nu\delta}=b_{\nu\delta},\ \pi\mid a_{\nu\delta}\cdot b_{\nu\delta}}1\Big). \tag{48}$$

We use the estimate

$$\sum_{a_{\nu\delta}=b_{\nu\delta},\ \pi/a_{\nu\delta}\cdot b_{\nu\delta}}1\ll\frac{Z^n}{\nu\cdot\delta\cdot\pi}\cdot\tau_{n^2}(\pi)\cdot\ln^{C_n}D. \tag{49}$$

The summation over q in (48) is bounded from above by a quantity depending on l:

$$q\leqslant\frac{Z^n-1}{D\nu\cdot\delta\cdot r_2\cdot\pi_3}-\frac{l}{\nu}.$$

But in the interval

$$\frac{Z^n-1}{D\cdot\nu\cdot\delta\cdot r_2\cdot\pi_3}-\frac{l}{\nu}\leqslant q_0\leqslant\frac{Z^n-1}{D\nu\cdot\delta\cdot r_2\cdot\pi_3}.$$

For any l there can be found no more than one value $q = q_0$. Therefore, if we replace the upper limit of the summation over q in (48) by $(Z^n - 1)/D \cdot \nu \cdot \delta \cdot r_2 \cdot \pi_3$ we thereby bring in a quantity of order no greater than

$$\frac{1}{r_2\pi_3}\sum_{l\leqslant\nu}\frac{1}{q_0\cdot\nu}\sum_{a_{\nu\delta}-b_{\nu\delta}\in\mathfrak{A}_\pi(Dq_0\cdot r_2\cdot\pi_3)}1.$$

But trivially we have

$$\sum_{a_{\nu\delta}-b_{\nu\delta}\in\mathfrak{A}_\pi(Dq_0\cdot r_2\cdot\pi_3)}1\ll\frac{D\cdot Z^n\cdot r_2\cdot q_0\cdot\pi_3}{\nu\cdot\delta\cdot\pi}\tau_{n^2}(\pi)\cdot\ln^{C_n}D. \tag{49a}$$

Therefore

$$\frac{1}{r_2\pi_3}\sum_{l\leqslant\nu}\frac{1}{q_0\cdot\nu}\sum_{a_{\nu\delta}-b_{\nu\delta}\in\mathfrak{A}_\pi(Dq_0\cdot r_2\cdot\pi_3)}1\ll\frac{D\cdot Z^n}{\nu\cdot\delta\cdot\pi}\tau_{n^2}(\pi)\cdot\ln^{C_n}D. \tag{50}$$

Taking into account estimates (49) and (50), we see that we have replaced our sum (45) by the quantity

$$\frac{1}{r_2\pi_3}\sum_{l=0}^{\nu-1}\sum_{\substack{q\geqslant 1\\ q\leqslant\frac{Z^n-1}{D\cdot\delta\cdot\nu\cdot r_2\pi_3}}}\frac{1}{q\cdot\nu+l}\sum_{a_{\nu\delta}-b_{\nu\delta}\in\mathfrak{A}_\pi(Dqr_2\pi_3)}1+O\Big(\frac{D\cdot Z^n\cdot\tau_{n^2}(\pi)}{\nu\cdot\delta\cdot\pi}\ln^{C_n}D\Big). \tag{51}$$

It remains to make the last step—to clear away l altogether.

We note that

$$\frac{1}{q\nu}-\frac{1}{q\nu+l}\leqslant\frac{l}{q^2\cdot\nu^2}\leqslant\frac{1}{q^2\nu},$$

so that sum (51) can be rewritten in the form

$$\frac{1}{r_2\pi_3}\sum_{q\leqslant\frac{Z^n-1}{D\delta\nu r_2\pi_3}}\frac{1}{q}\sum_{a_{\nu\delta}-b_{\nu\delta}\in\mathfrak{A}_\pi(Dqr_2\pi_3)}1$$

$$+O\Big(\frac{1}{r_2\pi_3}\sum_{q\leqslant\frac{Z^n-1}{D\delta\nu r_2\pi_3}}\frac{1}{q^2}\sum_{a_{\nu\delta}-b_{\nu\delta}\in\mathfrak{A}_\pi(Dqr_2\pi_3)}1\Big). \tag{52}$$

We estimate the quantity in the second line of (52). Using (49a) we obtain

$$\frac{1}{r_2\pi_3}\sum_{q\leqslant\frac{Z^n-1}{D\delta\nu\cdot\pi_3r_2}}\frac{1}{q^2}\sum_{a_{\nu\delta}-b_{\nu\delta}\in\mathfrak{A}_\pi(Dqr_2\pi_3)}1\ll\frac{D\cdot Z^n}{\nu\cdot\delta\cdot\pi}\tau_{n^2}(\pi)\cdot\ln^{C_n}D.$$

In this way (45) is finally transformed to the form

$$\frac{1}{r_2\pi_3}\sum_{q\leqslant\frac{Z^n-1}{D\delta\nu r_2\pi_3}}\frac{1}{q}\sum_{a_{\nu\delta}-b_{\nu\delta}\in\mathfrak{A}_\pi(Dqr_2\pi_3)}1+O\Big(\frac{D\cdot Z^n}{\nu\cdot\delta\cdot\pi}\tau_{n^2}(\pi)\cdot\ln^{C_n}D\Big).$$

Substituting this value of (45) into (44), we reduce (44) to the sum (42) in the form (43a) with the remainder we wanted. Thus we have established relation (32) under the assumption that (40) is fulfilled.

§5

To finally prove equation (32), it remains to derive relation (40). [1]

The condition $(a_\nu, b_\nu)=1$ and the equation $a_\nu-b_\nu=d\cdot q$ imply that $(a_\nu, q)=1$ and $(b_\nu, q)=1$. Therefore

$$\sum_{d=Q}^{2Q}\sum_{\substack{a_\nu-b_\nu=d\cdot q\\(a_\nu,b_\nu)=1}}1=\sum_{\substack{a_\nu\equiv b_\nu(q)\\Q\cdot q\leqslant a_\nu-b_\nu\leqslant 2Q\cdot q\\(a_\nu,b_\nu)=1}}1=\frac{1}{\varphi(q)}\sum_{\substack{Qq\leqslant a_\nu-b_\nu\leqslant 2Qq\\(a_\nu,b_\nu)=1}}\sum_{\chi_q}\chi(a_\nu)\cdot\overline{\chi}(b_\nu)$$

$$=\frac{1}{\varphi(q)}\sum_{a_\nu-b_\nu\in\mathfrak{A}(Qq,q)}1+\frac{1}{\varphi(q)}\sum_{\chi\neq\chi_0}\sum_{a_\nu-b_\nu\in\mathfrak{A}_1(Q\cdot q)}\frac{\chi(a_\nu)}{\chi(b_\nu)}. \tag{53}$$

For a nonprincipal character $\chi \pmod q$, we examine the sum

$$\sum_{\substack{Q\cdot q\leqslant a_\nu-b_\nu\leqslant 2Q\cdot q\\(a_\nu,b_\nu)=1}}\chi(a_\nu)\cdot\overline{\chi}(b_\nu)$$

1) *Editor's note*: See the correction at the end of the article.

considering it for a fixed collection $(\nu_1, \cdots, \nu_n)$ and $(\nu'_1, \cdots, \nu'_n)$. We write this sum in the form

$$\sum_{1 \leqslant b_\nu \leqslant \frac{Z^n}{\nu} - Q\cdot q} \bar{\chi}(b_\nu) \sum_{\substack{Y(b_\nu) \leqslant a_\nu \leqslant X(b_\nu) \\ (a_\nu, b_\nu)=1}} \chi(a_\nu), \tag{54}$$

where for fixed b_ν we select $Y(b_\nu)$ and $X(b_\nu)$ so that $Q \cdot q \leq a_\nu \to b_\nu \leq 2Q \cdot q$. We fix some b_ν and consider the sum

$$\sum_{\substack{Y(b_\nu) \leqslant a_\nu \leqslant X(b_\nu) \\ (a_\nu, b_\nu)=1}} \chi(a_\nu). \tag{55}$$

We write a_ν in the form $a_\nu = m_1 \cdot m_2 \cdot m_3 \cdot m_4 \cdot M_5$, where $M_5 = m_5 \cdots m_n$ (or 1 if $n = 4$). Here, by assumption, the m_i $(1 \leq i \leq n)$ run through the integers of the interval $(1, Z/\nu_i)$ which are relatively prime to b_ν. We write (55) in the form of a double sum

$$\sum_{\substack{U_1(b_\nu) \leqslant M_5 \leqslant U_2(b_\nu) \\ (M_5, b_\nu)=1}} \chi(M_5) \cdot \sum_{\substack{\frac{Y(b_\nu)}{M_5} \leqslant m_1 \cdot m_2 \cdot m_3 \cdot m_4 \leqslant \frac{X(b_\nu)}{M_5} \\ (m_i, b_\nu)=1}} \chi(m_1 \cdot m_2 \cdot m_3 \cdot m_4) \tag{56}$$

and for fixed M_5 we examine the sum

$$S_\chi(b_\nu, M_5) = \sum_{\substack{B \leqslant m_1 \cdot m_2 \cdot m_3 \cdot m_4 \leqslant A \\ (m_i, b_\nu)=1}} \chi(m_1 \cdot m_2 \cdot m_3 \cdot m_4),$$

where $A = X(b_\nu)/M_5$ and $B = Y(b_\nu)/M_5$. The isolation of this sum is precisely the central step of this method allowing us to estimate the high moments for $n \geq 4$ but which does not work for $n = 3$ or 2. From the further arguments of this section it will be clear why it is important to isolate the four variables m_i. In any case, we have estimates for A and B:

$$B \leqslant A \leqslant \frac{Z^4}{\nu_1 \cdot \nu_2 \cdot \nu_3 \cdot \nu_4}, \tag{57}$$

since $m_i \leq Z/\nu_i$. We represent the sum $S_\chi(b_\nu, M_5)$ as an integral:

$$S_\chi(b_\nu, M_5) = \frac{1}{2\pi i} \int_{1-iD^2}^{1+iD^2} \frac{A^s - B^s}{s} \prod_{i=1}^{4} \Bigg(\sum_{\substack{m_i \leqslant \frac{Z}{\nu_i} \\ (m_i, b_\nu)=1}} \frac{\chi(m_i)}{m_i^s} \Bigg) ds + O\left(\frac{1}{D}\right).$$

In addition, we have

$$\sum_{\substack{m_i \leqslant \frac{Z}{\nu_i} \\ (m_i, b_\nu)=1}} \frac{\chi(m_i)}{m_i^s} = \frac{1}{2\pi i} \int_{1-iD^2}^{1+iD^2} \frac{\left(\frac{Z}{\nu_i}\right)^{w_i}}{w_i} L(s + w_i, \chi) \cdot \prod_{p/b_\nu} \left(1 - \frac{\chi(p)}{p^{s+w_i}}\right) dw_i + O\left(\frac{1}{D^2}\right).$$

Consequently

$$S_\chi(b_\nu, M_5) = \frac{1}{(2\pi i)^5} \int_{1-iD^2}^{1+iD^2} \cdots \int \frac{(A^s - B^s) \cdot \prod_{i=1}^{4} \left(\frac{Z}{v_i}\right)^{w_i}}{s \cdot w_1 \cdot w_2 \cdot w_3 \cdot w_4} \cdot \prod_{i=1}^{4} L(s + w_i, \chi)$$

$$\times \prod_{i=1}^{4} \prod_{p \mid b_\nu} \left(1 - \frac{\chi(b_\nu)}{p^{s+w_i}}\right) ds \cdot dw_1 \ldots dw_4 + O\left(\frac{1}{D}\right).$$

For all four variables w_i $(1 \le i \le 4)$ we move the contour to the line $\mathrm{Re}\, w_i = 1/\ln D$, and the contour for s to the line

$$\mathrm{Re}\, s = \frac{1}{2} - \frac{1}{\ln D}.$$

Since χ is not the principal character, we do not pass through any singular points. But on these new lines of integration we obtain the estimate

$$|S_\chi(b_\nu, M_5)| \ll \frac{2\sqrt{A}}{(2\pi)^5} \int_{-D^2}^{D^2} \cdots \int \frac{\prod_{j=1}^{4} \left| L\left(\frac{1}{2} + i(t + v_j), \chi\right) \right| dt \cdot dv_1 \ldots dv_4}{\sqrt{\frac{1}{4} + t^2} \prod_{j=1}^{4} \sqrt{\frac{1}{\ln^2 D} + v_j^2}}$$

$$\times \prod_{p \mid b_\nu} \left(1 + \frac{1}{\sqrt{p}}\right)^4 + O\left(\frac{1}{D}\right). \tag{58}$$

The resulting five-fold integral depends on the character χ. Therefore, if we substitute (58) into (56) and into (54), then, considering estimate (57) and the estimates

$$M_5 \ll \frac{Z^{n-4}}{\nu_5 \ldots \nu_n}, \quad b_\nu \ll \frac{Z}{\nu}$$

we find that

$$\left| \sum_{\substack{Q \cdot q \ll a_\nu - b_\nu \ll 2Q \cdot q \\ (a_\nu, b_\nu) = 1}} \frac{\chi(a_\nu)}{\chi(b_\nu)} \right| \ll \frac{Z^{2n-2}}{\nu} \cdot I_5(\chi) \cdot \ln^{C_n} D.$$

Taking this estimate into account, we obtain

$$\left| \frac{1}{\varphi(q)} \sum_{\chi \neq \chi_0} \sum_{a_\nu - b_\nu \in \mathfrak{A}_1(Q \cdot q)} \frac{\chi(a_\nu)}{\chi(b_\nu)} \right| \ll \frac{Z^{2n-2}}{\nu} \cdot \ln^{C_n} D \cdot \frac{1}{\varphi(q)} \sum_{\chi \neq \chi_0} I_5(\chi), \tag{59}$$

where $I_5(\chi)$ is the five-fold integral in (58). We estimate the sum

$$\frac{1}{\varphi(q)} \sum_{\chi \neq \chi_0} I_5(\chi).$$

In the five-fold integral $I_5(\chi)$ we apply the inequality

$$\prod_{j=1}^{4} |L(s_j,\chi)| \ll C_0 \sum_{j=1}^{4} |L(s_j,\chi)|^4.$$

We have thereby actually reduced the problem to the average estimate of the fourth power of the Dirichlet L-series on the line at ½. But, thanks to the shortened functional equation (see [8]), it is well known that the fourth power of the Dirichlet L-series, averaged over all characters and along the imaginary axis, behaves like $\ln^{C_0} qt$ on the line at ½, which leads to the estimate

$$\frac{1}{\varphi(q)} \sum_{\chi \neq \chi_0} \int_{-T}^{T} \frac{\left|L\left(\frac{1}{2}+it,\chi\right)\right|^4}{\left|\frac{1}{2}+it\right|} dt \ll \ln^{C_0} qT.$$

Using this estimate, we trivially obtain

$$\frac{1}{\varphi(q)} \sum_{\chi \neq \chi_0} I_5(\chi) \ll \ln^{C_0} D. \tag{60}$$

It is precisely to get estimate (60), and together with it (59), that it is essential that $n \geq 4$, since only the fourth moment of the Dirichlet L-series gives us the minimally necessary gain of Z^2 in the reduction, which secures estimate (1). Since $q \leq Z$,

$$Z^{2n-2} \ll \frac{Z^{2n-1}}{q}.$$

Therefore

$$\left|\frac{1}{\varphi(q)} \sum_{\chi \neq \chi_0} \sum_{a_\nu - b_\nu \in \mathfrak{A}_1(Q\cdot q)} \frac{\chi(a_\nu)}{\chi(b_\nu)}\right| \ll \frac{Z^{2n-1}}{\nu\cdot q} \cdot \ln^{C_n} D. \tag{61}$$

But this is precisely the remainder whose validity was assumed in formula (40). Since the principal term in equation (58) was isolated precisely as in (40), it hence follows that relation (40) is actually valid. As was shown above, the validity of equation (32) follows from this.

If we try to operate with this method with $n = 3$, we can only isolate three variables m_i in sum (55); hence we must estimate the third power of the L-series on the line at ½, which gives us a reducing factor of $Z^{3/2}$ and a corresponding $Z^{2n-1/2}$ on the right in (61), which does not secure the remainder estimate required in (32).

§6

We calculate the sum in the right side of (32). We re-designate $u = q$. Then

$$\sum_{d=D}^{2D} \sum_{\substack{a_n - b_n = d\cdot q \\ (a_n\cdot b_n,\, q)=1}} 1 = 2 \sum_{q=1}^{\frac{Z^n-1}{D}} \sum_{d=D}^{2D} \sum_{\substack{a_n - b_n = d\cdot q \\ (a_n\cdot b_n,\, q)=1}} 1 + O(D\cdot Z^n \ln^{C_n} D). \quad (62)$$

But for fixed q the inner double sum in (62) can be represented in the form

$$\frac{1}{\varphi(q)} \sum_{\chi_q} \sum_{Dq \leqslant a_n - b_n \leqslant 2Dq} \frac{\chi(a_n)}{\chi(b_n)}.$$

Isolating the principal character, we obtain

$$\frac{1}{\varphi(q)} \sum_{a_n - b_n \in \mathfrak{A}(Dq,\, q)} 1 + \frac{1}{\varphi(q)} \sum_{\chi \neq \chi_0} \sum_{a_n - b_n \in \mathfrak{A}_1(Dq)} \chi(a_n)\cdot \overline{\chi}(b_n). \quad (63)$$

Applying to the sums over the nonprincipal characters in (63) the same device with the fourth power of the L-series that we used in getting (40), we obtain the estimate

$$\left| \frac{1}{\varphi(q)} \sum_{\chi \neq \chi_0} \sum_{a_n - b_n \in \mathfrak{A}_1(Dq)} \frac{\chi(a_n)}{\chi(b_n)} \right| \ll \frac{Z^{2n-1}}{q} \cdot \ln^{C_n} D.$$

Substituting this estimate into (63), and then into (62), we obtain

$$\sum_{d=D}^{2D} \sum_{\substack{a_n - b_n = d\cdot q \\ (a_n\cdot b_n,\, q)=1}} 1 = 2 \sum_{q \leqslant \frac{Z^n-1}{D}} \frac{1}{\varphi(q)} \sum_{a_n - b_n \in \mathfrak{A}(Dq,\, q)} 1 + O(D\cdot Z^n \cdot \ln^{C_n} D). \quad (64)$$

We split the double sum in the right side of (64) into three sums:

$$\sum_{q=1}^{\frac{Z^n-1}{2D}} \frac{1}{\varphi(q)} \sum_{\substack{b_n=1 \\ (b_n,\, q)=1}}^{Z^n-2Dq} \sum_{\substack{a_n=Dq+b_n \\ (a_n,\, q)=1}}^{2Dq+b_n} 1 + \sum_{q=1}^{\frac{Z^n-1}{2D}} \frac{1}{\varphi(q)} \sum_{\substack{b_n=Z^n-2Dq \\ (b_n,\, q)=1}}^{Z^n-Dq} \sum_{\substack{a_n=Dq+b_n \\ (a_n,\, q)=1}}^{Z^n} 1$$

$$+ \sum_{q=\frac{Z^n-1}{2D}}^{\frac{Z^n-1}{D}} \frac{1}{\varphi(q)} \sum_{\substack{b_n=1 \\ (b_n,\, q)=1}}^{Z^n-Dq} \sum_{\substack{a_n=Dq+b_n \\ (a_n,\, q)=1}}^{Z^n} 1.$$

We designate them by S_1, S_2 and S_3, respectively.

We have thus reduced the problem to calculating how many numbers a_n are in the entire intervals. We find asymptotics for these sums:

$$\sum_{\substack{a_n \leqslant x \\ (a_n, q)=1}} 1 = \frac{1}{(2\pi i)^{n+1}} \int_{1-iZ}^{1+iZ} \cdots \int \frac{x^s Z^{w_1+\cdots+w_n}}{s \cdot w_1 \cdot w_2 \ldots w_n} \cdot \prod_{i=1}^{n} \zeta(s+w_i)$$

$$\times \prod_{i=1}^{n} \prod_{p \mid q} \left(1 - \frac{1}{p^{s+w_i}}\right) ds \cdot dw_1 \ldots dw_n + O(Z^{n-1} \cdot \ln^{C_n} D). \qquad (65)$$

In the $(n+1)$-fold integral we move the contour of the variable s to the left onto the line $\operatorname{Re} s = 1 - 1/\ln D$; in doing so we do not pass through any singular points. We then move all the contours of the variables w_i in turn to the left onto the line

$$\operatorname{Re} w_i = \frac{1}{2 \ln D}.$$

We then pass through poles at the points $w_i = 1 - s$ with residues

$$\frac{Z^{1-s}}{1-s} \prod_{p/q} \left(1 - \frac{1}{p}\right).$$

After performing this operation, we obtain

$$\prod_{p/q} \left(1 - \frac{1}{p}\right)^n \cdot \frac{Z^n}{2\pi i} \int_{C_s} \frac{\left(\frac{x}{Z^n}\right)^s}{s \cdot (1-s)^n} ds$$

$$+ n \cdot \prod_{p/q} \left(1 - \frac{1}{p}\right)^{n-1} \frac{Z^{n-1}}{2\pi i} \int_{C_s} \int_{C_{w_1}} \frac{\left(\frac{x}{Z^{n-1}}\right)^s \cdot Z^{w_1} \cdot \zeta(s+w_1)}{w_1 \cdot s \cdot (1-s)^{n-1}} ds \cdot dw_1$$

$$+ \sum_{m=2}^{n} C_n^m \prod_{p/q} \left(1 - \frac{1}{p}\right)^{n-m} \frac{Z^{n-m}}{(2\pi i)^{m+1}} \int_{C_s} \int_{C_{w_1}} \cdots \int_{C_{w_m}} \frac{\left(\frac{x}{Z^{n-m}}\right)^s}{s \cdot (1-s)^{n-m}}$$

$$\times \frac{Z^{w_1+\ldots+w_m}}{w_1 \cdot w_2 \ldots w_m} \prod_{i=1}^{m} \zeta(s+w_i) \prod_{p/q} \left(1 - \frac{1}{p^{s+w_i}}\right) ds \cdot dw_1 \ldots dw_m, \qquad (66)$$

where

$$C_s \to \operatorname{Re} s = 1 - \frac{1}{\ln D}, \quad |\operatorname{Im} s| \leqslant Z,$$

$$C_{w_i} \to \operatorname{Re} w_i = \frac{1}{2 \ln D}, \quad |\operatorname{Im} w_i| \leqslant Z.$$

To estimate the second term in (66) we move the contour of the variable s

to the line $\operatorname{Re} s = 1/\ln D$, after which we transform $\zeta(s + w_1)$ by the functional equation, use Stirling's formula for $\Gamma(1 - s - w)$ and calculate the integral, not making the integrand coarse by taking a single modulus. The method of computation is the usual one in such cases, and we shall not dwell on it.

As a result, we obtain a quantity of order

$$\tau(q) \cdot Z^{n-1} \ln^{C_n} D. \tag{67}$$

We apply the following principle for integrals, depending on the summation index m. If $m = 2, 3$ or 4 we move the contour of the variable s onto the line $\operatorname{Re} s = \frac{1}{2}$ and we use the shortened functional equation for $\zeta(s + w_1)$ on the line at $\frac{1}{2}$ to compute the resulting integrals. We also get as a result that these integrals do not exceed the quantity (67).

If $m \geq 5$, we then proceed as follows. We move the contour for s onto the line $\operatorname{Re} s = \frac{1}{2}$. We leave the contours for w_1, w_2, w_3 and w_4 on the line $\operatorname{Re} w_i = 1/2 \ln D$ and we move the others to the right onto the line

$$\operatorname{Re} w_i = \frac{1}{2} - \frac{1}{\ln D}, \quad i \geqslant 5.$$

The problem has again been reduced to estimating the fourth moment of the zeta function on the line at $\frac{1}{2}$. We then obtain a quantity of order $Z^{n-2} \cdot \ln^{C_n} D$.

We thereby finally obtain the asymptotic formula

$$\sum_{\substack{a_n \leqslant x \\ (a_n, q)=1}} 1 = \prod_{p/q} \left(1 - \frac{1}{p}\right)^n \frac{Z^n}{2\pi i} \int_{C_s} \frac{\left(\frac{x}{Z^n}\right)^s}{s \cdot (1-s)^n} ds + O(Z^{n-1} \cdot \tau(q) \cdot \ln^{C_n} D).$$

We let

$$F(x) = \frac{Z^n}{2\pi i} \int_{1 - \frac{1}{\ln D} - i\infty}^{1 - \frac{1}{\ln D} + i\infty} \frac{\left(\frac{x}{Z^n}\right)^s}{s \cdot (1-s)^n} ds; \tag{68}$$

then

$$\sum_{\substack{a_n \leqslant x \\ (a_n, q)=1}} 1 = \prod_{p/q} \left(1 - \frac{1}{p}\right)^n \cdot F(x) + O(Z^{n-1} \cdot \tau(q) \cdot \ln^{C_n} D). \tag{69}$$

We note that for $x > 0$ the function $F(x)$ together with its derivative is continuous with respect to x. We shall use this remark later.

§7

Starting from formula (69), we obtain the following relations for the sums S_1, S_2 and S_3 introduced above:

$$S_1 = \sum_{q=1}^{\frac{Z^n-1}{2D}} \frac{1}{q} \prod_{p/q} \left(1-\frac{1}{p}\right)^{n-1} \sum_{b_n=1}^{Z^n-2Dq} [F(2Dq+b_n) - F(Dq+b_n)] + O(Z^{2n-1} \cdot \ln^{C_n} D),$$

$$S_2 = \sum_{q=1}^{\frac{Z^n-1}{2D}} \frac{1}{q} \prod_{p/q} \left(1-\frac{1}{p}\right)^{n-1} \sum_{b_n=Z^n-2Dq}^{Z^n-Dq} [F(Z^n) - F(Dq+b_n)] + O(Z^{2n-1} \cdot \ln^{C_n} D),$$

$$S_3 = \sum_{q=\frac{Z^n-1}{2D}}^{\frac{Z^n-1}{D}} \frac{1}{q} \prod_{p/q} \left(1-\frac{1}{p}\right)^{n-1} \sum_{b_n=1}^{Z^n-Dq} [F(Z^n) - F(Dq+b_n)] + O(Z^{2n-1} \cdot \ln^{C_n} D).$$

We shall later need an estimate of the derivatives of $F(x)$. From (68) we obtain

$$F'(x) = \frac{Z^n}{2\pi i} \int_{1-\frac{1}{\ln D}-i\infty}^{1-\frac{1}{\ln D}+i\infty} \frac{Z^{-ns} \cdot x^{s-1}}{(1-s)^n} ds, \tag{70}$$

from which it follows that $|F'(x)| \leq \ln^{C_n} D$ for $1 \leq x \leq Z^n$.

The following estimates are obtained from the same formula (70):

$$\left.\begin{aligned} &|F''(x)| \leqslant x^{-1} \cdot \ln^{C_n} D, \quad x \geqslant 1, \\ &|F'(2Dq+x) - F'(Dq+x)| < \frac{Dq}{Dq+x} \cdot \ln^{C_n} D, \\ &|F(Z^n) - F(Z^n - Dq)| < D \cdot q \cdot \ln^{C_n} D. \end{aligned}\right\} \tag{71}$$

In addition, if $x \leq Z^n$ we move the contour in (70) to the right to infinity to obtain

$$F(x) = x \cdot \sum_{m=0}^{n-1} \frac{1}{m!} \left(\ln \frac{Z^n}{x}\right)^m.$$

It follows from this relation that $F'(Z^n) = 0$.

Using partial summation over b_n, we transform each of the three sums S_1, S_2 and S_3 to the form

$$S_1 = \sum_{q=1}^{\frac{Z^n-1}{2D}} \prod_{p/q} \left(1-\frac{1}{p}\right)^{2n-1} \cdot \frac{\Phi_1(q)}{q} + O(Z^{2n-1}\cdot \ln^{C_n} D),$$

$$S_2 = \sum_{q=1}^{\frac{Z^n-1}{2D}} \prod_{p/q} \left(1-\frac{1}{p}\right)^{2n-1} \cdot \frac{\Phi_2(q)}{q} + O(Z^{2n-1}\cdot \ln^{C_n} D),$$

$$S_3 = \sum_{q=\frac{Z^n-1}{2D}}^{\frac{Z^n-1}{D}} \prod_{p/q} \left(1-\frac{1}{p}\right)^{2n-1} \cdot \frac{\Phi_3(q)}{q} + O(Z^{2n-1}\cdot \ln^{C_n} D),$$

where

$$\Phi_1(y) = F(Z^n - 2Dy)\cdot[F(Z^n) - F(Z^n - Dy)] - \int_1^{Z^n-2Dy} F(x)\cdot[F'(2Dy+x) - F'(Dy+x)]\,dx,$$

$$\Phi_2(y) = \int_{Z^n-2Dy}^{Z^n-Dy} [F(x) - F(Z^n - 2Dy)]\cdot F'(Dy+x)\,dx,$$

$$\Phi_3(y) = \int_1^{Z^n-Dy} F(x)\cdot F'(Dy+x)\,dx.$$

We now estimate the functions Φ_i $(i = 1, 2, 3)$ and their derivatives using inequalities (71). We first have

$$\left.\begin{aligned} |\Phi_1(y)| &\leqslant Z^n\cdot D\cdot y\cdot \ln^{C_n} D,\\ |\Phi_2(y)| &\leqslant D^2\cdot y^2 \ln^{C_n} D,\\ |\Phi_3(y)| &\leqslant Z^{2n}. \end{aligned}\right\} \tag{72}$$

We estimate the derivatives with respect to y:

$$\Phi_1'(y) = -2D\cdot F'(Z^n - 2Dy)\cdot[F(Z^n) - F(Z^n - Dy)] - 2D\cdot F(Z^n - 2Dy)\cdot F'(Z^n - D\cdot y) - D\cdot \int_1^{Z^n-2Dy} F(x)\cdot[2\cdot F''(2Dy+x) - F''(Dy+x)]\,dx.$$

From this formula and estimates (71), we obtain

$$|\Phi_1'(y)| \leqslant D\cdot Z^n\cdot \ln^{C^n} D.$$

Further,

$$\Phi_2'(y) = -D \cdot \int_{Z^n - 2Dy}^{Z^n - Dy} \{[F(x) - F(Z^n - 2Dy)] \cdot F''(Dy + x) + 2F'(Z^n - 2Dy) \cdot F'(Dy + x)\}\, dx.$$

Consequently

$$|\Phi_2'(y)| \leqslant D^2 \cdot y \cdot \ln^{C_n} D.$$

Finally,

$$\Phi_3'(y) = D \cdot \int_1^{Z^n - Dy} \cdot F(x) \cdot F''(Dy + x)\, dx.$$

Thus for the derivatives we have

$$\left.\begin{aligned} |\Phi_1'(y)| &\leqslant D \cdot Z^n \cdot \ln^{C_n} D, \\ |\Phi_2'(y)| &\leqslant D^2 \cdot y \cdot \ln^{C_n} D, \\ |\Phi_3'(y)| &\leqslant D \cdot Z^n \cdot \ln^{C_n} D. \end{aligned}\right\} \tag{73}$$

We apply partial summation over q to S_1, S_2 and S_3, where we shall sum $\Pi_{p/q}(1 - 1/p)^{2n-1}$; for $i = 1$ and 2 we then obtain

$$S_i = -\int_1^{\frac{Z^n - 1}{2D}} S(y) \cdot \left[\frac{\Phi_i'(y)}{y} - \frac{\Phi_i(y)}{y^2}\right] dy + S\left(\frac{Z^n - 1}{2D}\right) \cdot \frac{2D}{Z^n - 1} \cdot \Phi_i\left(\frac{Z^n - 1}{2D}\right) + O(Z^{2n-1} \cdot \ln^{C_n} D);$$

$$S_3 = -\int_{\frac{Z^n - 1}{2D}}^{\frac{Z^n - 1}{D}} \left[S(y) - S\left(\frac{Z^n - 1}{2D}\right)\right] \cdot \left[\frac{\Phi_3'(y)}{y^2} - \frac{\Phi_3(y)}{y^2}\right] dy$$

$$+ \left[S\left(\frac{Z^n - 1}{D}\right) - S\left(\frac{Z^n - 1}{2D}\right)\right] \cdot \frac{D}{Z^n - 1} \cdot \Phi_3\left(\frac{Z^n - 1}{D}\right) + O(Z^{2n-1} \cdot \ln^{C_n} D).$$

But

$$S(y) = \sum_{q \leqslant y} \prod_{p/q} \left(1 - \frac{1}{p}\right)^{2n-1} = \frac{1}{2\pi i} \int_{2-iy}^{2+iy} \frac{y^s}{s} \cdot \mathfrak{S}(s) \cdot \zeta(s) \cdot ds + O(1) = \mathfrak{S}(1) \cdot y + O(\ln^{C_n} D).$$

Hence for $i = 1$ or 2 we have

$$S_i = -\mathfrak{S}(1)\cdot \int_1^{\frac{Z^n-1}{2D}} \left[\Phi_i'(y) - \frac{\Phi_i(y)}{y}\right] dy$$
$$+\mathfrak{S}(1)\cdot\Phi_i\left(\frac{Z^n-1}{2D}\right) + O\left(Z^{2n-1}\cdot \ln^{C_n} D\right)$$
$$+O\left\{\int_1^{\frac{Z^n-1}{2D}} \left(\frac{|\Phi_i'(y)|}{y} + \frac{|\Phi_i(y)|}{y^2}\right) dy \cdot \ln^{C_n} D\right\};$$

$$S_3 = -\mathfrak{S}(1) \int_{\frac{Z^n-1}{2D}}^{\frac{Z^n-1}{D}} \left[\Phi_3'(y) - \frac{\Phi_3(y)}{y}\right] dy + \mathfrak{S}(1)\cdot\Phi_3\left(\frac{Z^n-1}{D}\right)$$
$$-\mathfrak{S}(1)\cdot\Phi_3\left(\frac{Z^n-1}{2D}\right) + O\left(Z^{2n-1}\cdot\ln^{C_n} D\right)$$
$$+O\left\{\int_{\frac{Z^n-1}{2D}}^{\frac{Z^n-1}{D}} \left(\frac{|\Phi_3'(y)|}{y} + \frac{|\Phi_3(y)|}{y^2}\right) dy\cdot \ln^{C_n} D\right\}.$$

Using estimates (72) and (73), we find that all quantities under the O signs have order of magnitude $D \cdot Z^n \cdot \ln^{C_n} D$. Thus for $i = 1$ or 2 we have

$$S_i = -\mathfrak{S}(1)\cdot \int_1^{\frac{Z^n-1}{2D}} \left[\Phi_i'(y) - \frac{\Phi_i(y)}{y}\right] dy$$
$$+\mathfrak{S}(1)\,\Phi_i\left(\frac{Z^n-1}{2D}\right) + O(D\cdot Z^n\cdot \ln^{C_n} D);$$

$$S_3 = -\mathfrak{S}(1)\cdot \int_{\frac{Z^n-1}{2D}}^{\frac{Z^n-1}{D}} \left[\Phi_3'(y) - \frac{\Phi_3(y)}{y}\right] dy$$
$$+\mathfrak{S}(1)\cdot\left[\Phi_3\left(\frac{Z^n-1}{D}\right) - \Phi_3\left(\frac{Z^n-1}{2D}\right)\right] + O(D\cdot Z^n\cdot\ln^{C_n} D).$$

On removing the derivatives from under the integral sign using the equations

$$\int_B^A \Phi_i'(y)\, dy = \Phi_i(A) - \Phi_i(B)$$

we obtain, for $i = 1$ or 2,

$$S_i = \mathfrak{S}(1) \cdot \int_1^{\frac{Z^n-1}{2D}} \frac{\Phi_i(y)}{y} dy + O(D \cdot Z^n \ln^{C_n} D);$$

$$S_3 = \mathfrak{S}(1) \cdot \int_{\frac{Z^n-1}{2D}}^{\frac{Z^n-1}{D}} \frac{\Phi_3(y)}{y} dy + O(D \cdot Z^n \cdot \ln^{C_n} D).$$

We write $\Phi_2(y)$ in the form

$$\Phi_2(y) =$$

$$\int_{Z^n-2Dy}^{Z^n-Dy} F(x) \cdot F'(Dy + x)\, dx - F(Z^n - 2Dy)[F(Z^n) - F(Z^n - Dy)].$$

We note that

$$\Phi_1(y) = \int_1^{Z^n-2Dy} F(x) \cdot [F'(2Dy + x) - F'(Dy + x)]\, dx$$
$$+ F(Z^n - 2Dy) \cdot [F(Z^n) - F(Z^n - Dy)].$$

Hence, when adding $S_1 + S_2$ terms with opposite signs cancel. We obtain

$$S_1 + S_2 = \mathfrak{S}(1) \cdot \int_1^{\frac{Z^n-1}{2D}} \frac{\Phi_1(y) + \Phi_2(y)}{y} dy + O(D \cdot Z^n \cdot \ln^{C_n} D)$$

$$= \mathfrak{S}(1) \cdot \int_1^{\frac{Z^n-1}{2D}} \frac{1}{y} \int_1^{Z^n-Dy} F(x) \cdot F'(Dy + x)\, dx \cdot dy$$

$$- \mathfrak{S}(1) \cdot \int_1^{\frac{Z^n-1}{2D}} \frac{1}{y} \int_1^{Z^n-2Dy} F(x) \cdot F'(2Dy + x)\, dx \cdot dy + O(D \cdot Z^n \cdot \ln^{C_n} D).$$

Adding S_3 we find that

$$S_1 + S_2 + S_3 = \mathfrak{S}(1) \cdot \int_1^{\frac{Z^n-1}{D}} \frac{1}{y} \int_1^{Z^n - Dy} F(x) \cdot F'(Dy + x)\, dx \cdot dy$$

$$- \mathfrak{S}(1) \cdot \int_1^{\frac{Z^n-1}{2D}} \frac{1}{y} \int_1^{Z^n - D2y} F(x) \cdot F'(2Dy + x) \cdot dx \cdot dy + O(D \cdot Z^n \cdot \ln^{C_n} D).$$

We make the substitution $2y \to y$ in the second double integral; then

$$\int_1^{\frac{Z^n-1}{2D}} \frac{1}{y} \int_1^{Z^n - 2Dy} F(x) \cdot F'(2Dy + x) \cdot dx \cdot dy$$

$$= \int_2^{\frac{Z^n-1}{D}} \frac{1}{y} \int_1^{Z^n - Dy} F(x) \cdot F'(Dy + x)\, dx \cdot dy.$$

But this integral differs from the first in that its lower limit of integration with respect to y equals two, while it is one in the first. Therefore

$$S_1 + S_2 + S_3 = \mathfrak{S}(1) \cdot \int_1^{2} \frac{1}{y} \int_1^{Z^n - Dy} F(x) \cdot F'(Dy + x)\, dx \cdot dy$$

$$+ O(D \cdot Z^n \cdot \ln^{C_n} D).$$

From the estimate

$$| F'(Dy + x) - F'(x) | < \frac{Dy}{x} \cdot \ln^{C_n} D$$

we find

$$F'(Dy + x) = F'(x) + O\left(\frac{D}{x} \cdot \ln^{C_n} D\right) \quad (y \leqslant 2),$$

and since $|F(x)/x| \leq \ln^{C_n} D$ it follows that

$$S_1 + S_2 + S_3$$

$$= \mathfrak{S}(1) \cdot \int_1^{2} \frac{1}{y} \int_1^{Z^n - Dy} F(x) \cdot F'(x) \cdot dx \cdot dy + O(D \cdot Z^n \cdot \ln^{C_n} D).$$

In addition, we note that

$$\int_{Z^n-Dy}^{Z^n} F(x)\cdot F'(x)\cdot dx \leqslant D\cdot Z^n\cdot \ln^{C_n} D \qquad (y \leqslant 2).$$

Hence we finally have

$$S_1+S_2+S_3=\mathfrak{S}(1)\cdot\ln 2\cdot\int_1^{Z^n} F(x)\cdot F'(x)\,dx + O(D\cdot Z^n\cdot \ln^{C_n} D).$$

We consequently obtain the following expression for the sum (62), and along with it for our basic sum in the left side of (61):

$$\sum_{d=D}^{2D}\ \sum_{\substack{a_n-b_n=d\cdot u\\ (a_n\cdot b_n,\, d)=1}} 1 = 2\cdot\mathfrak{S}(1)\cdot\ln 2\cdot\int_1^{Z^n}\cdot F(x)\cdot F'(x)\cdot dx + O(D\cdot Z^n\cdot\ln^{C_n} D). \tag{74}$$

We take up the integral in the right side of (74). Formula (68) gives

$$F(x)\cdot F'(x)=\frac{Z^{2n}}{(2\pi i)^2}\int_{L_1}\int_{L_2}\frac{Z^{-n(s_1+s_2)}\cdot x^{s_1+s_2-1}}{s_1(1-s_1)^n\cdot(1-s_2)^n}\,ds_1\cdot ds_2.$$

Both contours go along the line $\operatorname{Re} s = 1 - 1/\ln D$. Since $n \geq 4$ the integration under the double integral sign is possible, and so

$$\int_1^{Z^n} F(x)\cdot F'(x)\,dx=\frac{Z^{2n}}{(2\pi i)^2}\cdot\int_{L_1}\int_{L_2}\frac{ds_1\cdot ds_2}{s_1\cdot(1-s_1)^n\cdot(1-s_2)^n\cdot(s_1+s_2)}+O(\ln^{C_n} D).$$

We examine the double integral obtained on the right. We move the contour for s_1 to the left to infinity; we then pass two poles, at the points $s_1 = 0$ and $s_1 = -s_2$ with residues

$$\frac{1}{2\pi i}\int_{L_2}\frac{ds_2}{s_2(1-s_2)^n}-\frac{1}{2\pi i}\int_{L_2}\frac{ds_2}{s_2\cdot(1+s_2)^n\cdot(1-s_2)^n}\,.$$

But the first integral is easily evaluated by moving the contour to the left to infinity:

$$\frac{1}{2\pi i}\int_{L_2}\frac{ds_2}{s_2(1-s_2)^n}=1.$$

By moving the contour in the second integral to the right to infinity and evaluating its residue at the point $s = 1$, we reduce it to the numerical quantity

$$\frac{1}{2\pi i}\int\limits_{L_2}\frac{ds_2}{s_2(1+s_2)^n\cdot(1-s_2)^n}=\frac{1}{2^n}\sum_{m=0}^{n-1}\frac{C_{n+m-1}^m}{2^m},$$

where

$$C_{n+m-1}^m=\frac{n\cdot(n+1)\cdots(n+m-1)}{m!}.$$

In this way, we finally have

$$\int\limits_1^{Z^n}F(x)\cdot F'(x)\cdot dx=Z^{2n}\cdot\left(1-\frac{1}{2^n}\sum_{m=0}^{n-1}\frac{C_{n+m-1}^m}{2^m}\right)+O(\ln^{C_n}D).$$

It remains to show that

$$S_n=\sum_{m=0}^{n-1}\frac{C_{n+m-1}^m}{2^m}=2^{n-1}.$$

To do this, we consider the polynomial

$$f(x)=\left(\frac{1}{2}+x\right)^{n-1}+\left(\frac{1}{2}+x\right)^n+\ldots+\left(\frac{1}{2}+x\right)^{2n-2}=\sum_{\nu=0}^{2n-2}c_\nu\cdot x^\nu.$$

We note that the $(n-1)$th coefficient is actually the sum we need:

$$c_{n-1}=\sum_{m=0}^{n-1}\frac{C_{n+m-1}^m}{2^m}=S_n.$$

But, on the other hand, the polynomial $f(x)$ can be transformed as a geometric progression:

$$f(x)=\left(\frac{1}{2}+x\right)^{n-1}\cdot\frac{\left(\frac{1}{2}+x\right)^n-1}{x-\frac{1}{2}}.$$

Multiplying both sides by $x - \frac{1}{2}$, we obtain

$$f(x)\cdot\left(x-\frac{1}{2}\right)=\left(\frac{1}{2}+x\right)^{2n-1}-\left(\frac{1}{2}+x\right)^{n-1}. \qquad (75)$$

But if we write

$$f(x)\cdot\left(x-\frac{1}{2}\right)=\sum_{\nu=0}^{2n-1} a_\nu\cdot x^\nu,$$

then, by equating coefficients, we find that $a_n = c_{n-1} - c_n{}^{1/2}$. We note that

$$c_n = S_{n+1} - \frac{C_{2n-1}^{n-1}}{2^{n-1}} - \frac{C_{2n}^{n}}{2^n},$$

where S_{n+1} is the sum we need for $(n+1)$. Consequently

$$a_n = S_n - \frac{1}{2}S_{n+1} + \frac{1}{2}\left(\frac{C_{2n-1}^{n-1}}{2^{n-1}} + \frac{C_{2n}^{n}}{2^n}\right).$$

On the other hand, the coefficient of the nth power in the right side of equation (75) is $C_{2n-1}^{n-1}/2^{n-1}$. Equating coefficients of x^n, we obtain

$$\frac{C_{2n-1}^{n-1}}{2^{n-1}}.$$

This means that $S_{n+1} = 2\cdot S_n$. But $S_1 = 2^0$, and hence $S_n = 2^{n-1}$.

This gives us the equation for our integral:

$$S_n - \frac{1}{2}S_{n+1} = \frac{C_{2n-1}^{n-1}}{2^n} - \frac{C_{2n}^{n}}{2^{n+1}} = 0$$

from which it follows that our sum actually has the asymptotics we need:

$$\int_1^{Z^n} F(x)\cdot F'(x)\,dx = \frac{1}{2}Z^{2n} + O(\ln^{C_n} D).$$

This finishes the proof of estimate (1), and of Theorem 1 as well.

We conclude by noting that to obtain a new proof of Ju. V. Linnik's theorem on the solubility of the equation $n = p + x^2 + y^2$ (a problem of Hardy and Littlewood [5]) along the lines of C. Hooley [9], we must have in Theorem 1 a quantity ϵ tending to zero, as D grows, at a sufficiently fast rate:

$$\varepsilon \leqslant \frac{1}{(\ln D)^{1-\delta_0+\eta}}, \qquad (76)$$

where δ_0 is the lowering power of the logarithm which appears in Hooley's paper; $\eta > 0$ is an arbitrarily small, but fixed quantity.

A detailed analysis of our proof shows that $\epsilon \leq 1/(\ln D)^{1/2-\eta}$ (since ϵ^2 is involved), where $\eta > 0$ is an arbitrarily small fixed number. Whether it is possible to get a quantity ϵ of order (76) by exploiting all the possibilities of this method is still unclear.

Note. For the case $n = 2$, instead of estimate (1) we can obtain the stronger estimate

$$\sum_{\chi_d \neq \chi_0} \left| \sum_{m \leqslant Z} \chi_1(m) \right|^4 \leqslant d \cdot Z^2 \cdot \ln^{C_n} D, \tag{77}$$

from which (1) follows trivially by summing over d (as before $D \leq d \leq 2D$ and $\sqrt{d} \leq Z \leq d$).

Estimate (77) is easily obtained from the "inversion" principle for the sum of Dirichlet characters through the point $\sqrt{d}$.

In fact, using I. M. Vinogradov's "cups" [10], we obtain the equations

$$\sum_{m \leqslant Z} \chi(m) = \varepsilon(\chi) \cdot \sum_{m=1}^{\infty} \frac{a_m}{m} \chi(m) + \sum_{m=1}^{\sqrt{d}} c_m \cdot \chi(-m) + \sum_{m=Z}^{Z+\sqrt{d}} c'_m \cdot \chi(m), \tag{78}$$

where $\epsilon(\chi)$ is a gaussian sum, and a_m, c_m and c'_m are coefficients determined in detail in [10]. We split the infinite series into three sums:

$$\sum_{m=1}^{\frac{d}{Z}} + \sum_{m=\frac{d}{Z}}^{\sqrt{d}} + \sum_{m \geqslant \sqrt{d}}^{\infty}$$

and designate them by

$$S_\chi\left(\frac{d}{Z}\right), \quad S_\chi\left(\frac{d}{Z}, \sqrt{d}\right), \quad S_\alpha(\sqrt{d}),$$

respectively. We use the inequality $|a_m|/m \leq c_0 Z/d$ to estimate the sum $S_\chi(d/Z)$; we then obtain

$$\sum_{\chi} \left| S_\chi\left(\frac{d}{Z}\right) \right|^4 \leqslant \frac{Z^2}{d} \cdot \ln^{C_0} d.$$

To estimate the sums $S_\chi(d/Z, \sqrt{d})$ and $S_\chi(\sqrt{d})$ we use the inequality

$$|a_m| \leqslant \left(\frac{\sin \Delta m}{\Delta m}\right)^r,$$

where $\Delta = d^{-1/2}$ and $r \geq 2$. We find that

$$\sum_{\chi} \left| S_{\chi}\left(\frac{d}{Z}, \sqrt{d}\right) \right|^4 \leqslant \frac{Z^2}{d} \cdot \ln^{C_0} d,$$

$$\sum_{\chi} | S_{\chi}(\sqrt{d}) |^4 \leqslant \ln^{C_0} d.$$

Moreover, $|\epsilon(\chi)|^4 = d^2$, so that

$$\sum_{\chi} \left| \varepsilon(\chi) \cdot \sum_{m=1}^{\infty} \frac{a_m}{m} \chi(m) \right|^4 \leqslant d \cdot Z^2 \cdot \ln^{C_0} d.$$

The fourth moments of the two sums remaining in the right side of (78) are estimated trivially. Their order also does not exceed $d \cdot Z^2 \cdot \ln^{C_0} d$. Consequently, inequality (77) is truly valid.

CORRECTION [1]

Some conditions on the relative primeness of the indices of summation were omitted in the proof of equation (32). To restore accuracy, the sentence just before (36) must be deleted. The divisibility arithmetic of the number a_n and b_n by the number ν has another form and is given by the following rule.

Let $a_n = m_1 \cdot m_2 \cdots m_n \equiv 0 \pmod{\nu}$. We define ν_1 by the equation $(m_1, \nu) = \nu_1$, from which it follows that $(m_1/\nu_1, \nu/\nu_1) = 1$. If we are already given the numbers $\nu_1, \cdots, \nu_{i-1}$ in the decomposition $\nu = \nu_1 \cdots \nu_n$, then the number ν_i is defined by the equation $(m_i, \nu/\nu_1 \cdots \nu_{i-1}) = \nu_i$, from which it follows that

$$\left(\frac{m_i}{\nu_i}, \frac{\nu}{\nu_1 \cdot \nu_2 \ldots \nu_i}\right) = 1, \quad 1 \leqslant i \leqslant n. \tag{a}$$

The divisibility arithmetic of the number b_n by ν is obtained by this same rule.

In this connection, we must impose on the numbers a_ν and b_ν the additional conditions (a) of relative primeness which the components m_i/ν_i (m_i'/ν_i') in the decomposition

$$a_\nu = \frac{m_1}{\nu_1} \cdot \frac{m_2}{\nu_2} \cdots \frac{m_n}{\nu_n} \quad \left(b_\nu = \frac{m_1'}{\nu_1'} \cdot \frac{m_2'}{\nu_2'} \cdots \frac{m_n'}{\nu_n'}\right)$$

must obey. These conditions must be taken into account in several places,

---

1) *Editor's note*: Translated from Izv. Akad. Nauk SSSR 30 (1966), 719–720.

for example, in the decomposition $r = r_1 \cdot r_2$ in (43) and (44), and in the decomposition $\pi = \pi_1 \cdot \pi_2 \cdot \pi_3$ in (43a) and (44a). We note that in these cases conditions (a) are imposed on the numbers q and q_1 in the last analysis so that they run through the entire intervals. This makes it possible not to introduce yet another discontinuous factor.

Conditions (a) must also be taken into account in the derivation of equation (40) in §5.

BIBLIOGRAPHY

[1] M. B. Barban, *The "density" of the zeroes of Dirichlet L-series and the problem of the primes and "near primes"*, Mat. Sb. **61** (103) (1963), 418–425. (Russian) MR **30** #1992.

[2] A. I. Vinogradov, *On the density hypothesis and the quasi-Riemann hypothesis*, Dokl. Akad. Nauk SSSR **158** (1964), 1014–1017 = Soviet Math. Dokl. 5 (1964), 1339–1342. MR 30 #68.

[3] Y. Wang, *On the representation of large integer as a sum of a prime and an almost prime*, Acta. Math. Sinica **10** (1960), 168–181 = Chinese Math. 1 (1962), 181–195. MR 27 #5733.

[4] B. V. Levin, *Distribution of "near primes" in polynomial sequences*, Mat. Sb. **61** (103) (1963), 389–407. (Russian) MR 30 #1991.

[5] Ju. V. Linnik, *An asymptotic formula in an additive problem of Hardy and Littlewood*, Izv. Akad. Nauk SSSR Ser. Mat. 24 (1960), 629–706; English transl., Amer. Math. Soc. Transl. (2) 46 (1965), 65–148. MR 23 #A130.

[6] ———, *Divisor problems and related binary additive problems*, Proc. Internat. Math. Congress (Edinburg, 1958), Cambridge Univ. Press, New York, 1960, pp. 313–321. (Russian) MR **25** #3012.

[7] ——— *The dispersion method in binary additive problems*, Izdat. Leningrad. Univ., Leningrad, 1961, Chapter VIII; English transl., Transl. Math. Monographs, vol. 4, Amer. Math. Soc., Providence, R. I., 1963. MR 25 #3920.

[8] ———, *All large numbers are sums of a prime and two squares (A problem of Hardy and Littlewood)*. II, Mat. Sb. **53** (95) (1961), 3–38; English transl., Amer. Math. Soc. Transl. (2) **37** (1964), 197–240. MR 22 #10964.

[9] C. Hooley, *On the representation of a number as the sum of two squares and a prime*, Acta. Math. 97 (1957), 189–210. MR 19, 532.

[10] I. M. Vinogradov, *Selected works*, Izdat. Akad. Nauk SSSR, Moscow, 1952. (Russian) MR 14, 610.

[11] E. C. Titchmarsch, *A divisor problem*, Rend. Circ. Mat. Palermo 54 (1930), 414–429.

Translated by:
N. Koblitz

A FUNCTIONAL EQUATION FOR DIRICHLET L-SERIES AND THE PROBLEM OF DIVISORS IN ARITHMETIC PROGRESSIONS

UDC 511

A. F. LAVRIK

A new form of functional equation for Dirichlet L-series is introduced and the problem of divisors in intervals of an arithmetic progression is examined for growing differences.

§1. Introduction

In this paper we continue the investigations of Ju. V. Linnik [1], [2] in the direction of a shortened functional equation for Dirichlet L-series and the problem of divisors in intervals of arithmetic progressions of growing difference.

A new form of functional equation for $L(s, \chi_d)$ is established here. Approximate functional equations with uniform estimates in d and $|s|$ are derived from it for the critical strip. Then, based on this and using D. A. Burgess' latest estimate [3] for the sum of characters, we examine the question of asymptotics for the sums

$$\sum_{dn+l\leqslant x} \tau_k(dn+l), \tag{1}$$

where $\tau_k(m)$ is the number of decompositions of m into k positive factors.

We note that the method applied to get the fundamental equation (2) leads to analogous functional equations for powers of Dirichlet L-functions, Dedekind zeta-functions, L-series of algebraic number fields and, in general, of functions representable by Dirichlet series and satisfying functional equations of the Riemann type. The value of this type of equation is that it leads to approximate equations with uniform estimates in the basic arguments.

§2. Statements of results

Theorem 1. *If χ is a primitive character of the modulus $d > 1$, $z \neq 0$, and $|\arg z| \leq \pi/2$, then the following equation holds in the whole s-plane:*

$$\begin{aligned}\Gamma\left(\frac{s+a}{2}\right)L(s,\chi) &= \sum_{n=1}^{\infty}\frac{\chi(n)}{n^s}\Gamma\left(\frac{s+a}{2}, \frac{\pi n^2 z}{d}\right)\\ &+ \varepsilon(\chi)\left(\frac{d}{\pi}\right)^{\frac{1}{2}-s}\sum_{n=1}^{\infty}\frac{\bar{\chi}(n)}{n^{1-s}}\Gamma\left(\frac{1-s+a}{2}, \frac{\pi n^2}{dz}\right),\end{aligned} \tag{2}$$

where $\epsilon(\chi)$ *has the same meaning as in the equation for* $L(s, \chi)$, $a = \frac{1}{2}[1 - \chi(-1)]$, $\Gamma(\alpha)$ *is the gamma-function, and*

$$\Gamma(\alpha, u) = \int_u^{\infty \arg u} e^{-\xi} \xi^{\alpha-1} \, d\xi.$$

The information about $L(s, \chi)$ contained in its well-known functional equation (see [4]) is contained in (2), since, by the symmetry of the right side, this equation is a trivial consequence of (2). Moreover, by its very essence equation (2) is shortened. Namely, if we take, for example,

$$z = \exp i \operatorname{sgn} \operatorname{Im} s \left(\frac{\pi}{2} - \frac{1}{|\operatorname{Im} s| + 1}\right),$$

then the left side of (2) is determined with great precision by the intervals of the series of the right side with $n \leq N$, where

$$N = \sqrt{d(|t| + 1)} \ln d(|t| + 1)$$

and here we take the factor quenching the growth of the function $\Gamma^{-1}((s + a)/2)$ as $|t| \longrightarrow \infty$ out of each summand.

More precisely, we have the following result.

Theorem 2. *If* χ *is a primitive character of the modulus* $d > 1$, $s = \sigma + it$, $N = \sqrt{d(t + 1)}$, $0 < \sigma < 1$, *and* t *is a real number, then*

$$L(s, \chi) = \sum_{n \leqslant N \ln N} \frac{\chi(n)}{n^s} C_n(s, d, a)$$

$$+ \varepsilon(\chi) \left(\frac{d}{\pi}\right)^{\frac{1}{2} - s} \frac{\Gamma\left(\frac{1 - s + a}{2}\right)}{\Gamma\left(\frac{s + a}{2}\right)} \sum_{n \leqslant N \ln N} \frac{\bar{\chi}(n)}{n^{1-s}} C_n(1 - s, d, a) + O(N^{-b}),$$

where $C_n(\omega, d, a)$ *is a quantity bounded in all the arguments by an absolute constant and* b *is any positive constant.*

Remark. The exact value of $C_n(\omega, d, a)$ is determined by the formula

$$C_n(\omega, d, a) = \frac{\Gamma\left(\frac{\omega + a}{2}, \frac{\pi n^2}{d} z_\omega\right)}{\Gamma\left(\frac{\omega + a}{2}\right)},$$

$$z_\omega = \begin{cases} z, & \text{if} \quad \omega = s, \\ \bar{z}, & \text{if} \quad \omega = 1 - s, \end{cases}$$

where z has the value defined above and $\bar{z}$ is its conjugate.

From Theorem 1 follows

Theorem 3. *Let m be an integer, $N = d(|t| + 1)$, and*

$$V_m = \sqrt{\frac{d|t|}{2\pi}} + m\sqrt{\frac{d}{2\pi}}, \quad T = \sqrt{|t|+1}\ln N.$$

Then for a primitive character χ of the modulus $d > 1$ and for $s = \sigma + it$ $(0 < \sigma < 1$, t real) we have

$$L(s, \chi) = A \sum_{|m| \leqslant T} \left[\max_{V \leqslant V_m} \left| \sum_{\substack{V_{m-1} < n \leqslant V \\ n > 0}} \frac{\chi(n)}{n^s} \right| \right.$$
$$\left. + BN^{\frac{1}{2}-\sigma} \max_{u \leqslant V_m} \left| \sum_{\substack{V_{m-1} < n \leqslant u \\ n > 0}} \frac{\bar{\chi}(n)}{n^{1-s}} \right| \right] + O(N^{-b}),$$

where A and B are quantities bounded by absolute constants and b is any positive constant.

This expression has an especially simple form on the line $\sigma = 1/2$. Namely, for $s = 1/2 + it$,

$$L(s, \chi) = A \sum_{|m| \leqslant T} \max_{V \leqslant V_m} \left| \sum_{\substack{V_{m-1} < n < V \\ n > 0}} \frac{\chi(n)}{n^s} \right| + O(N^{-b}).$$

Thus, for $|t| \geq 1$, instead of the entire sum over $n \leq \sqrt{d|t|}\ \ln d|t|$, we have here $T = \sqrt{|t|}\ \ln d|t|$ nonoverlapping intervals, each of length $< \sqrt{d}$.

Concerning the sums (1) we have

Theorem 4. *For integral $k \geq 4$, $4 \leq 2m \leq k$, l and d mutually prime uniformly in d and $0 < l \leq d$, we have*

$$\sum_{\substack{n \leqslant x \\ n \equiv l \pmod d}} \tau_k(n) = \frac{x}{\varphi(d)} \sum_{\delta | d^k} \sum_{\delta = \delta_1 \dots \delta_k} \frac{\mu(\delta_1) \dots \mu(\delta_k)}{\delta} P_k\left(\ln \frac{x}{\delta}\right) + R, \quad (3)$$

where

$$R \ll \frac{1}{\varphi(d)} x^{1 - \frac{1}{2^\nu m}} d^{\frac{3k+2m}{2^{\nu+3} m} + \varepsilon_0} \ln^b x,$$

ϕ is Euler's function, μ is the Möbius function, P_k is a polynomial of degree $k-1$ defined by the equation

$$P_k\left(\ln \frac{x}{\delta}\right) = \frac{\delta}{(k-1)!\, x} \lim_{s \to 1} \frac{\partial^{k-1}}{\partial s^{k-1}} \left\{ \frac{\zeta^k(s)(s-1)^k}{s} \left(\frac{x}{\delta}\right)^s \right\},$$

ζ is the Riemann zeta-function, $\nu \geq (2k - 3m)/m$ is an integer, $\epsilon_0 = 0$ if $2m = k$ and ϵ_0 equals any arbitrarily small $\epsilon > 0$ if $2m < k$, and b is a constant depending only on k.

As the main growth term is the quantity of order $x \ln^{k-1} x/\phi(d)$, (3) gives a bound for the asymptotics of the sums (1) in the form

$$d \leqslant \frac{x^\gamma}{\ln^b x}, \quad \gamma = \frac{8}{3k + 2m} - \varepsilon_0 \quad (4 \leqslant 2m \leqslant k)$$

and a corresponding lowering of $x^{-\beta}$ in the remainder term, with

$$\beta = \frac{1}{2^\nu m}\left[1 - \gamma\left(\frac{3k + 2m}{8} + \varepsilon_0\right)\right], \quad \nu \geqslant \frac{2k - 3m}{m}.$$

It is hence evident that the best bound for d is reached for $m = 2$ with $\nu = 8/(3k + 4) - \epsilon$, and here the lowering of the remainder as k grows decreases exponentially ($\nu = k - 3$). In the other extreme case, when $2m = k$, we have a bound for d with $\nu = 2/k$, but the lowering of the remainder as k grows decreases as $x^{-\beta}$, $\beta = 1/k - \nu/2$. In some applications, the latter gives more than the former, which is why the parameter is introduced.

The problem of asymptotics for the sums (1) was examined earlier in papers [5], [6], where less precise results were obtained.

The case of primitive arithmetic progressions we examined in Theorem 4 is fundamental, since every other case can be reduced to it using the following theorem.

Theorem 5. *Let $(l, d) > 1$. Set*

$$(l, d) = d_1 d_2, \quad (d_1, d_2) = 1, \quad \left(d_1, \frac{d}{(l, d)}\right) = 1,$$

where (n, m) is the greatest common divisor of n and m and for $d_1 > 1$ every prime dividing d_1 also divides $d/(l, d)$, while $d_2 = p_1^{\nu_1} \cdots p_\lambda^{\nu_\lambda}$ for $d_2 > 1$, where p_j are distinct prime numbers and $\nu_j \geq 1$. Then, regardless of the values of $d_1 \geq 1$ and $d_2 \geq 1$, we have

$$\sum_{\substack{n \leqslant x \\ n \equiv l \pmod{d}}} \tau_k(n) = \tau^{k-1}(d_1) \sum_{\substack{n \leqslant x/d_1 \\ n \equiv \frac{l}{d_1} \left(\mathrm{mod}\ \frac{d}{d_1}\right)}} \tau_k(n),$$

and if $d_1 \geq 1$ but $d_2 > 1$, then

$$\sum_{\substack{n \leqslant x \\ n \equiv l \pmod d}} \tau_k(n) = \tau^{k-1}(d_1) \sum_{s_1=\nu_1} \cdots \sum_{s_\lambda=\nu_\lambda} \sum_{t_1=0}^{k} \cdots \sum_{t_\lambda=0}^{k} (-1)^{t_1+\cdots+t_\lambda}$$

$$\times \prod_{i=1}^{\lambda} C_k^{t_i} (1+s_i)^{k-1} \sum_{\substack{n \leqslant x/d'd_1 \\ n \equiv l' \left(\mathrm{mod}\ \frac{d}{(l,d)}\right)}} \tau_k(n),$$

where $\tau(d_1)$ *is the number of divisors of* d_1; *the summing over* s_i *proceeds until the first term equal to zero; and*

$$d' = \prod_{i=1}^{\lambda} p_i^{s_i+t_i}, \quad l' = \frac{l}{d_1} \prod_{i=1}^{\lambda} p_i^{-s_i-t_i},$$

where $p_\nu^{-s_\nu-t_\nu}$ *designates the number inverse to* $p_\nu^{s_\nu+t_\nu}$ *modulo*

$$d\left(d_1 \prod_{i=1}^{\nu} p_i^{\nu_i}\right)^{-1}, \quad C_k^t = \frac{k!}{(k-t)!\,t!}.$$

We note that since the numbers l/d_1, d/d_1 and l', $d/(l, d)$ will be relatively prime for $d_2 = 1$ and for $d_2 > 1$, respectively, it follows that the problem of asymptotics for the sums (1) in the general case reduces to that for progressions with $(l, d) = 1$.

§3. Derivation of the functional equation

By the principle of analytic continuation, it suffices to prove Theorem 1 only for the critical strip.

We set

$$\xi(\omega, \chi) = \left(\frac{d}{\pi}\right)^{\frac{\omega}{2}} \Gamma\left(\frac{\omega+a}{2}\right) L(\omega, \chi), \tag{4}$$

$$F_\eta(\omega, \chi) = \eta^{\frac{1}{2}-\omega} \xi(\omega, \chi), \tag{5}$$

where $\eta \neq 0$ and $|\arg \eta| \leq \pi/2$.

For $s = \sigma + it$, where $0 < \sigma < 1$ and t is any real number, we let

$$I = \frac{1}{2\pi i} \int_{2-i\infty}^{2+i\infty} F_\eta(\omega, \chi) \frac{d\omega}{s-\omega}. \tag{6}$$

We take the integral of the function $F_\eta(\omega, \chi)(s-\omega)^{-1}$ over the rectangle with vertices at the points

$$A_1 = 2 - iT, \quad A_2 = 2 + iT, \quad A_3 = \alpha + iT, \quad A_4 = \alpha - iT,$$

where $0 < \alpha < \sigma$ and $T > 0$ is chosen so that the point s is inside the contour.

If χ is a primitive character of modulus $d > 1$, then $F_\eta(\omega, \chi)$ is an entire function, and the residue theorem gives

$$-F_\eta(s, \chi) = \frac{1}{2\pi i}\left\{\int_{A_1A_2} + \int_{A_2A_3} + \int_{A_3A_4} + \int_{A_4A_1}\right\} F_\eta(\omega, \chi)\frac{d\omega}{s-\omega}.$$

As $|\operatorname{Im}\omega|$ grows, $|\Gamma(\omega)|$ decreases exponentially, while $|L(\omega, \chi)|$ grows no faster than a power, so that the integrals over the horizontal segments of the contour tend to zero as $T \longrightarrow \infty$, and consequently

$$I = -F_\eta(s, \chi) + \frac{1}{2\pi i}\int_{\alpha - i\infty}^{\alpha + i\infty} F_\eta(\omega, \chi)\frac{d\omega}{s-\omega}. \tag{7}$$

We designate the integral in (7) by I_1. The transformation $z = s - \omega$ takes it from the line $\operatorname{Re}\omega = \alpha$ in the ω-plane to the line $\operatorname{Re} z = \sigma - \alpha = \beta$ in the z-plane, so that

$$I_1 = \frac{1}{2\pi i}\int_{\beta - i\infty}^{\beta + i\infty} F_\eta(s - z, \chi)\frac{dz}{z}.$$

We apply the functional equation of $L(\omega, \chi)$ (see [4]) in the form

$$\xi(\omega, \chi) = \varepsilon(\chi)\xi(1 - \omega, \bar{\chi}).$$

By (4) and (5) we then have

$$I_1 = \frac{\varepsilon(\chi)}{2\pi i}\int_{\beta - i\infty}^{\beta + i\infty} \eta^{\frac{1}{2}+z-s}\left(\frac{d}{\pi}\right)^{\frac{1-s+z}{2}}\Gamma\left(\frac{1-s+z+a}{2}\right)\sum_{n=1}^{\infty}\frac{\bar{\chi}(n)}{n^{1-s+z}}\frac{dz}{z}. \tag{8}$$

To further transform (8), for $0 < \gamma \le 2 - \operatorname{Re}\rho$, $x \ne 0$, $|\arg x| \le \pi/4$, $\rho = s$ and $\rho = 1 - s$ we examine the function

$$P(\rho, x, \gamma) = \frac{1}{2\pi i}\int_{\gamma - i\infty}^{\gamma + i\infty} \Gamma\left(\frac{z+\rho+a}{2}\right)\frac{x^{-z}}{z}\,dz$$

$$= \frac{1}{2\pi i}\int_{\frac{1}{2}(\gamma + \operatorname{Re}\rho + a) - i\infty}^{\frac{1}{2}(\gamma + \operatorname{Re}\rho + a) + i\infty} \Gamma(\omega)\, x^{-2\omega+\rho+a}\frac{2d\omega}{2\omega - \rho - a}. \tag{9}$$

We perform the integration along the rectangle with vertices

$$\frac{1}{2}(\gamma+\operatorname{Re}\rho+a)-iT, \quad \frac{1}{2}(\gamma+\operatorname{Re}\rho+a)+iT,$$
$$-m-\frac{1}{2}+iT, \quad -m-\frac{1}{2}-iT,$$

where m is a positive integer and

$$T>\max(|\operatorname{Im}\rho|, 2|x|^2).$$

We let $P_{m,T}(\rho, x, \gamma)$ designate this integral. By the residue theorem,

$$P_{m,T}(\rho, x, \gamma)=\Gamma\left(\frac{\rho+a}{2}\right)-\sum_{\nu=0}^{m}\operatorname*{res}_{\omega=-\nu}\left\{x^{-2\omega+\rho+a}\frac{2\Gamma(\omega)}{2\omega-\rho-a}\right\}$$
$$=\Gamma\left(\frac{\rho+a}{2}\right)-2\sum_{\nu=0}^{m}\frac{(-1)^\nu}{\nu!}\frac{x^{2\nu+\rho+a}}{2\nu+\rho+a}. \tag{10}$$

We estimate the integral entering into $P_{m,T}(\rho, x, \gamma)$ along the horizontal lines of the contour. Here the integrand is a quantity of order no greater than

$$C_m\exp\left[-\frac{\pi T}{2}-\operatorname{Im}(-2\omega+\rho)\arg x\right]T^{\delta-\frac{3}{2}}|x|^{-2\delta+\operatorname{Re}\omega+a},$$

where C_m is a quantity depending only on m, and

$$-m-\frac{1}{2}\leqslant\delta\leqslant\frac{1}{2}(\gamma+\operatorname{Re}\rho+a).$$

Consequently, the integral along the upper segment of the contour can be estimated as

$$\ll C_m\exp\left[-\frac{\pi T}{2}+\operatorname{Im}(2\omega-\rho)\arg x\right]\frac{|x|^{\operatorname{Re}\omega+a}}{\ln(T/|x|^2)}T^{-\frac{3}{2}}\left(\frac{T}{|x|}\right)^{\frac{1}{2}(\gamma+\operatorname{Re}\rho+a)},$$

and if $|\arg x|\leq\pi/4$ and $0<\gamma\leq 2-\operatorname{Re}\rho$, then this integral tends to zero as $T\longrightarrow\infty$ for every fixed m. Understandably, the same holds for the integral along the lower segment of the contour.

It remains to examine the integral along the line $\operatorname{Re}\omega=-m-1/2$.

We have (m is an integer ≥ 1)

$$\left|\Gamma\left(-m-\frac{1}{2}+iu\right)\right|\ll e^{-\frac{\pi|u|}{2}}\frac{1}{(m-1)!\,(|u|^2+0,01)}.$$

On the line in question the integrand is estimated as

$$\ll \exp\left[-2\varphi u - \frac{\pi |u|}{2} + \varphi \operatorname{Im} \rho\right] \frac{|x|^{2m+1+\operatorname{Re}\rho+a}}{m!\,(|u|^2+0,01)}.$$

Therefore, if $|\phi| = |\arg x| \leq \pi/4$, then the integral along the line $\operatorname{Re}\omega = -m - 1/2$ tends to zero as $m \longrightarrow \infty$.

As a result we obtain

$$\lim_{m\to\infty} \lim_{T\to\infty} P_{m,T}(\rho, x, \gamma) = P(\rho, x, \gamma).$$

By (9) and (10) it follows that

$$P(\rho, x, \gamma) = \Gamma\left(\frac{\rho+a}{2}\right) - 2\sum_{\nu=0}^{\infty} \frac{(-1)^\nu}{\nu!} \frac{x^{2\nu+\rho+a}}{2\nu+\rho+a}.$$

This latter expression is the familiar decomposition of the incomplete gamma-function.

Thus, if

$$0 < \operatorname{Re}\rho < 1, \quad x \neq 0, \quad |\arg x| \leqslant \frac{\pi}{4}, \quad 0 < \gamma \leqslant 2 - \operatorname{Re}\rho,$$

then

$$P(\rho, x, \gamma) = \Gamma\left(\frac{\rho+a}{2}, x^2\right). \tag{11}$$

We return to equation (8). Using equation (11) with

$$\gamma = \sigma - \alpha, \quad \rho = 1 - s, \quad x^2 = \frac{\pi n^2}{d\eta}, \quad \eta \neq 0, \quad |\arg \eta| \leqslant \frac{\pi}{2},$$

under the condition that we can justify changing the order of summation and integration, we obtain

$$I_1 = \varepsilon(\chi)\left(\frac{d}{\pi}\right)^{\frac{1-s+a}{2}} \eta^{\frac{1}{2}-s} \sum_{n=1}^{\infty} \frac{\bar{\chi}(n)}{n^{1-s}} \Gamma\left(\frac{1-s+a}{2}, \frac{\pi n^2}{d\eta}\right). \tag{12}$$

It is immediately evident from the definition that $\Gamma(\alpha, u)$ is bounded for $0 < \operatorname{Re}\alpha < 1$. Moreover, we know that if $|\arg u| < 3\pi/2$, then

$$\Gamma(\alpha, u) \sim e^{-u} u^{\alpha-1}$$

as $|u| \longrightarrow \infty$. Therefore the series in the right side of (12) converges (absolutely if $\operatorname{Re}\eta > 0$), and hence our reversing the order of operations is legitimate.

We again examine the intergal I from (6). Using equation (11) with

$$\gamma = 2 - \sigma, \quad \rho = s, \quad x^2 = \frac{\pi n^2}{d}\eta, \quad \eta \neq 0, \quad |\arg \eta| \leqslant \frac{\pi}{2},$$

along with (4) and (5), as before we find that

$$I = -\frac{1}{2\pi i}\int_{2-\sigma-i\infty}^{2-\sigma+i\infty} F_\eta(s+z, \chi)\frac{dz}{z}$$

$$= -\left(\frac{d}{\pi}\right)^{\frac{s+a}{2}} \eta^{\frac{1}{2}-s} \sum_{n=1}^{\infty} \frac{\chi(n)}{n^s} \Gamma\left(\frac{s+a}{2}, \frac{\pi n^2}{d}\eta\right). \tag{13}$$

Uniting relations (4), (7), (12) and (13) completes the derivation of Theorem 1 for $0 < \operatorname{Re} s < 1$ and, by the principle of analytic continuation, for the whole s-plane as well.

Equation (2) can be differentiated with respect to z. Because

$$\frac{\partial}{\partial x}\Gamma(\alpha, x) = -e^{-x}x^{\alpha-1},$$

we have

$$z^{\frac{1}{2}+a}\sum_{n=1}^{\infty}\chi(n)\,n^a e^{-\frac{\pi n^2 z}{a}} = \varepsilon(\chi)\sum_{n=1}^{\infty}\bar{\chi}(n)\,n^a e^{-\frac{\pi n^2}{dz}}.$$

This is E. Landau's identity, which forms the basis for R. O. Kuz'min's derivation of his functional equation for $L(s, \chi)$.

§4. Derivation of the approximate equations

We introduce additional notation to prove Theorems 2 and 3.

Let ω take only the two values s and $1-s$, where $s = \sigma + it$, $0 < \sigma < 1$, and t is any real number.

We set

$$z = \exp\left[i \operatorname{sign} t\left(\frac{\pi}{2} - \frac{1}{|t|+1}\right)\right], \quad z_\omega = \begin{cases} z, & \text{if } \omega = s, \\ \bar{z}, & \text{if } \omega = 1-s, \end{cases} \tag{14}$$

where sign is Kronecker's symbol and $\bar{y}$ is the complex conjugate of y.

We show that under these conditions the function

$$C_n(\omega, d, a) = \Gamma\left(\frac{\omega+a}{2}, \frac{\pi n^2}{d}z_\omega\right)\Big/\Gamma\left(\frac{\omega+a}{2}\right) \tag{15}$$

is bounded by an absolute constant in all its arguments.

We first note that since

$$\Gamma\left(\frac{\omega+a}{2}, \frac{\pi n^2}{d} z_\omega\right) = \overline{\Gamma\left(\frac{\bar\omega+a}{2}, \frac{\pi n^2}{d} z_{\bar\omega}\right)},$$

$$\Gamma\left(\frac{\omega+a}{2}\right) = \overline{\Gamma\left(\frac{\bar\omega+a}{2}\right)},$$

we can limit ourselves to nonnegative values of t.

First let $t \geq 1$. We set

$$\Delta(x) = \exp\left(-\frac{\pi n^2 x}{d} z_\omega + \frac{\omega+a-2}{2} \ln x\right), \quad T = \sqrt{\frac{dt}{2\pi}},$$

$$I_1(\alpha) = \int_\alpha^\infty \Delta(x)\,dx, \quad I_2(\beta) = \int_0^\beta \Delta(x)\,dx, \tag{16}$$

where

$$\alpha = \begin{cases} 1, & \text{if } n > 2T, \\ 1 + t^{-\frac{1}{2}}, & \text{if } T \leqslant n \leqslant 2T, \end{cases}$$

$$\beta = \begin{cases} 1, & \text{if } 1 \leqslant n \leqslant \frac{1}{2} T, \\ 1 - t^{-\frac{1}{2}}, & \text{if } \frac{1}{2} T < n < T. \end{cases} \tag{17}$$

We note that while the integrals (16) are more complicated than those considered in [1], however the concepts of the saddle-point method are applicable to them.

For $\operatorname{Re}\omega > 0$, using the substitution

$$x = \exp \frac{2 \ln u}{\operatorname{Re}\omega + a}$$

we obtain

$$I_1(\alpha) = \frac{2}{\operatorname{Re}\omega + a} \int_\gamma^\infty \exp\left(-\frac{\pi n^2}{d} \operatorname{Re} z_\omega \exp \frac{2 \ln u}{\operatorname{Re}\omega + a}\right)$$

$$\times \exp\left(i\left(\frac{\operatorname{Im}\omega \cdot \ln u}{\operatorname{Re}\omega + a} - \frac{\pi n^2}{d} \operatorname{Im} z_\omega \exp \frac{2 \ln u}{\operatorname{Re}\omega + a}\right)\right) du,$$

where

$$\gamma = \exp \frac{\operatorname{Re}\omega + a}{2} \ln \alpha.$$

We introduce the functions

$$p(u) = \frac{\pi n^2}{d} \cos \frac{1}{t+1} \exp \frac{2 \ln u}{\operatorname{Re}\omega + a} - \frac{t \ln u}{\operatorname{Re}\omega + a},$$

$$h(u) = \frac{\pi n^2}{d} \sin \frac{1}{t+1} \exp \frac{2 \ln u}{\operatorname{Re}\omega + a}, \qquad g(u) = h(u) \mp ip(u),$$

where the minus sign is taken if $\omega = s$ and the plus sign if $\omega = 1 - s$.

Taking into consideration that for $t \geq 1$

$$z = \exp i\left(\frac{\pi}{2} - \frac{1}{t+1}\right), \qquad \operatorname{Re} z_\omega = \sin\frac{1}{t+1}$$

and for $\omega = 1 - s$ and for $\omega = s$

$$\operatorname{Im} z_\omega = -\cos\frac{1}{t+1} \text{ and } \cos\frac{1}{t+1}, \qquad \operatorname{Im}\omega = -t \text{ and } t,$$

respectively, we find that

$$I_1(\alpha) = \frac{2}{\operatorname{Re}\omega + a}\int_\gamma^\infty e^{-g(u)}\,du. \tag{18}$$

We show that the saddle-point of the function $p(u)$ lies outside the interval of integration.

If $p'(u_0) = 0$, it follows that

$$u_0 = \left(\frac{2\pi n^2}{dt}\cos\frac{1}{t+1}\right)^{-\frac{\operatorname{Re}\omega + a}{2}}.$$

Let $n \geq T$; then, using the estimate

$$\cos\varphi \geqslant \frac{2}{\pi}\left(\frac{\pi}{2} - \varphi\right) \quad \left(0 \leqslant \varphi \leqslant \frac{\pi}{2}\right)$$

we obtain

$$u_0 = \left(1 - \frac{2}{\pi(t+1)}\right)^{-\frac{\operatorname{Re}\omega + a}{2}} < \left(1 + \frac{1}{\sqrt{t}}\right)^{\frac{\operatorname{Re}\omega + a}{2}} = \gamma.$$

But if $n \geq 2T$ and $\operatorname{Re}\omega > 0$, then

$$u_0 \leqslant \left(4\cos\frac{1}{t+1}\right)^{-\frac{\operatorname{Re}\omega + a}{2}} < 1.$$

Thus, if α is defined by formula (17), then $p'(u) \neq 0$ for $u \geq \gamma$ and the expression (18) can be written in the form

$$I_1(\alpha) = \frac{2}{\operatorname{Re}\omega + a}\int_{u=\gamma}^\infty e^{-g(u)}\frac{dp(u)}{p'(u)}.$$

It follows that

$$\frac{\operatorname{Re}\omega + a}{2} I_1(\alpha) = \mp i \left.\frac{e^{-g(u)}}{p'(u)}\right|_{\gamma}^{\infty} \mp i \int_{\gamma}^{\infty} e^{-g(u)} \frac{p''(u)}{[p'(u)]^2} du$$

$$\pm i \int_{\gamma}^{\infty} e^{-g(u)} \frac{h'(u)}{p'(u)} du = \mp i \left.\frac{e^{-g(u)}}{p'(u)}\right|_{\gamma}^{\infty} \mp i I_1^{(1)}(\alpha) \pm i I_1^{(2)}(\alpha). \quad (19)$$

We estimate the integrals in the right side of (19). We have

$$p''(u) = \frac{u^{-2}}{\operatorname{Re}\omega + a}\left\{\left(\frac{2}{\operatorname{Re}\omega + a} - 1\right)\frac{\pi n^2}{d}\cos\frac{1}{t+1}\exp\frac{2\ln u}{\operatorname{Re}\omega + a} + t\right\}$$

and hence $p''(u) > 0$ for $0 < \operatorname{Re}\omega < 1$ and $u \geq \gamma$.

Further, the functions $p'(u)$, $h(u)$ and $h'(u)$ are positive and monotonically increasing. Consequently,

$$|I_1^{(1)}(\alpha)| \leqslant e^{-h(\gamma)} \int_{u=\gamma}^{\infty} \frac{dp'(u)}{[p'(u)]^2} = \frac{e^{-h(\gamma)}}{p'(\gamma)},$$

$$|I_1^{(2)}(\alpha)| \leqslant \frac{1}{p'(\gamma)} \int_{u=\gamma}^{\infty} e^{-h(n)}\, dh(u) = \frac{e^{-h(\gamma)}}{p'(\gamma)}.$$

Substituting these estimates into (19), we have

$$|I_1(\alpha)| \ll \frac{2}{\operatorname{Re}\omega + a}\frac{e^{-h(\gamma)}}{p'(\gamma)}. \quad (20)$$

Let $n \geq T$. Then $\alpha = 1 + t^{-1/2}$ and

$$\frac{1}{p'(\gamma)} = (\operatorname{Re}\omega + a)\left(1 + \frac{1}{\sqrt{t}}\right)^{\frac{\operatorname{Re}\omega + a}{2}} \left\{\frac{2\pi n^2}{d}\cos\frac{1}{t+1}\exp\ln\left(1 + \frac{1}{\sqrt{t}}\right) - t\right\}^{-1}$$

$$\leqslant \frac{2(\operatorname{Re}\omega + a)}{t}\left\{\left(1 + \frac{1}{\sqrt{t}}\right)\cos\frac{1}{t+1} - 1\right\}^{-1}$$

$$\leqslant \frac{2(\operatorname{Re}\omega + a)}{t}\left\{\left(1 + \frac{1}{\sqrt{t}}\right)\left(1 - \frac{2}{\pi(t+1)}\right) - 1\right\}^{-1} \leqslant \frac{4(\operatorname{Re}\omega + a)}{\sqrt{t}}.$$

But if $n \geq 2T$, then $\gamma = 1$ and

$$\frac{1}{p'(u)} = \frac{\operatorname{Re}\omega + a}{\frac{2\pi n^2}{d}\cos\frac{1}{t+1}} \leqslant \frac{\operatorname{Re}\omega + a}{\frac{\pi n^2}{d}}.$$

These estimates and relations (20) give

$$|I_1(\alpha)| \ll \begin{cases} t^{-\frac{1}{2}}, & \text{if } T \leqslant n < 2T, \\ \frac{d}{n^2}, & \text{if } n \geqslant 2T. \end{cases} \quad (21)$$

We can now obtain the assertion about $C_n(\omega, d, a)$ in the cases $n \geq T$ and $t \geq 1$.

In the earlier notation,

$$\Gamma\left(\frac{\omega+a}{2}, \frac{\pi n^2}{d} z_\omega\right) = \left(\frac{\pi n^2}{d} z_\omega\right)^{\frac{\omega+a}{2}} \int_1^\infty \Delta(x)\, dx$$
$$= \left(\frac{\pi n^2}{d} z_\omega\right)^{\frac{\omega+a}{2}} \left\{I_1(\alpha) + \int_1^\alpha \Delta(x)\, dx\right\}. \tag{22}$$

If $n \geq 2T$, then $\alpha = 1$ and, by estimate (21), the absolute value of the right side of (22) does not exceed the quantity

$$\sigma e^{-\frac{\pi t}{4}} t^{\frac{\operatorname{Re}\omega+a}{2}-1}.$$

If $T \leq n < 2T$, then $\alpha = t^{-1/2}$ and the integral along the segment $[1, \alpha]$ obviously is no larger than $t^{-1/2}$. We therefore have the estimate

$$\ll e^{-\frac{\pi t}{4}} t^{\frac{\operatorname{Re}\omega+a-1}{2}}.$$

Consequently, for $n \geq T$ and $t \geq 1$,

$$\left|\Gamma\left(\frac{\omega+a}{2}, \frac{\pi n^2}{d} z_\omega\right)\right| \ll e^{-\frac{\pi t}{4}} t^{\frac{\operatorname{Re}\omega+a-1}{2}}.$$

Hence, by the well-known asymptotics for the Γ-function, it follows that for $n \geq T$ and $t \geq 1$ we have

$$|C_n(\omega, d, a)| = \left|\frac{\Gamma\left(\frac{\omega+a}{2}, \frac{\pi n^2}{d} z_\omega\right)}{\Gamma\left(\frac{\omega+a}{2}\right)}\right| < C, \tag{23}$$

where C is an absolute constant.

We turn to the derivation of the analogous estimate for $n \leq T$ and $t \geq 1$. We have

$$\Gamma\left(\frac{\omega+a}{2}, \frac{\pi n^2}{d} z_\omega\right) = \Gamma\left(\frac{\omega+a}{2}\right) - \gamma\left(\frac{\omega+a}{2}, \frac{\pi n^2}{d} z_\omega\right), \tag{24}$$

where $y = \pi n^2 z_\omega / d$ for $\operatorname{Re}\omega > 0$ and, by definition,

$$\gamma\left(\frac{\omega+a}{2}, \frac{\pi n^2}{d} z_\omega\right) = \int_0^y e^{-\xi} \xi^{\frac{\omega+a}{2}-1}\, d\xi. \tag{25}$$

By the same token, our problem reduces to estimating the integral $I_2(\beta)$ defined by formulas (16) and (17).

We introduce the functions $\bar{p}(u)$ and $\bar{g}(u)$ as follows:

$$\bar{p}(u) = -p(u), \quad \bar{g}(u) = h(u) \pm i\bar{p}(u).$$

Setting

$$\delta = \exp \frac{\operatorname{Re}\omega + a}{2} \ln \beta$$

under the condition that $\bar{p}'(u)$ does not vanish in the interval $(0, \delta)$, we obtain

$$I_2(\beta) = \frac{2}{\operatorname{Re}\omega + a} \int_0^\delta e^{-\bar{g}(u)} du = \frac{2}{\operatorname{Re}\omega + a} \int_{u=0}^{u=\delta} e^{-\bar{g}(u)} \frac{d\bar{p}(u)}{\bar{p}'(u)}$$

$$= \frac{2}{\operatorname{Re}\omega + a} \left\{ \pm i \frac{e^{-\bar{g}(u)}}{\bar{p}'(u)} \Bigg|_0^\delta \pm i \int_0^\delta e^{-\bar{g}(u)} \frac{\bar{p}''(u)}{[\bar{p}'(u)]^2} du \mp i \int_0^\delta e^{-\bar{g}(u)} \frac{h'(u)}{\bar{p}'(u)} du \right\}$$

$$= \frac{\pm 2i}{\operatorname{Re}\omega + a} \left\{ \frac{e^{-\bar{g}(u)}}{\bar{p}'(u)} \Bigg|_0^\delta + I_2^{(1)}(\beta) - I_2^{(2)}(\beta) \right\}. \tag{26}$$

We have

$$\bar{p}'(u) = \frac{1}{u(\operatorname{Re}\omega + a)} \left\{ t - \frac{\pi n^2}{d} \cos \frac{1}{t+1} \exp \frac{2 \ln u}{\operatorname{Re}\omega + a} \right\}. \tag{27}$$

The point u_0 for which $\bar{p}'(u) = 0$ is the same as for $p'(u)$. We show that it is located outside the segment $[0, \delta]$.

Let $2/T < n \leq T$; then $\beta = 1 - t^{1/2}$ and

$$u_0 = \left(\frac{2\pi n^2}{dt} \cos \frac{1}{t+1} \right)^{-\frac{\operatorname{Re}\omega + a}{2}} \geqslant \left(\cos \frac{1}{t+1} \right)^{-\frac{\operatorname{Re}\omega + a}{2}} > \delta.$$

If $1 \leq n \leq 2/T$, then $\beta = 1$ and it is obvious that

$$u_0 \geqslant \left(\frac{4}{\cos \frac{1}{t+2}} \right)^{\frac{\operatorname{Re}\omega + a}{2}} > 1.$$

Hence $\bar{p}'(u) \neq 0$ for $0 \leq u \leq \beta$ and the transformations performed in (26) are legitimate.

Further, it follows from (27) that $\bar{p}'(u) > 0$ and $\bar{p}''(u) < 0$, i.e. $\bar{p}'(u)$ decreases monotonically, reaching its least value at the point $u = \delta$. Therefore, proceeding as before, we obtain

$$|I_2(\beta)| \ll \frac{1}{\sqrt{t}}. \tag{28}$$

It follows from (25) and (28) that for $n \leq T$ and $t \geq 1$ we have

$$\gamma\left(\frac{\omega+a}{2}, \frac{\pi n^2}{d} z_\omega\right) = \left(\frac{\pi n^2}{d} z_\omega\right)^{\frac{\omega+a}{2}} \left\{I_2(\beta) + \int_\beta^1 \Delta(x)\,dx\right\}$$

and

$$\left|\gamma\left(\frac{\omega+a}{2}, \frac{\pi n^2}{d} z_\omega\right)\right| \ll e^{-\frac{\pi t}{4}} t^{\frac{\operatorname{Re}\omega+a-1}{2}}. \tag{29}$$

Hence, by (24) for $n \leq T$ and $t \geq 1$ we obtain, as before,

$$|C_n(\omega, d, a)| < C,$$

where C is an absolute constant.

It remains to examine the estimate of $C_n(\omega, d, a)$ for $0 \leq t \leq 1$. The function $\Gamma^{-1}((\omega+a)/2)$ is obviously bounded for $0 < \operatorname{Re}\omega < 1$, while $C_n(\omega, d, a)$ is bounded for the incomplete gamma-functions for $n \leq T$ by relations (24) and (25) and for $n > T$ by relation (22). The assertion of Theorem 2 about the quantity $C_n(\omega, d, a)$ is thus completely proved.

For the values of z_ω indicated in (14), we examine the function

$$R(\omega, N) = \sum_{n > N \ln N} \frac{\chi(n)}{n^\omega} \Gamma\left(\frac{\omega+a}{2}, \frac{\pi n^2}{d} z_\omega\right). \tag{30}$$

We have

$$\Gamma\left(\frac{\omega+a}{2}, \frac{\pi n^2}{d} z_\omega\right) = \left(\frac{\pi n^2}{d} z_\omega\right)^{\frac{\omega+a}{2}} \int_1^\infty \Delta(x)\,dx.$$

Since $0 < \operatorname{Re}\omega < 1$ and

$$\operatorname{Re} z_\omega = \cos\left[\operatorname{sign} t\left(\frac{\pi}{2} - \frac{1}{|t|+1}\right)\right], \quad \sin\frac{1}{|t|+1} > \frac{2}{\pi(|t|+1)},$$

this latter integral is estimated as

$$\ll \frac{d(|t|+1)}{n^2} \exp\left(-\frac{2n^2}{d(|t|+1)}\right).$$

Therefore

$$|R(\omega, N)| \ll \exp\left(-\frac{\pi|t|}{4} - \ln^2 N\right).$$

This estimate can also be written in the form

$$|R(\omega, N)| \ll \left|\Gamma\left(\frac{s+a}{2}\right)\right| [d(|t|+1)]^{-A} \tag{31}$$

where A is any positive constant.

Applying Theorem 1 with the z from (14) and using formulas (15), (30) and (31), we obtain the assertion of Theorem 2.

We turn to the proof of Theorem 3. Because

$$L(s,\chi) = \overline{L(\bar{s},\bar{\chi})} \quad (\mathrm{Re}\, s > 0), \tag{32}$$

it suffices to prove the theorem only for the case $t \geq 0$.

We first examine the case $t \geq 1$. Keeping the values for z_ω indicated in (14), we introduce additional notation. Let

$$\gamma_m = 1 + \frac{m}{\sqrt{t}}, \quad N_m = \gamma_m \sqrt{\frac{dt}{2\pi}} \, .$$

For integral $m \geq 1$ we set

$$F_m(\omega,\chi) = \sum_{N_{m-1}<n\leqslant N_m} \frac{\chi(n)}{n^\omega} \Gamma\left(\frac{\omega+a}{2}, \frac{\pi n^2}{d} z_\omega\right), \tag{33}$$

and for integral $m \leq 0$ we set

$$f_m(\omega,\chi) = \sum_{\substack{N_{m-1}<n\leqslant N_m \\ n>0}} \frac{\chi(n)}{n^\omega} \gamma\left(\frac{\omega+a}{2}, \frac{\pi n^2}{d} z_\omega\right). \tag{34}$$

We examine $F_m(\omega, \chi)$. Since

$$\frac{\partial}{\partial x}\Gamma(\alpha, x) = -e^{-x}x^{\alpha-1},$$

we obtain

$$F_m(\omega,\chi) = 2\left(\frac{\pi z_\omega}{d}\right)^{\frac{\omega+a}{2}} \int_{N_{m-1}}^{N_m} \sum_{N_{m-1}<n\leqslant V} \frac{\chi(n)}{n^\omega} \exp\left(-\frac{\pi V^2}{d} z_\omega\right)$$

$$\times V^{\omega+a-1}\,dV + \Gamma\left(\frac{\omega+a}{2}, \frac{\pi N_m^2}{d} z_\omega\right) \sum_{N_{m-1}<n\leqslant N_m} \frac{\chi(n)}{n^\omega}$$

$$= F'_m(\omega,\chi) + F''_m(\omega,\chi). \tag{35}$$

We further find that

$$|F'_m(\omega,\chi)| \ll |z_\omega^{\frac{\omega}{2}}|\, d^{-\frac{\operatorname{Re}\omega+a}{2}} \max_{V\le N_m}\Big|\sum_{N_{m-1}<n\le V}\frac{\chi(n)}{n^\omega}\Big|$$

$$\times \int_{N_{m-1}}^{N_m} \exp\Big(-\frac{\pi V^2}{d}\operatorname{Re} z_\omega\Big) V^{\operatorname{Re}\omega+a-1}\,dV$$

$$\ll e^{-\frac{\pi t}{4}}\frac{N^{\frac{\operatorname{Re}\omega+a}{2}}}{\sqrt{t}}\, d^{-\frac{\operatorname{Re}\omega+a}{2}} \max_{V\le N_m}\Big|\sum_{N_{m-1}<n\le V}\frac{\chi(n)}{n^\omega}\Big|,$$

which can be rewritten in the form

$$|F'_m(\omega,\chi)| \ll \Big|\Gamma\Big(\frac{\omega+a}{2}\Big)\Big| \max_{V\le N_m}\Big|\sum_{N_{m-1}<n\le V}\frac{\chi(n)}{n^\omega}\Big|. \tag{36}$$

For $F''_m(\omega,\chi)$ it immediately follows from estimate (23) that

$$|F''_m(\omega,\chi)| \ll \Big|\Gamma\Big(\frac{\omega+a}{2}\Big)\Big|\cdot\Big|\sum_{N_{m-1}<n\le N_m}\frac{\chi(n)}{n^\omega}\Big|. \tag{37}$$

Similarly, taking into consideration that

$$\frac{\partial}{\partial x}\gamma(\alpha,x) = -\frac{\partial}{\partial x}\Gamma(\alpha,x),$$

we find by estimate (29) that for $t\ge 1$ we have

$$|f_m(\omega,\chi)| \ll \Big|\Gamma\Big(\frac{\omega+a}{2}\Big)\Big| \max_{V\le N_m}\Big|\sum_{\substack{N_{m-1}<n\le V\\ n>0}}\frac{\chi(n)}{n^\omega}\Big|. \tag{38}$$

We write the functional equation (2) in the form

$$\Gamma\Big(\frac{s+a}{2}\Big)L(s,\chi) = Q(s,\chi) + \varepsilon(\chi)\Big(\frac{d}{\pi}\Big)^{\frac{1}{2}-s} Q(1-s,\chi).$$

By (24), (30), (33) and (34), for $\omega = s$ and $\omega = 1-s$ we then have

$$Q(\omega,\chi) = \Gamma\Big(\frac{\omega+a}{2}\Big)\sum_{n\le\sqrt{\frac{dt}{2\pi}}}\frac{\chi(n)}{n^\omega} - \sum_{m=0}^{\sqrt{t}+1} f_{-m}(\omega,\chi)$$

$$+\sum_{m=1}^{\sqrt{t}\ln N} F_m(\omega,\chi) + O(N^{-b}).$$

It follows from (35)–(38) that

$$|Q(\omega, \chi)| \ll \left|\Gamma\left(\frac{\omega+a}{2}\right)\right| \sum_{|m|=0}^{\sqrt{t}\ln N} \max_{V \leq N_m} \left| \sum_{\substack{N_{m-1}<n\leq V \\ n>0}} \frac{\chi(n)}{n^\omega} \right| + O(N^{-b}).$$

Taking this estimate for $\omega = s$, and then for $\omega = 1 - s$, and in the latter case replacing χ by $\bar{\chi}$, in view of (39) we obtain Theorem 3 for $t \geq 1$.

It remains to examine the case $0 \leq t \leq 1$. Using Theorem 1 with $z = 1$ and estimate (30), we find

$$Q(\omega, \chi) = \sum_{n \leq \sqrt{d}\ln d} \frac{\chi(n)}{n^\omega} \Gamma\left(\frac{\omega+a}{2}, \frac{\pi n^2}{d}\right) + O(d^{-b}).$$

Expressing this sum in the form of an integral and taking into consideration that the function $\Gamma(\alpha, x)$ is bounded, we obtain, as before,

$$|Q(\omega, \chi)| \ll \max_{V \leq \sqrt{d}\ln d} \left| \sum_{n \leq V} \frac{\chi(n)}{n^\omega} \right| + O(d^{-b}).$$

This estimate with $\omega = s$ and $\omega = 1 - s$, together with estimate (39) and consideration of the boundedness of the function $\Gamma^{-1}((s+a)/2)$ for $0 \leq t \leq 1$ proves Theorem 3 for the case $0 \leq t \leq 1$.

Theorem 4 can be proved using Theorem 2 and an estimate for the sum of characters recently obtained by D. A. Burgess [3]. The proof does not essentially require new ideas, and we shall not bring it in here.

Theorem 5 can be proved using several tedious but completely elementary sieve considerations.

BIBLIOGRAPHY

[1] Ju. V. Linnik, *All large numbers are sums of a prime and two squares (A problem of Hardy and Littlewood)*. II, Mat. Sb. 53 (95) (1961), 3–38; English transl., Amer. Math. Soc. Transl. (2) 37 (1964), 197–240. MR 22 #10964.

[2] ———, *An asymptotic formula in an additive problem of Hardy and Littlewood*, Izv. Akad. Nauk SSSR Ser. Mat. 24 (1960), 629–706; English transl., Amer. Math. Soc. Transl. (2) 46 (1965), 65–148. MR 23 #A130.

[3] D. A. Burgess, *On character sums on L-series*. II, Proc. London Math.

Soc. (3) 13 (1963), 524–536. MR 26 #6133.

[4] N. G. Čudakov, *Introduction to the theory of Dirichlet's L-functions*, OGIZ, Moscow, 1947. (Russian) MR 11, 234.

[5] A. F. Lavrik, *The sum over the characters of powers of the modulus of the Dirichlet L-function in the critical strip*, Dokl. Akad. Nauk SSSR 154 (1964), 34–37 = Soviet Math. Dokl. 5 (1964), 28–31. MR 28 #68.

[6] O. Saparnijazov, *A character sum of powers of the modulus of the Dirichlet L-functions in the critical strip*, Izv. Akad. Nauk UzSSR, Ser. Fiz.-Mat. 1964, no. 4, 13–19. (Russian) MR 30 #1101.

Translated by:
N. Koblitz

UPPER BOUNDS AND NUMERICAL CALCULATION OF THE NUMBER OF IDEAL CLASSES OF REAL QUADRATIC FIELDS

I. Š. SLAVUTSKIĬ

1. It is well known that the formulas for the number $h(d)$ of ideal classes of the quadratic field $R(\sqrt{d})$ can be written in the form

$$h(d) = \begin{cases} \dfrac{\sqrt{d}}{2 \ln E_1} L(1 \mid \chi), & d > 0, \\ \dfrac{\sqrt{|d|}}{\pi} L(1 \mid \chi), & d < -4, \end{cases} \tag{1}$$

where d is the discriminant of the field $R(\sqrt{d})$, E_1 is the fundamental unit of the field, $L(1|\chi) = \Sigma_{n=1}^{\infty} \chi(n)/n$, L is a Dirichlet series, and $\chi(n) = (d/n)$ is the Kronecker symbol. Using results of Pólya and Schur, Landau [1] proved that the estimate $h(d) = O(\sqrt{d})$ holds in the case of a real quadratic field. Ankeny and Chowla [2] noticed that assuming the extended Riemann hypothesis leads to a more precise estimate of the order of growth: $h(d) = O(\sqrt{d} \ln \ln d / \ln d)$.

On the other hand, Ankeny, Brauer, and Chowla [3], in particular, proved that for any $\epsilon > 0$ there exist infinitely many real quadratic fields for which $h(d) > d^{1/2 - \epsilon}$. Yet in many cases involving an upper estimate for $h(d)$, the estimate reduces to special cases and is too excessive ([2], [4]–[6]). The purpose of this note is to prove the inequalities

$$h(d) < \sqrt{d} \quad \text{for } d > 0, \tag{2}$$

and

$$h(d) < \frac{1}{3} \sqrt{|d|} \ln |d| \quad \text{for } d < -4, \tag{3}$$

and also to indicate a method of numerical calculation of the number of ideal classes of a real quadratic field based on some congruences found by A. A. Kiselev and the author.

2. Let $d > 0$. Then, setting $k = \phi(d)/2$ and taking into account that $|\Sigma_{n=a}^{b} \chi(n)| \le k$, we shall estimate L, which is the Dirichlet series $L(1|\chi) = \Sigma_{n=1}^{\infty} \chi(n)/n$. Setting $S_n = \Sigma_{r=1}^{n} \chi(r)$, we obtain

$$|L(1 \mid \chi)| = \left| \sum_{n=1}^{\infty} \frac{S_n - S_{n-1}}{n} \right| = \left| \sum_{n=1}^{\infty} \frac{S_n}{n(n+1)} \right| <$$

$$< \sum_{n=1}^{k-1} \frac{n}{n(n+1)} + \sum_{n=k}^{\infty} \frac{k}{n(n+1)} = \sum_{n=2}^{k} \frac{1}{n} + k \sum_{n=k}^{\infty} \left(\frac{1}{n} - \frac{1}{n+1} \right)$$

$$= \sum_{n=1}^{k} \frac{1}{n} < \ln k + C + \frac{1}{k} - \frac{1}{2(k+1)},$$

where Euler's constant $C = 0.5772\cdots$ appears in the well-known estimate from above for a segment of the harmonic series (see for example [7], Russian p. 126). Since $\phi(d) \le d - 1$, we further have

$$|L(1|\chi)| < \ln(d-1) - \ln 2 + C + \frac{\varphi(d)+4}{\varphi(d)(\varphi(d)+2)}$$

or

$$|L(1|\chi)| < \ln(d-1) - 0.69 + 0.58 + \frac{\varphi(d)+4}{\varphi(d)(\varphi(d)+2)}.$$

Let $d \ge 60$. Then $\phi(d) \ge 16$, $(\phi(d)+4)/\phi(d)(\phi(d)+2) < 0.07$, and

$$|L(1|\chi)| < \ln(d-1) - 0.04. \tag{4}$$

We estimate the fundamental unit $E_1 = (t + u\sqrt{d})/2$. It is well known that $t^2 - u^2 d = \pm 4$; hence in all cases $t \ge u\sqrt{d-4}$, and, consequently,

$$E_1 > \frac{1}{2} u (\sqrt{d} + \sqrt{d-4}) > \sqrt{d-4}. \tag{5}$$

Thus, in the case of a real quadratic field, (1), (4) and (5) lead to

$$h(d) < \sqrt{d} \frac{\ln(d-1) - 0.04}{\ln(d-4)},$$

where $(\ln(d-1) - 0.04)/\ln(d-4) < 1$ for $d \ge 80$, and, because $\ln((d-1)/(d-4)) < 0.04$, we finally have $h(d) < \sqrt{d}$ for $d \ge 80$. For $d < 80$ inequality (2) can be easily checked directly. In the case of an imaginary quadratic field with $d < -4$, analogous calculations lead to the inequality $|L(1|\chi)| < \ln|d|$; hence $h(d) < \sqrt{|d|}\ln|d|/3$.

3. The estimate for $h(d)$ in the case of a real quadratic field allows the class number to be calculated purely arithmetically (without tables and transcendental functions) from congruences following from the Dirichlet formulas of [7]–[9]:

$$h(d) \frac{U_l}{p^{l-1}} \equiv -\frac{T_l}{2\varepsilon m} \sum_{n=1}^{u} \left(\frac{\varepsilon n}{u}\right) B_m\left(\frac{u}{n}\right) \pmod{p^l}, \quad d = np, \quad n \geqslant 1; \tag{6}$$

and

$$h(d)\,\frac{\overline{U}_l}{p^l} \equiv -\frac{\overline{T}_l}{4dm}\sum_{u=1}^{d}\left(\frac{d}{u}\right)B_{2m}\left(\frac{u}{d}\right) \pmod{p^l}, \qquad p\nmid d; \tag{7}$$

where $T_l + U_l\sqrt{d} = E_1^{p^{l-1}}$, $\overline{T}_l + \overline{U}_l\sqrt{d} = E_1^{(p-(d/p))p^{l-1}}$, $E_1 = T_1 + U_1\sqrt{d}$ is the fundamental unit of $R(\sqrt{d})$, $m = ((p-1)/2)p^{l-1}$, $\epsilon = (-1)^{(p-1)/2}$, p is an odd prime, $l \geq 1$ is a natural number, $(d/4)$ is the Kronecker symbol, and $B_m(x) = \sum_{k=0}^m C_m^k B_k x^{m-k}$ is the Bernoulli polynomial, where Bernoulli numbers are defined by the symbolic identity $(B+1)^k = B^k$, $k = 2, 3, \cdots$; $B_0 = 1$. With this aim we first note that we have

Lemma. *In our notation, U_l/p^{l-1} and U_1 are divisible by the same power (zero included) of the prime $p > 3$, and $\overline{U}_l/p^l$ and $\overline{U}_1/p$ by the same power of any odd prime.*

Let $p|d$ and $d = np$, $n \geq 1$. Then

$$U_l = \sum_{\substack{k=1\\(2,k)=1}}^{p^{l-1}} C_{p^{l-1}}^k T_1^{p^{l-1}-k} U_1^k d^{\frac{k-1}{2}}$$

and

$$\frac{U_l}{p^{l-1}} = U_1\left\{T_1^{p^{l-1}-1} + d\sum_{\substack{k=3\\(k,2)=1}}^{p^{l-1}} \frac{1}{k} C_{p^{l-1}-1}^{k-1} T_1^{p^{l-1}-k} U_1^{k-1} d^{\frac{k-3}{2}}\cdot\right\}.$$

Since $k-3/k$ is p-integral for $p > 3$ and $k > 3$ (because if $p^r \| k$, obviously $(k-3)/2 \geq (p^r-3)/2 \geq r$)[1] and $T_1 \not\equiv 0 \pmod{p}$, it follows that the number in braces is p-integral and relatively prime with p. The first part of the lemma is proved. Now let $(p, d) = 1$. We first show that $\overline{U}_1 \equiv 0 \pmod{p}$. In fact, when $(d/p) = -1$, we have $\overline{U}_1 \equiv T_1^p U_1 + T_1 U_1^p d^{(p-1)/2} \equiv 0 \pmod{p}$. But if $(d/p) = +1$, then

$$\overline{T}_1 \equiv T_1^{p-1} + T_1^{p-3}U_1^2 d + \ldots + U_1^{p-1} d^{\frac{p-1}{2}} \equiv \frac{(T_1^2)^{\frac{p+1}{2}} - (U_1^2 d)^{\frac{p+1}{2}}}{T_1^2 - U_1^2 d}$$

$$\equiv 1 \pmod{p},$$

and, since $\overline{T}_1^2 - \overline{U}_1^2 d = +1$, again $\overline{U}_1 \equiv 0 \pmod{p}$. But then

$$\frac{\overline{U}_l}{p^l} = \frac{\overline{U}_1}{p}\left\{\overline{T}_1^{p^{l-1}} + \overline{U}_1 \sum_{\substack{k=3\\(k,2)=1}}^{p^{l-1}} C_{p^{l-1}}^{k-1}\overline{T}_1^{p^{l-1}-k} d^{\frac{k-1}{2}} \frac{\overline{U}_1^{k-2}}{k}\right\},$$

1) Here and below $p^r \| k$ means that $p^r | k$ but $p^{r+1} \nmid k$.

Therefore, because $\overline{T}_1^2 \equiv 1 \pmod p$, and, along with $\overline{U}_1^{k-2}/k$, the number is p-integral (if $p^r \| k$, then $k-2 \geq p^r - 2 \geq r$ for any p and r), the lemma is completely proved.[1)]

Now let p^k be the highest power of p dividing U_1 or $\overline{U}_1/p$. By (2) we can calculate $h(d)$ from (6) or (7) if we choose the modulus p^l of the congruence so that $p^{l-k} > \sqrt{d}$ and, using the lemma, solve the congruence of the first degree in $h(d)$, taking into account that $0 < h(d) < p^{l-k}$.

For the specific calculation it helps to restrict the interval of summation in (6) and (7):

$$h(d)\,\frac{U_l}{p^{l-1}} \equiv -\frac{T_l}{nm} \sum_{1 \leqslant u \leqslant \frac{n}{2}} \left(\frac{\varepsilon n}{u}\right) B_m\left(\frac{u}{n}\right) \pmod{p^l},\quad d = np,\ n \geqslant 1; \tag{6'}$$

$$h(d)\,\frac{\overline{U}_l}{p^l} \equiv -\frac{\overline{T}_l}{2dm} \sum_{1 \leqslant u \leqslant \frac{d}{2}} \left(\frac{d}{u}\right) B_{2m}\left(\frac{u}{d}\right) \pmod{p^l},\quad p \nmid d. \tag{7'}$$

Here, for example in (6′), we use that

$$\left(\frac{\varepsilon n}{n-u}\right) B_m\left(\frac{n-u}{n}\right) = \left(\frac{\varepsilon n}{u}\right) B_m\left(\frac{u}{n}\right),$$

since, if $p \equiv 1 \pmod 4$, then $\epsilon = 1$ and $B_m((n-u)/n) = B_m(u/n)$, while in the case $p \equiv 3 \pmod 4$ we have $\epsilon = -1$ and $B_m((n-u)/n) = -B_m(u/n)$.

This method of calculating $h(d)$ is more convenient than the arithmetic calculation of the class number presented in papers [11] and [12] by Hasse and Bergström. Let $d = 65$; then $h(d) < \sqrt{65} < 13$ and $E_1 = 8 + \sqrt{65}$ so that, setting $l = 1$ in (6′), we get

$$h(65) \equiv 2\,\frac{8}{5}\left(\frac{5}{13}\right) \sum_{1 \leqslant u \leqslant \frac{5}{2}} \left(\frac{5}{u}\right) B_6\left(\frac{u}{5}\right) \pmod B$$

or

$$h \equiv -\frac{3}{5}\left\{15\,\frac{1}{5^2}\,B_4 + 15\,\frac{1}{5^4}\,B_2 + 6\,\frac{1}{5^5}\,B_1 + \frac{1}{5^6} - 15\left(\frac{2}{5}\right)^2 B_4\right.$$
$$\left.-15\left(\frac{2}{5}\right)^4 B_2 - 6\left(\frac{2}{5}\right)^5 B_1 - \left(\frac{2}{5}\right)^6\right\} \equiv 2\left\{(-2+8)\left(-\frac{1}{30}\right) + (2-6)\,\frac{1}{6} +\right.$$

---

1) Some remarks similar to the first part of the lemma are found in Dirichlet's paper [10], while the congruence $\overline{U}_1 \equiv 0 \pmod p$ was pointed out by A. A. Kiselev [6].

$$+\left(\frac{6}{5}+\frac{3}{5}\right)\left(-\frac{1}{2}\right)+31\Big\}\equiv 2\left\{-\frac{1}{5}-\frac{2}{3}-\frac{9}{10}-2\right\}$$
$$\equiv -3+3+6-4\equiv 2 \pmod{13}.$$

Thus, $h(65) = 2$.

Using a generalized congruence of Voronoĭ from [6] and [8], congruences (6) and (7) can be given another form, which avoids the rapidly growing Bernoulli numbers:

$$h(d)\frac{U_l}{p^{l-1}}\equiv -\frac{T_l}{2}\sum_{0<\nu<np^l}\left(\frac{}{\nu}\right)\frac{1}{n\nu}\left[\frac{\nu}{p^l}\right] \pmod{p^l},\ n>1,\ p\neq 3 \text{ for } l=1,$$

$$h(d)\frac{U_l}{p^{l-1}}\equiv -\frac{T_l}{4}\sum_{0<\nu<p^l}\left(\frac{\nu}{p}\right)\frac{1}{g\nu}\left[\frac{g\nu}{p^l}\right] \pmod{p^l},\ d=p,\ p\equiv 1 \pmod 4,$$
$$\left(\frac{g}{p}\right)=-1,$$

$$h(d)\frac{\overline{U}_l}{p^l}\equiv -\frac{\overline{T}_l}{2}\sum_{0<\nu<dp^l}\left(\frac{d}{\nu}\right)\left(\frac{\nu}{p}\right)^2\frac{1}{d\nu}\left[\frac{\nu}{p^l}\right] \pmod{p^l},\ p\nmid d,$$

where $[z]$ is the integral part of z.

BIBLIOGRAPHY

[1] E. Landau, *Abschätzungen von Charaktersummen, Einheit und Klassenzahl*, Nachr. Akad. Wiss. Göttingen. Mat.-Phys. Kl. 1 (1918), 79–97.

[2] N. C. Ankeny and S. Chowla, *A note on the class number of real quadratic fields*, Acta Arith. 6 (1960), 145–147. MR 22 #6780.

[3] N. C. Ankeny, R. Brauer and C. Chowla, *A note on the class numbers of algebraic number fields*, Amer. J. Math. 78 (1956), 51–61. MR 18, 565.

[4] L. Carlitz, *Note on the class number of real quadratic fields*, Proc. Amer. Math. Soc. 4 (1953), 535–537. MR 15, 104.

[5] H.-W. Leopldt, *Über Klassenzahlprimteiler reeller abelscher Zahlkörper als Primteiler verallgemeinerter Bernoullischer Zahlen*, Abh. Math. Sem. Univ. Hamburg 23 (1959), 36–47. MR 21 #1967.

[6] A. A. Kiselev and I. Š. Slavutskiĭ, *On the number of classes of ideals of a quadratic field and its rings*, Dokl. Akad. Nauk SSSR 126 (1959), 1191–1194. (Russian) MR 23 #A141.

[7] A. O. Gel'fond, *The calculus of finite differences*, GITTL, Moscow, 1952; French transl., Coll. Univ. Math., XII, Dunod, Paris, 1963. MR 14, 759; MR 28 #376.

[8] I. Š. Slavutskiĭ, *On the ideal class number of a real quadratic field*, Izv. Vysš. Učebn. Zaved. Matematika 1960, no. 4 (17), 173–177. (Russian) MR 24 #A1905.

[9] ———, *On the ideal class number of a real quadratic field with prime discriminant*, Leningrad. Gos. Ped. Inst. Učen. Zap. 218 (1961), 179–189. (Russian)

[10] L. Dirichlet, "Sur une propriété des formes quadratiques a determinant positif " in *Mathematische Werke*. Band II, Berlin, 1897, pp. 189–194.

[11] H. Hasse, *Productformel für verallgemeinerte Gaussche Summen und ihre Anwendung auf die Klassenzahlformel für reelle quadratische Zahlkörper*, Math. Z. 46 (1940), 303–314. MR 2, 39.

[12] H. Bergström, *Die Klassenzahlformel für reelle quadratische Zahlkörper mit zusammengesetzter Diskriminante als Produkt verallgemeinerter Gausscher Summen*, J. Reine Angew. Math. 186 (1944), 91–115. MR 7, 148.

Translated by:
N. Koblitz

ON SYSTEMS OF CONGRUENCES

UDC 511

A. A. KARACUBA

The article investigates the question of estimating the number of solutions of systems of congruence of a definite form.

Introduction

We consider the system of congruences

$$\left.\begin{array}{l} x_1 + \ldots + x_k - y_1 - \ldots - y_k \equiv \lambda_1 \\ \ldots\ldots\ldots\ldots\ldots\ldots\ldots\ldots \\ x_1^n + \ldots + x_k^n - y_1^n - \ldots - y_k^n \equiv \lambda_n \end{array}\right\} \pmod{q} \qquad (1)$$

$$1 \leqslant x_i, y_i \leqslant P, \quad i = 1, 2, \ldots, k,$$

and designate the number of solutions of this system by $N_k(\lambda_1, \cdots, \lambda_n)$. Let r be an integer satisfying the conditions $2 \leq r \leq n (n \geq 2)$ and $A_1 P^r \leq q \leq A_2 P^r$, where A_1 and A_2 are constants depending on n and k.

To estimate rational trigonometric sums it is necessary to know quite precisely a bound from above for the quantity $N_k = N_k(0, \cdots, 0) = N_k(q)$. We trivially obtain

$$N_k(\lambda_1, \ldots, \lambda_n) = q^{-n} \sum_{a_1, \ldots, a_n = 1}^{q} \left| \sum_{x=1}^{P} \exp \frac{2\pi i}{q} (a_1 x + \ldots + a_n x^n) \right|^{2k}$$
$$\times \exp \frac{-2\pi i}{q} (a_1 \lambda_1 + \ldots + a_n \lambda_n) \leqslant N_k.$$

N_k is easily estimated from below. In fact, we note that in the system (1) the first r congruences are equations. Therefore λ_ν will vary within the limits

$$-kP^\nu \leqslant \lambda_\nu \leqslant kP^\nu, \quad \nu \leqslant r; \quad 0 \leqslant \lambda_\nu \leqslant q - 1, \quad \nu \geqslant r + 1.$$

Thus we have

$$P^{2k} = \sum_{\lambda_1, \ldots, \lambda_n} N_k(\lambda_1, \ldots, \lambda_n) \leqslant G N_k,$$

where

$$G = \sum_{\lambda_1, \ldots, \lambda_n} 1 \leqslant (2k)^r A_2^{n-r} P^{rn - \frac{r(r-1)}{2}}$$

Consequently

$$N_k \geqslant (2k)^{-r} A_2^{-n+r} P^{2k-rn+\frac{r(r-1)}{2}}.$$

It is natural to expect the following estimate from above for the quanity N_k for suitable k:

$$N_k \leqslant c\,(n, k)\, P^{2k-rn+\frac{r(r-1)}{2}} \tag{2}$$

In fact, from I. M. Vinogradov's theorem on the mean [1] estimate (2) easily follows for $k \geq [4n^2 \ln n]$. However, to refine the estimates of rational trigonometric sums with denominator q it is important to obtain estimate (2) for possibly small k (for a statement of the problem, see H. M. Korobov's report [2] to the Fourth All-Union Mathematical Congress in Leningrad).

In this article we obtain a class of congruences (1) for which estimate (2) is valid for $k \geq 6rn \ln n$. In addition, we investigate the possibility of obtaining estimate (2) for small $k \leq n^2$ for systems (1) with arbitrary q. The method used in solving the problem is a further development of ideas of the author first used in [4].

We introduce the necessary notation: n and r are integers, $n \geq 2$, $2 \leq r \leq n$, $\tau = [5r \ln n]$; p_i are prime numbers, $1 \leq i \leq \tau$, where

$$0{,}25 p_i^{1-\frac{1}{r}} \leqslant p_{i+1} \leqslant 0{,}5 p_i^{1-\frac{1}{r}}, \qquad p_1 = p > \exp crn^5,$$

$$q_1 = p_1^r \ldots p_\tau^r; \qquad A_3 q^{\frac{1}{r}} \leqslant P \leqslant A_4 q^{\frac{1}{r}};$$

$c, c_1, c_2, \cdots$ are absolute (positive) constants, and $A, A_1, A_2, \cdots$ are positive constants depending on n and k.

§1

We prove the following preliminary lemma.

Lemma. *Consider the system of congruences*

$$\begin{cases} x_1 + \ldots + x_n \equiv y_1 + \ldots + y_n + \lambda_1 \pmod{p}, \\ \ldots\ldots\ldots\ldots\ldots\ldots\ldots\ldots\ldots\ldots \\ x_1^{r-1} + \ldots + x_n^{r-1} \equiv y_1^{r-1} + \ldots + y_n^{r-1} + \lambda_{r-1} \pmod{p^{r-1}}, \\ x_1^r + \ldots + x_n^r \equiv y_1^r + \ldots + y_n^r + \lambda_r \pmod{p^r}, \\ \ldots\ldots\ldots\ldots\ldots\ldots\ldots\ldots\ldots\ldots \\ x_1^n + \ldots + x_n^n \equiv y_1^n + \ldots + y_n^n + \lambda_n \pmod{p^r}, \end{cases}$$

$$0 \leqslant x_i, y_i \leqslant Np^r - 1, \quad i = 1, 2, \ldots, n,$$

$$(N \text{ is a nonnegative integer}) \tag{3}$$

$$x_i \neq x_j \pmod p, \quad i \neq j;\ p \geqslant n^2.$$

If $T_n(\lambda_1, \cdots, \lambda_n)$ *is the number of solutions of this system, then*[1) 2)]

$$\frac{1}{4} N^{2n} p^{rn + \frac{r(r-1)}{2}} \leqslant T_n = T_n(0, \ldots, 0) \leqslant n!\, N^{2n} p^{rn + \frac{r(r-1)}{2}}. \tag{4}$$

Proof. We first prove the right side of the inequality. There are precisely $N^n p^{rn}$ possible collections of the variables $y_1, \cdots, y_n$. Therefore we need to obtain the corresponding estimate of the number of solutions of congruences of the form (3), where in the right side instead of the forms $y_1^\nu + \cdots + y_n^\nu$, $\nu = 1, 2, \cdots, n$, we have certain integers $N_1, \cdots, N_n$.

Let

$$x_i = x_{i1} + p x_{i2} + \ldots + p^{r-1} x_{ir} + p^r x_{ir+1}$$

be the p-adic decomposition of x_i, $i = 1, 2, \cdots, n$. It follows from the conditions on x_i that

$$x_{i1} \neq x_{j1}, \quad i \neq j,$$
$$0 \leqslant x_{i\nu} \leqslant p - 1, \quad 1 \leqslant \nu \leqslant r; \quad 0 \leqslant x_{ir+1} \leqslant N - 1.$$

For every $2 \leq s \leq r$ we consider the following system of congruences T''_s, where the variables $x_{1s}, \cdots, x_{ns}$ are unknowns:

$$\left.\begin{array}{l} (x_{11} + p x_{12} + \ldots + p^{s-1} x_{1s})^s + \ldots \\ \ldots + (x_{n1} + p x_{n2} + \ldots + p^{s-1} x_{ns})^s \equiv N_s \\ \ldots\ldots\ldots\ldots\ldots\ldots\ldots\ldots \\ (x_{11} + p x_{12} + \ldots + p^{s-1} x_{1s})^n + \ldots \\ \ldots + (x_{n1} + p x_{n2} + \ldots + p^{s-1} x_{ns})^n \equiv N_n \end{array}\right\} \pmod{p^s}.$$

Let T'_s be the number of solutions of T''_s. Introducting the notation

$$x_i = x_{i1} + p x_{i2} + \ldots + p^{s-2} x_{is-1}$$

for brevity, after obvious simplifications we arrive at the system:

---

1) The condition $p \geq n^2$ is necessary for the proof of the left part of inequality (4). The right part is derived for $p > n$.

2) For $r = n$ we obtain a lemma of Ju. V. Linnik [3].

$$\left.\begin{array}{l}(x_1')^s + \ldots + (x_n')^s + sp^{s-1}((x_1')^{s-1}x_{1s} + \ldots + (x_n')^{s-1}x_{ns}) \equiv N_s \\ \ldots\ldots\ldots\ldots\ldots\ldots\ldots\ldots\ldots\ldots\ldots\ldots \\ (x_1')^n + \ldots + (x_n')^n + np^{s-1}((x_1')^{n-1}x_{1s} + \ldots + (x_n')^{n-1}x_{ns}) \equiv N_n\end{array}\right\} \pmod{p^s}. \quad (5)$$

But the quantities $x_1', \cdots, x_n'$ satisfy the system of congruences

$$\left.\begin{array}{l}(x_1')^s + \ldots + (x_n')^s \equiv N_s \\ \ldots\ldots\ldots\ldots\ldots\ldots \\ (x_1')^n + \ldots + (x_n')^n \equiv N_n\end{array}\right\} \pmod{p^{s-1}}.$$

Therefore, cancelling p^{s-1} in system (5), we obtain

$$\left.\begin{array}{l}s((x_1')^{s-1}x_{1s} + \ldots + (x_n')^{s-1}x_{ns}) \equiv N_s \\ \ldots\ldots\ldots\ldots\ldots\ldots\ldots\ldots \\ n((x_1')^{n-1}x_{1s} + \ldots + (x_n')^{n-1}x_{ns}) \equiv N_n'\end{array}\right\} \pmod{p}.$$

Since $x_{i_1} \neq x_{j_1}$ for $i \neq j$, we can let, say, the variables $x_{i_1}, \cdots, x_{n-j_1}$ differ from zero. Then

$$\Delta_s = n(n-1)\ldots s \begin{vmatrix} (x_1')^{s-1} & \ldots & (x_{n-s+1}')^{s-1} \\ \ldots & \ldots & \ldots \\ (x_1')^{n-1} & \ldots & (x_{n-s+1}')^{n-1}\end{vmatrix} \not\equiv 0 \pmod{p}.$$

Hence for fixed $x_{n-s+2,s}, \cdots, x_{ns}$ the variables $x_{1s}, \cdots, x_{n-s+1,s}$ are uniquely determined, i.e. $T_s' \leq p^{s-1}$. In addition, we know that $T_1' \leq n!$ and it trivially follows that $T_{r+1}' \leq N^n$. We hence obtain the first assertion of the lemma:

$$T_n \leqslant N^n p^{rn} T_1' \ldots T_{r+1}' \leqslant n!\, N^{2n} p^{rn + \frac{r(r-1)}{2}}.$$

We now prove the left part of inequality (4). The number of systems $x_{11}, \cdots, x_{n1}$ such that $x_{i1} \neq x_{j1}$ equals

$$p(p-1)\cdots(p-n+1) \geq p^n - n^2 p^{n-1}/2 \geq p^n/2$$

(we recall that $p \geq n^2$). We carry on from here just as we did before.

Let $T_n'(\lambda_1, \cdots, \lambda_n)$ be the number of solutions of (3) under the condition that the unknowns $y_1, \cdots, y_s$ vary just as $x_1, \cdots, x_n$. Then

$$\sum_{\lambda_1,\ldots,\lambda_n} T_n'(\lambda_1, \ldots, \lambda_n) \geqslant N^{2n} p^{2rn} \frac{1}{4}.$$

On the other hand, we have

$$\sum_{\lambda_1,\ldots,\lambda_n} T_n'(\lambda_1, \ldots, \lambda_n) \leqslant T_n'(0, \ldots, 0) \sum_{\lambda_1,\ldots,\lambda_n} 1 \leqslant T_n p^{rn - \frac{r(r-1)}{2}}.$$

Uniting the last two inequalities, we obtain the second assertion of the lemma.

Theorem 1. *Let* $N_k(\lambda_1, \cdots, \lambda_n)$ *be the number of solutions of system* (1) *with* $q = q_1$. *Then the following estimate holds for* $k \geq 6rn \ln n$:

$$N_k(\lambda_1, \ldots, \lambda_n) \leqslant N_k(0, \ldots, 0) = N_k \leqslant AP^{2k - rn + \frac{r(r-1)}{2} + \delta},$$

where

$$A = \exp c_1 k(\tau + \ln A_4), \qquad \delta = 4n\tau \left(1 - \frac{1}{r}\right)^{\tau} \leqslant \frac{c_3 \ln n}{n^3}.$$

Proof. 1. We define the numbers $P_1, P_2, \cdots, P_\tau$ as follows:

$$P_\nu = [P_{\nu-1} p_\nu^{-1}] + 1, \quad P_0 = P, \quad \nu = 1, 2, \ldots, \tau.$$

From the definition of P_ν we have $P_{\nu-1} < P_\nu p_\nu$. Let N'_k be the number of solutions of the system of congruences

$$\left.\begin{array}{l} (x_1 + p_1\bar{x}_1) + \ldots + (x_k + p_1\bar{x}_k) \equiv \\ \equiv (y_1 + p_1\bar{y}_1) + \ldots + (y_k + p_1\bar{y}_k) \\ \ldots\ldots\ldots\ldots\ldots\ldots\ldots\ldots \\ (x_1 + p_1\bar{x}_1)^n + \ldots + (x_k + p_1\bar{x}_k)^n \equiv \\ \equiv (y_1 + p_1\bar{y}_1)^n + \ldots + (y_k + p_1\bar{y}_k)^n \end{array}\right\} \pmod{q_1}, \qquad (6)$$

$$0 \leqslant x_i, y_j \leqslant p_1 - 1,$$

$$0 \leqslant \bar{x}_i, \bar{y}_i \leqslant P_1 - 1, \quad i = 1, 2, \ldots, k. \qquad (7)$$

We obviously have the inequality $N_k \leq N'_k$. We divide all collections of numbers $x_1, \cdots, x_k$ into two sets A and B. We put the collections $x_1, \cdots, x_n$ in which n different quantities x_i can be found in the set A. We put all others in B. Then we have

$$N'_k = N_k(A) + N_k(B),$$

where $N_k(A)$ is the number of solutions of system (6) with the condition that $x_1, \cdots, x_k$ and $y_1, \cdots, y_k$ are in A, and $N_k(B)$ is the number of solutions of system (6) with the condition that either $x_1, \cdots, x_k$ or $y_1, \cdots, y_k$ are in B.

2. We estimate $N_k(A)$. The n different quantities $x_i(y_i)$ can be situated in the set $x_1, \cdots, x_k$ (in $y_1, \cdots, y_k$) in C_k^n different ways. Clearly,

$$N_k(A) \leqslant (C_k^n)^2 N'_k(A),$$

where $N'_k(A)$ is the number of solutions of the system:

$$\left.\begin{array}{c}(x_1+p_1\bar{x}_1)+\ldots-(y_k+p_1\bar{y}_k)\equiv 0\\ \ldots\ldots\ldots\ldots\ldots\ldots\\ (x_1+p_1\bar{x}_1)^n+\ldots-(y_k+p_1\bar{y}_k)^n\equiv 0\end{array}\right\}\pmod{q_1}, \tag{8}$$

while $x_i\neq x_j$, $y_i\neq y_j$, $i\neq j$, $1\leq i,\ j\leq n$; and x_ν, y_ν, $\nu\geq n+1$, and $\bar{x}_s, \bar{y}_s$, $s=1,2,\cdots,k$, satisfy inequalities (7). For later arguments it helps to have an analytic formula for the quantity $N'_k(A)$. Let

$$f(x)=a_1x+\ldots+a_nx^n$$

and

$$S(x)=\sum_{y=0}^{p_1-1}\exp\frac{2\pi i}{q_1}f(x+p_1y).$$

Then we have

$$N'_k(A)=q_1^{-n}\sum_{a_1,\ldots,a_n=1}^{q_1}\left|\sum_{x_1,\ldots,x_n}S(x_1)\ldots S(x_n)\right|^2\left|\sum_{x=0}^{p_1-1}S(x)\right|^{2k-2n}$$

$$\leqslant p_1^{2k-2n-1}\sum_{x=0}^{p_1-1}q_1^{-n}\sum_{a_1,\ldots,a_n=1}^{q_1}\left|\sum_{x_1,\ldots,x_n}S(x_1)\ldots S(x_n)\right|^2|S(x)|^{2k-2n}$$

$$=p_1^{2k-2n-1}\sum_{x=0}^{p_1-1}\sigma(x)\leqslant p_1^{2k-2n}\max_{0\leqslant x\leqslant p_1-1}\sigma(x),$$

where $\sigma(x)$ is the analytic formula for the number of solutions of system (8) with the condition that $x_{n+1}=\cdots=y_k=x$. The solutions of system (8) have the following notable property: *If* $(x_1,\bar{x}_1,\cdots,y_k,\bar{y}_k)$ *is a solution of* (8), *then* $(x_1+x,\bar{x}_1,\cdots,y_k+x,\bar{y}_k)$ *is a solution of* (8) *as well, for any integer* x.

This assertion is verified by direct substitution.

We thereby pass from $\sigma(x)$ to $\sigma'(x)$, the number of solutions of the system

$$\left.\begin{array}{c}(x_1-x+p_1\bar{x}_1)+\ldots-(y_n-x+p_1\bar{y}_n)\equiv p_1(\bar{x}_{n+1}+\ldots-\bar{y}_k)\\ \ldots\ldots\ldots\ldots\ldots\ldots\\ (x_1-x+p_1\bar{x}_1)^n+\ldots-(y_n-x+p_1\bar{y}_n)^n\equiv p_1^n(\bar{x}_{n+1}^n+\ldots-\bar{y}_k^n)\end{array}\right\}\pmod{q_1},$$

where x is an integer.

Along with this system we consider two more systems of congruences:

$$\left.\begin{array}{l}\bar{x}_{n+1} + \ldots - \bar{y}_k \equiv \mu_1 \\ \ldots\ldots\ldots\ldots\ldots \\ \bar{x}^n_{n+1} + \ldots - \bar{y}^n_k \equiv \mu_n\end{array}\right\} \pmod{q_1 p_1^{-r}},$$

$$0 \leqslant \bar{x}_\nu, \bar{y}_\nu \leqslant P_1 - 1, \quad \nu = n+1, \ldots, k,$$

and

$$\left.\begin{array}{l}(x_1 - x + p_1\bar{x}_1) + \ldots - (y_n - x + p_1\bar{y}_n) \equiv \mu_1 p_1 \\ \ldots\ldots\ldots\ldots\ldots\ldots\ldots\ldots\ldots\ldots \\ (x_1 - x + p_1\bar{x}_1)^n + \ldots - (y_n - x + p_1\bar{y}_n)^n \equiv \mu_n p_1^n\end{array}\right\} \pmod{q_1},$$

$$0 \leqslant x_i, y_i \leqslant p_1 - 1; \quad x_i \neq x_j, \quad y_i \neq y_j, \quad i \neq j, \quad 1 \leqslant i, j \leqslant n;$$

$$0 \leqslant \bar{x}_i, \bar{y}_i \leqslant P_1 - 1, \quad i = 1, 2, \ldots, n.$$

Let $\overline{N}_{k-n}(\mu_1, \cdots, \mu_n)$ and $\overline{\overline{N}}(\mu_1 p_1, \cdots, \mu_n p_1^n)$ be the numbers of solutions of these two systems. We then obtain

$$\sigma'(x) = \sum_{\mu_1, \ldots, \mu_n} \overline{N}_{k-n}(\mu_1, \ldots, \mu_n) \overline{\overline{N}}(\mu_1 p_1, \ldots, \mu_n p_1^n)$$

$$\leqslant \overline{N}_{k-n} \sum_{\mu_1, \ldots, \mu_n} \overline{\overline{N}}(\mu_1 p_1, \ldots, \mu_n p_1^n),$$

where the summation is over all possible collections $\mu_1, \cdots, \mu_n$. But this last sum is the number of solutions of the system of congruences of the lemma. Applying the estimate of our lemma, we obtain

$$N_k(A) \leqslant 4^{n\tau} n! (C_k^n)^2 p_1^{2k - rn + \frac{r(r-1)}{2}} P_1^{2n} p_1^{2rn\left(1 - \frac{1}{r}\right)^r} \overline{N}_{k-n}.$$

3. We estimate $N_k(B)$. As in step 2, we have

$$N_k(B) = q_1^{-n} \sum_{a_1, \ldots, a_n = 1}^{q_1} \left[\sum_{x_1, \ldots, y_k} S(x_1) \ldots \overline{S(y_1)}\right],$$

where the variables $x_1, \cdots, y_k$ vary so that either the collection $x_1, \cdots, x_k$ or the collection $y_1, \cdots, y_k$ belongs to B. We calculate the number of collections in B. By definition, the collection $x_1, \cdots, x_k$ belongs to B if there are no more than $n-1$ different x_i. Let the different values be already chosen (there are no more than p_1^{n-1} of them). Then the collection $x_1, \cdots, x_k$ consists of these values and ones coinciding with them; in other words, $x_1, \cdots, x_k$ corresponds to a k-digit number in the $(n-1)$-adic number system. Consequently, for the different $x_1, \cdots, x_{n-1}$ fixed, there exist

$(n-1)^k$ collections $x_1, \cdots, x_k$ of B, and so there are no more than $n^k p_1^{n-1}$ collections in B. Using the fact that the geometric mean of non-negative numbers does not exceed the arithmetic mean, we arrive at the inequality

$$N_k(B) \leqslant q_1^{-n} n^k p_1^{k+n-1} \sum_{x=0}^{p_1-1} |S(x)|^{2k}.$$

Since $|S(x)| \leq P_1$ and since the systems under consideration have property α, we see that

$$N_k(B) \leqslant q_1^{-n} n^k p_1^{k+n-1} P_1^{2n} \sum_{x=0}^{p_1-1} |S(x)|^{2k-2n} = n^k p_1^{k-n} P_1^{2n} \overline{N}_{k-n}.$$

Bringing together the estimates for $N_k(A)$ and $N_k(B)$, we obtain

$$N_k \leqslant N_k(A) + N_k(B)$$
$$\leqslant p_1^{2k-rn+\frac{r(r-1)}{2}} P_1^{2n} p_1^{2rn\left(1-\frac{1}{r}\right)^\tau} \overline{N}_{k-n} [4^{n\tau} n! (C_k^n)^2 + n^k p_1^{-k+rn+n}]$$
$$\leqslant e^{c_4 k} P_1^{2n} p_1^{2k-rn+\frac{r(r-1)}{2}} p_1^{2rn\left(1-\frac{1}{r}\right)^\tau} \overline{N}_{k-n},$$

since

$$k \geqslant 6rn \ln n, \quad C_k^n \leqslant 2^k, \quad p_1 > \exp crn^5.$$

We recall that $\overline{N}_{k-n}$ is the number of solutions of system (1) with $\lambda_1 = \cdots = \lambda_n = 0$ and with P replaced by P_1, q_1 by $q_1 p_1^{-r}$, and k by $k-n$.

4. We proceed analogously with the quantity $\overline{N}_{k-n}$. Continuing this process τ times, we obtain the inequality

$$N_k \leqslant e^{c_4 k\tau} (P_1 \cdots P_\tau)^{2n} (p_1 \cdots p_\tau)^{2k-rn+\frac{r(r-1)}{2}} p_2^{-2n} \cdots p_\tau^{-2n(\tau-1)} P_\tau^{2k-2n\tau}$$
$$\times p_1^{2rn(1-1/r)^\tau} p_2^{2rn(1-1/r)^{\tau-1}} \cdots p_\tau^{2rn(1-1/r)}.$$

The assertion of the theorem is obtained from the following simple calculations:

$$P_\tau \leqslant P_\tau;\ P_{\tau-1} \leqslant P_\tau p_\tau;\ P_{\tau-2} \leqslant P_{\tau-1} p_{\tau-1} \leqslant P_\tau p_{\tau-1} p_\tau;\ \ldots;$$
$$P_1 \leqslant P_2 p_2 \leqslant \ldots \leqslant P_\tau p_2 \ldots p_\tau;$$
$$P_1 \ldots P_\tau \leqslant p_2 p_3^2 \ldots p_\tau^{\tau-1} P_\tau^\tau; \quad P_\tau < P_{\tau-1} p_\tau^{-1} + 1 < P_{\tau-2} p_{\tau-1}^{-1} p_\tau^{-1} + 1 + p_\tau^{-1}$$
$$< \ldots < P p_1^{-1} \ldots p_\tau^{-1} + 1 + p_\tau^{-1} + \ldots + p_2^{-1} \ldots p_\tau^{-1} \leqslant 2A_4.$$

Remark 1. We note that for somewhat different relationships among the quantities p_i we can obtain an estimate for N_k with fewer factors depending on n.

Remark 2. I. M. Vinogradov's theorem on the mean follows from this theorem for $r = n$.

§2

We apply our theorem to estimate a rational trigonometric sum.

Theorem 2. *Let* $f(x) = a_1 x + \cdots + a_{n+1} x^{n+1}$, $(a_{n+1}, q_1) = 1$, *and*

$$S = \sum_{x \leqslant P} \exp \frac{2\pi i}{q_1} f(x).$$

Then

$$|S| \leqslant e^{c_5 (r \ln n + \ln A_4)} P^{1 - \frac{\gamma}{rn \ln n}},$$

where c_5 *and* γ *are absolute (positive) constants.*

Proof. We take $P_1 = [P^{1-\gamma/rn \ln n}]$. It is obvious that

$$|S| \leqslant P_1^{-1} \sum_{y=1}^{P_1} \Big| \sum_{x=1}^{P} \exp \frac{2\pi i}{q_1} f(x+y) \Big| + 2P^{1 - \frac{\gamma}{rn \ln n}}.$$

We estimate

$$S_1 = P_1^{-1} \sum_{y=1}^{P_1} \Big| \sum_{x=1}^{P} \exp \frac{2\pi i}{q_1} f(x+y) \Big|.$$

We raise this equation to the power $2k$ and apply Hölder's inequality:

$$S_1^{2k} = P_1^{-1} \sum_{y=1}^{P_1} \Big| \sum_{x=1}^{P} \exp \frac{2\pi i}{q_1} [\varphi_1(y) x + \ldots + \varphi_{n+1}(y) x^{n+1}] \Big|^{2k}$$

$$= P_1^{-1} \sum_{y=1}^{P_1} \sum_{\lambda_1, \ldots, \lambda_{n+1}} N_k(\lambda_1, \ldots, \lambda_{n+1}) \exp \frac{2\pi i}{q_1} [\varphi_1(y) \lambda_1 + \ldots + \varphi_{n-1}(y) \lambda_{n+1}],$$

where $\phi_\nu(y) = f^{(\nu)}(y)/\nu!$, $\nu = 1, 2, \cdots, n+1$; in particular, $\phi_{n+1}(y) = a_{n+1}$. Using the fact that $\Sigma_{\lambda_{n+1}} N_k(\lambda_1, \cdots, \lambda_{n+1}) = N_k(\lambda_1, \cdots, \lambda_n)$, we obtain

$$S_1^{2k} \leqslant P_1^{-1} \sum_{\lambda_1, \ldots, \lambda_n} N_k(\lambda_1, \ldots, \lambda_n) \Big| \sum_{y=1}^{P_1} \exp \frac{2\pi i}{q_1} [\varphi_1(y) \lambda_1 + \ldots + \varphi_n(y) \lambda_n] \Big|.$$

We square this inequality:

$$S_1^{4k} \ll P_1^{-2} P^{2k} \sum_{\lambda_1,\ldots,\lambda_n} N_k(\lambda_1,\ldots,\lambda_n) \sum_{y,y_1} \exp\frac{2\pi i}{q_1}[(\varphi_1(y)-\varphi_1(y_1))\lambda_1+\ldots]$$

$$\ll P^{2k+rn-\frac{r(r+1)}{2}} P_1^{-1} N_k q_1,$$

i.e., for $k = c_6 rn \ln n$, we have

$$S_1 \ll e^{c_7(r\ln n + \ln A_4)} P^{1-\frac{\gamma}{rn\ln n}},$$

which implies the assertion of the thoerem.

We note that, somewhat complicating the proof, we can free ourselves from $\ln n$ in the lowering factor $\gamma/rn\ln n$.

Theorem 1 can be refined. We have

Theorem 3. *The following estimate holds for* N_k *with* $k \geq 6rn$:

$$N_k \ll A_5 P^{2k-rn+\frac{r(r-1)}{2}}. \qquad (*)$$

This theorem is proved just as Lemma 7 in [1] (Russian p. 418), except that Theorem 1 must be used in the appropriate place. Moreover, in this case we can obtain an asymptotic formula for N_k.

The following theorem shows that the results of Theorems 1 and 3 do not hold for arbitrary q.

Theorem 4. *Let* $q_2 = p_1^{\alpha_1}\cdots p_s^{\alpha_s}$, $\alpha_1 = \max_{1\leq\nu\leq s}\alpha_\nu > r$ *and* $p_1 \to \infty$ *as* $q_2 \to \infty$. *Then estimate* $(*)$ *cannot be obtained for*

$$k < \frac{1}{2}\alpha n - \frac{1}{4}\alpha(\alpha-1),$$

where $\alpha = \min(\alpha_1, n)$.[1]

Proof. We consider system (1) with $q = q_2$ and $\lambda_1 = \cdots = \lambda_n = 0$. Since

$$P \gg A_3 q^{\frac{1}{r}} > A_3 p_1^{\frac{\alpha}{r}} \gg A_3 p_1^{1+\frac{1}{r}},$$

we have $Pp_1^{-1} > A_3 p_1^{1/r} \to \infty$ as $q_2 \to \infty$. We shall estimate from below the number of solutions of our system of congruences of the form $(p_1\bar{x}_1,\cdots,p_1\bar{y}_k)$. In other words, we consider the system of congruences:

1) N. M. Korobov proved this theorem for $q = p^s$ and $s > n$.

$$\left.\begin{array}{l} \bar{x}_1 + \ldots - \bar{y}_k = 0 \\ \ldots\ldots\ldots\ldots\ldots \\ \bar{x}_1^r + \ldots - \bar{y}_k^r = 0 \\ \left.\begin{array}{l} p_1^{r+1}(\bar{x}_1^{r+1} + \ldots - \bar{y}_1^{r+1}) \equiv 0 \\ \ldots\ldots\ldots\ldots\ldots \\ p_1^n(\bar{x}_1^n + \ldots - \bar{y}_k^n) \equiv 0 \end{array}\right\} \pmod{q_2} \end{array}\right\} \qquad (9)$$

$$1 \leqslant \bar{x}_\nu, \bar{y}_\nu \leqslant Pp_1^{-1}, \quad \nu = 1, 2, \ldots, k.$$

Now let $q_{r+1} = q_2 p_1^{-r-1}, \cdots, q_\alpha = q_2 p_1^{-\alpha}$, $q_\nu = q_2 p_1^{-\alpha}$, $n \geq \nu > \alpha$. Then for $\lambda_1 = \cdots = \lambda_n = 0$ system (9) is equivalent to the system

$$\left\{\begin{array}{l} \bar{x}_1 + \ldots - \bar{y}_k = \lambda_1 \\ \ldots\ldots\ldots\ldots\ldots \\ \bar{x}_1^r + \ldots - \bar{y}_k^r = \lambda_r \\ \bar{x}_1^{r+1} + \ldots - \bar{y}_k^{r+1} \equiv \lambda_{r+1} \pmod{q_{r+1}} \\ \ldots\ldots\ldots\ldots\ldots\ldots\ldots \\ \bar{x}_1^n + \ldots - \bar{y}_k^n \equiv \lambda_n \pmod{q_n} \end{array}\right.$$

$$1 \leqslant \bar{x}_\nu, \bar{y}_\nu \leqslant Pp_1^{-1}, \quad \nu = 1, 2, \ldots, k.$$

We let $N_k'(\lambda_1, \cdots, \lambda_n)$ designate the number of solutions of this system. It follows from what was said above that $N_k(q_2) > N_k'$. We now proceed just as at the beginning of the article when estimating N_k from below:

$$P^{2k} p_1^{-2k} = \sum_{\lambda_1, \ldots, \lambda_n} N_k'(\lambda_1, \ldots, \lambda_n) \leqslant k^n (Pp_1^{-1})^{\frac{r(r+1)}{2}} \prod_{\nu=r+1}^{\alpha} q_2 p_1^{-\nu} \prod_{\nu=\alpha+1}^{n} q_2 p_1^{-\alpha} N_k'.$$

From this we obtain

$$N_k(q_2) > N_k' > A_6 P^{2k - rn + \frac{r(r-1)}{2}} p_1^{-2k + \alpha n - \frac{\alpha(\alpha-1)}{2}},$$

which proves the theorem.

Remark 1. Theorems 1, 2, and 3 can also be proved for other q similar in structure to q_1.

Remark 2. Evidently, the following assertion holds: *let* $q_3 = q_4 p_1^{\alpha_1} \cdots p_s^{\alpha_s}$, *where* $q_4 \leq c_8$ *and* $p_i \to \infty$, $1 \leq i \leq s$, *as* $q_3 \to \infty$. *Then if*

$$\max_{1 \leqslant \nu \leqslant s} \alpha_\nu \leqslant r,$$

it follows that for $k = c_9 m \ln n$ *we have*

$$N_k(q_3) \leqslant A_7 P^{2k-rn+\frac{r(r-1)}{2}}.$$

Proving this assertion would allow the results of Theorem 2 to be carried over to arbitrary moduli q.

BIBLIOGRAPHY

[1] I. M. Vinogradov, *Selected works,* Izdat. Akad. Nauk SSSR, Moscow, 1952. (Russian) MR 14, 610.

[2] N. M. Korobov, *Weyl sum estimates and their applications,* Proc. Fourth All-Union Math. Congress (Leningrad, 1961), vol. II, "Nauka", Leningrad, 1964, pp. 112–116. (Russian) MR 36 #2570.

[3] Ju. V. Linnik, *On Weyl's sums,* Dokl. Akad. Nauk SSSR **34** (1942), 184–186. (Russian) MR 4, 211.

[4] A. A. Karacuba, *Waring's problem for a congruence modulo the power of a prime,* Vestnik. Moskov. Univ. Ser. I Mat. Meh. **1962**, no. 4, 28–38. (Russian) MR 26 #97.

Translated by:
N. Koblitz

ON THE REPRESENTATION OF NUMBERS BY POSITIVE BINARY DIAGONAL QUADRATIC FORMS

G. A. LOMADZE

§1

It is well known that it is possible to obtain by elementary methods the formulas for the number of representations of natural numbers by a positive binary quadratic form whose genus consists of one class. Concerning the formulas for the number of representations by a binary quadratic form whose genus contains more than one class, the only result, as far as we know, is in the paper by van der Blij [7] where explicit formulas are obtained simultaneously for the number of representations of integers by forms $x_1^2 + x_1x_2 + 6x_2^2$, $2x_1^2 + x_1x_2 + 3x_2^2$ and $2x_1^2 - x_1x_2 + 3x_2^2$, belonging to the same genus. However, these results are of a particular nature and are obtained by a method which does not allow generalization.

Paper [3] gives an indication of a proof of explicit formulas for the number of representations of natural numbers by forms $x_1^2 + 11x_2^2$, $x_1^2 + 17x_2^2$ and $4x_1^2 + 5x_2^2$, each belonging to a genus containing more than one class. In this proof, similarly to the papers [18], [17] and [2], we have employed the apparatus of integral modular forms belonging to the principal congruence group $\Gamma(N)$, which involves laborious, though quite elementary calculations.

In the present paper, presenting the details of the formulas of our note [3], we give a general approach for obtaining the explicit formulas for the number of representations of a number by means of an arbitrary positive binary quadratic form. Moreover, as in the author's paper [4] we use the apparatus of integral modular forms adjoined to the congruence subgroup $\Gamma_0(N)$ which enables us to decrease considerably the amount of calculations.

Further, the method is illustrated by the deduction of the explicit formulas for representation of numbers by forms $x_1^2 + 11x_2^2$, $x_1^2 + 17x_2^2$, $x_1^2 + 20x_2^2$ and $4x_1^2 + 5x_2^2$. Formulas for other concrete forms may be obtained similarly.

§2

In the present paper the following notation is used: N, a, d, k, n, q, r, λ, ω denote natural numbers (except for §3, where q may be any integer); s is a natural number not less than 2; b and u are odd natural numbers, p denotes a prime; κ and l are nonnegative integers; H, c, f, g, h, j, m, x, y, α, β, γ, δ are integers; A, μ, ν, σ, t, w are real numbers; i is $\sqrt{-1}$;

z, ζ, τ, A, C are complex quantities where $\operatorname{Im} \tau > 0$. K denotes a positive number, not necessarily the same for every case and possibly dependent on parameters. These letters, when necessary, are accompanied by indices and strokes.

(h, m) denotes the greatest common divisor of h and m. $d \mid m$ means that d divides m, $d \nmid m$ means d does not divide m, $p^\lambda \| m$ means that $p^\lambda \mid m$ but $p^{\lambda+1} \nmid m$ (for $m = 0$ assume $\lambda = \infty$, i.e. that λ is greater than any given number); $p^\circ \| m$ means $p \nmid m$. The symbol (h/u) is the generalized Jacobi symbol; (h/c), $c > 0$ denotes the Kronecker symbol if h is not a square and $h \equiv 0$ or $1 \pmod 4$.

Further, take $I(u) = i^{\frac{1}{4}(u-1)^2}$; $e(z) = e^{2\pi i z}$, and let $\Sigma_{h \bmod q}$ and $\Sigma'_{h \bmod q}$ denote sums where h runs through a complete and a reduced system of residues modulo q respectively. Empty sums are assumed equal to zero, and empty products are equal to one.

For convenience of reference, we now give necessary definitions and quote some known results as lemmas.

Definition 1. If $z \neq 0$ we set

$$z^{\frac{k}{2}} = (z^{\frac{1}{2}})^k, \quad -\frac{\pi}{2} < \arg z^{\frac{1}{2}} \leqslant \frac{\pi}{2}.$$

Consequently

$$0 < \arg (\mu\tau + \nu)^{\frac{1}{2}} < \frac{\pi}{2} \quad \text{for } \mu > 0,$$

$$-\frac{\pi}{2} < \arg (\mu\tau + \nu)^{\frac{1}{2}} < 0 \quad \text{for } \mu < 0.$$

Further, for $\mu \neq 0$

$$\{-(\mu\tau + \nu)\}^{\frac{1}{2}} = i^{-\operatorname{sgn} \mu} (\mu\tau + \nu)^{\frac{1}{2}}. \tag{2.1}$$

Let Γ be a modular group, i.e. the group of linear substitutions $\tau' = (\alpha\tau + \beta)/(\gamma\tau + \delta)$ where $\alpha\delta - \beta\gamma = 1$. Further, let $\Gamma_0(N)$ denote the subgroup of Γ whose substitutions satisfy the congruence $\gamma \equiv 0 \pmod N$.

Definition 2 (see, for example, [11], p. 716 or [12], pp. 808–809). An analytic function $F(\tau)$ is called an *integral modular form* of dimension $-r$, adjoined to the subgroup $\Gamma_0(N)$, if it satisfies the following conditions:

1) $F(\tau)$ is regular and invariant in the upper halfplane.

2) For every substitution of the group $\Gamma_0(N)$

$$F\left(\frac{\alpha\tau+\beta}{\gamma\tau+\delta}\right)=\varepsilon(\delta)(\gamma\tau+\delta)^r F(\tau), \quad |\varepsilon(\delta)|=1.$$

3) In the neighborhood of $\tau=\infty$ the following expansion is valid:

$$F(\tau)=\sum_{m=0}^{\infty} C_m e\left(\frac{m\tau}{N}\right). \tag{2.2}$$

4) In the neighborhood of $\tau=-\delta/\gamma$ $(\gamma\neq 0,\ (\gamma,\delta)=1)$ the following expansion holds:

$$(\gamma\tau+\delta)^r F(\tau)=\sum_{m=0}^{\infty} C'_m e\left(\frac{m}{N}\frac{\alpha\tau+\beta}{\gamma\tau+\delta}\right), \tag{2.3}$$

where $\alpha\delta-\beta\gamma=1$.

Definition 3 (see for example [12], pp. 808 and 811). An integral modular form $F(\tau)$ is called a *function of divisor* N if all exponents m which occur in the expansion (2.2) are divisible by N. Then (2.2) becomes

$$F(\tau)=\sum_{m=0}^{\infty} A_m e(m\tau). \tag{2.4}$$

Lemma 1 ([12], pp. 811 and 853). *An integral modular form $F(\tau)$ of dimension $-r$, adjoined to the subgroup $\Gamma_0(N)$ and of divisor N, is identically zero if in the expansion* (2.4)

$$A_m=0 \quad \textit{for all} \quad m\leqslant\frac{r}{12}N\prod_{p\mid N}\left(1+\frac{1}{p}\right).$$

Lemma 2. *Let $S(h,q)=\sum_{j \bmod q} e(hj^2/q)$. Then $S(rh,rq)=rS(h,q)$ for $(h,q)=1$.*

Lemma 3 ([1], p. 10, Lemma 1). *Let $(h,q)=1$. Then $|S(h,q)|\leq(2q)^{1/2}$.*

Lemma 4 ([1], p. 11, Lemma 2). *If h is odd then*

$$S(h,2^\lambda)=\begin{cases} 0 & \textit{for } \lambda=1,\\ (1+i^h)2^{\frac{\lambda}{2}} & \textit{for even } \lambda,\\ e\left(\frac{h}{8}\right)2^{\frac{1}{2}(\lambda+1)} & \textit{for odd } \lambda>1.\end{cases}$$

Lemma 5 (see [1], p. 15, Lemma 8). *Let $(h,u)=1$. Then $S(h,u)=(h/u)I(u)u^{1/2}$.*

Lemma 6 ([1], p. 177, formula (20)). *Let $q=p^\lambda$ and $p^\kappa \| h$. Then*

$$c(h,q)=\sum_{j \bmod q}{}' e\left(\frac{hj}{q}\right)=\begin{cases}0 & \text{for } \varkappa<\lambda-1,\\ -p^{\lambda-1} & \text{for } \varkappa=\lambda-1,\\ p^{\lambda-1}(p-1) & \text{for } \varkappa>\lambda-1.\end{cases}$$

Lemma 7 ([8], p. 27, Lemma 6). *Let* $p>2$, $p^{\kappa}\| h$. *Then*

$$\sum_{j \bmod p^{\lambda}}{}' \left(\frac{j}{p}\right) e\left(\frac{hj}{p^{\lambda}}\right)=\begin{cases}0 & \text{for } \varkappa\neq\lambda-1,\\ p^{\lambda-\frac{1}{2}}\left(\dfrac{p^{-\varkappa}h}{p}\right) I(p) & \text{for } \varkappa=\lambda-1.\end{cases}$$

Lemma 8 ([16], p. 172, Theorem 213). *Every* $h\equiv 0$ *or* $1 \pmod 4$ *which is not a square is uniquely representable in the form* $h=fm^2$ *where* $m>0$ *and* f *is the fundamental discriminant.*

Lemma 9 ([5], p. 148 and 155). *The Kronecker symbol* (h/q) *is a character with leading modulus*

$$k^*=\begin{cases}|f|, & \text{if } f\equiv 1 \pmod 4,\\ 4\,|f|, & \text{if } f\not\equiv 1 \pmod 4,\end{cases}$$

where f *is a square-free kernel of number* h.

Lemma 10 ([6], p. 33, Theorem 12). *If* h *is the fundamental discriminant, then the Kronecker symbol* (h/c) *for all natural* c *coincides with the real primitive character* $\chi(c)$ *whose modulus* k *is equal to* $|h|$, *where* h *and* k *satisfy the following relation:* $h=\chi(-1)\cdot k$.

Lemma 11 ([18], p. 31, Lemma 31, or [10], pp. 345 and 392–394). *For* $\operatorname{Re}(z+\zeta)>1$ *the integral*

$$U(z;\zeta,\tau,\nu)=\int_{-\infty}^{\infty}\frac{e(-\nu t)\,dt}{(t+\tau)^{\zeta}|t+\tau|^{z}}$$

is absolutely convergent. The function $U(z;\zeta,\tau,\nu)$ *has the following properties:*

1) *For* $\nu\neq 0$ *it may be continued over the entire* z*-plane.*

2) $U(z;\zeta,\tau,0)$ *is regular in the halfplane* $\operatorname{Re}(z+\zeta)>1$.

3) *If* D *is a bounded domain in the* z*-plane and if* $\nu\neq 0$, *then the following inequality holds in* D:

$$|U(z;\zeta,\tau,\nu)|<Ke^{-K'|\nu|},$$

where K *and* K' *are independent of* ν *and* z; *if* $\nu=0$, *then this inequality holds for* $\operatorname{Re} z>1-\operatorname{Re}\zeta$.

$$4)\ U(0;\zeta,\tau,\nu)=\begin{cases}0 & g\ \textit{for}\ \nu<0,\\ \dfrac{(2\pi)^{\zeta}\nu^{\zeta-1}}{\Gamma(\zeta)\,e\left(\frac{\zeta}{4}\right)}\,e(\tau\nu) & \textit{for}\ \nu>0,\ \zeta\neq 0,-1,-2,\ldots.\end{cases}$$

Lemma 12 ([16], p. 204, Theorem 237 and [15], p. 449). *Let* $L(z,\chi)=\sum_{n=1}^{\infty}\chi(n)n^{-z}$. *Then for* Re $z>1$ *we have*

$$\prod_{p}(1-\chi(p)p^{-z})=\frac{1}{L(z,\chi)}.$$

If χ *is a nonprincipal character this relation holds also for* $z=1$.

Lemma 13 ([6], p. 91, Theorem 33). *If* χ *is a primitive character modulo* k, *satisfying* $\chi(-1)=-1$ *then in the whole* z*-plane apart from isolated points* $z=0,-1,-2,\cdots$ *the following functional equation holds:*

$$L(1-z,\chi)=\frac{2k^{z-\frac{1}{2}}\Gamma(z)}{(2\pi)^{z}}\sin\frac{\pi}{2}zL(z,\overline{\chi}).$$

Lemma 14 ([6], p. 111). *Let* χ *be a character modulo* k, *let* k^{*} *be its leading modulus, and let* χ^{*} *be the corresponding primitive character modulo* k^{*}. *Then between* $L(z,\chi)$ *and* $L(z,\chi^{*})$ *there exists a relation* $L(z,\chi)=\prod_{p|k,p\nmid k^{*}}(1-\chi^{*}(p)p^{-z})L(z,\chi^{*})$.

Lemma 15 ([9], p. 489–491, formulas (d), (e), (f), (g)). *Let*

$$L(k,m)=\sum_{u=1}^{\infty}\left(\frac{m}{u}\right)\frac{1}{u^{k}} \tag{2.5}$$

and let ω *be a square-free number. Then*

$$\begin{aligned}L(1,-\omega)&=\frac{\pi}{4}\quad \textit{for}\ \omega=1,\\ &=\frac{\pi}{\omega^{\frac{1}{2}}}\sum_{1\leqslant h\leqslant\frac{\omega}{4}}\left(\frac{h}{\omega}\right)\quad \textit{for}\ \omega\equiv 1\ (\mathrm{mod}\ 4),\ \omega>1,\\ &=\frac{\pi}{2\omega^{\frac{1}{2}}}\sum_{1\leqslant h\leqslant\frac{\omega}{2}}\left(\frac{h}{\omega}\right)\quad \textit{for}\ \omega\equiv 3\ (\mathrm{mod}\ 4),\\ &=\frac{\pi}{2^{\frac{3}{2}}}\quad \textit{for}\ \omega=2,\end{aligned}$$

$$= \frac{\pi}{\omega^{\frac{1}{2}}} \left\{ \sum_{1 \leq h \leq \frac{\omega}{16}} \left(\frac{h}{\frac{1}{2}\omega} \right) - \sum_{\frac{3\omega}{16} < h \leq \frac{\omega}{4}} \left(\frac{h}{\frac{1}{2}\omega} \right) \right\} \text{ for } \omega \equiv 2 \pmod 8), \omega > 2,$$

$$= \frac{\pi}{\omega^{\frac{1}{2}}} \sum_{\frac{\omega}{16} < h \leq \frac{3\omega}{16}} \left(\frac{h}{\frac{1}{2}\omega} \right) \text{ for } \omega \equiv 6 \pmod 8).$$

§3

From now on we assume that the coefficients of the form $F = a_1x_1^2 + a_2x_2^2 + \cdots + a_sx_s^2$ are relatively prime. The determinant of F will be denoted by Δ, and the least common multiple of all coefficients a_k will be denoted by a.

Consider the function

$$\Psi_s(\tau, z) = \frac{\left(\frac{i}{2}\right)^{\frac{s}{2}}}{2\Delta^{\frac{1}{2}}} \sum_{\substack{q, H=-\infty \\ (H,q)=1}}^{\infty}{}' i^{\frac{s}{2}(\operatorname{sgn} q - 1)} \frac{\prod_{k=1}^{s} S(-a_kH \operatorname{sgn} q, |q|)}{|q|^{\frac{s}{2}} (q\tau + H)^{\frac{s}{2}}} \cdot \frac{1}{|q\tau + H|^z}, \tag{3.1}$$

where the prime signifies that the terms with $q = 0$ are absent. By Lemma 3 the function $\Psi_s(\tau, z)$ is regular for a fixed τ and $\operatorname{Re} z > 2 - s/2$.

Lemma 16. *Let*

$$A_q\left(\frac{m}{4a}\right) = q^{-s} \sum_{h \bmod q}{}' e\left(-\frac{mh}{4aq}\right) \prod_{k=1}^{s} S(a_kh, q). \tag{3.2}$$

Then, for $\operatorname{Re} z > 2 - s/2$ *we have*

$$\Psi_s(\tau, z) = 1 + \frac{i^{\frac{s}{2}}}{2^{\frac{3s}{2} - 2 + 2z} \cdot \Delta^{\frac{1}{2}} \cdot a^{\frac{s}{2} - 1 + z}} \sum_{\substack{m=-\infty \\ 4a|m}}^{\infty} T_s(m, z) V_s\left(\frac{m}{4a}, z\right), \tag{3.3}$$

where

$$T_s(m, z) = \int_{-\infty}^{\infty} \frac{e(-mt)\, dt}{\left(\frac{\tau}{4a} + t\right)^{\frac{s}{2}} \left|\frac{\tau}{4a} + t\right|^z}, \tag{3.4}$$

$$V_s\left(\frac{m}{4a}, z\right) = \sum_{q=1}^{\infty} A_q\left(\frac{m}{4a}\right) q^{-z}. \tag{3.5}$$

Proof. For $\operatorname{Re} z > 2 - s/2$ it follows from (3.1) and (2.1) that

$$\Psi_s(\tau, z) = 1 + \frac{\left(\frac{i}{2}\right)^{\frac{s}{2}}}{\Delta^{\frac{1}{2}}} \sum_{q=1}^{\infty} \sum_{\substack{H=-\infty \\ (H,q)=1}}^{\infty} \frac{\prod_{k=1}^{s} S(-a_k H, q)}{q^{\frac{s}{2}} (q\tau + H)^{\frac{s}{2}}} \cdot \frac{1}{|q\tau + H|^z}.$$

Further, by reasoning similar to that used in the proof of Lemma 20 in [2] (p. 115) we obtain the required statement.

Lemma 17. *Let*

$$\mathrm{X}_p\left(\frac{m}{4a}, z\right) = 1 + \sum_{\lambda=1}^{\infty} A_{p^\lambda}\left(\frac{m}{4a}\right) p^{-\lambda z}. \tag{3.6}$$

Then for $\operatorname{Re} z > 2 - s/2$ *we have*

$$V_s\left(\frac{m}{4a}, z\right) = \prod_p \mathrm{X}_p\left(\frac{m}{4a}, z\right). \tag{3.7}$$

Proof. Let $(q_1, q_2) = 1$. Then from (3.2) we obtain

$$A_{q_1}\left(\frac{m}{4a}\right) A_{q_2}\left(\frac{m}{4a}\right)$$

$$= (q_1 q_2)^{-s} \sum_{h_1 \bmod q_1}' \sum_{h_2 \bmod q_2}' e\left(-\frac{Hm}{4aq_1q_2}\right) \prod_{k=1}^{s} \{S(a_k h_1, q_1) S(a_k h_2, q_2)\},$$

where $H = h_1 q_2 + h_2 q_1$ runs through a reduced system of residues modulo $q_1 q_2$. It is known ([13], p. 149, formula (2.4.13)) that $S(a_k h_1, q_1) S(a_k h_2, q_2) = S(a_k H, q_1 q_2)$. Therefore $A_q(m/4a)$ is multiplicative with respect to q.

Again, according to Lemma 3, the series $\sum_{q=1}^{\infty} A_q(m/4a) q^{-z}$ is absolutely convergent for $\operatorname{Re} z > 2 - s/2$. Consequently, for $\operatorname{Re} z > 2 - s/2$ we have

$$\sum_{q=1}^{\infty} A_q\left(\frac{m}{4a}\right) q^{-z} = \prod_p \mathrm{X}_p\left(\frac{m}{4a}, z\right).$$

Lemma 18. *If* $\operatorname{Re} z > 2 - s/2$, *then for every substitution of the group* $\Gamma_0(4a)$ *the following relation holds:*

$$\Psi_s\left(\frac{\alpha\tau + \beta}{\gamma\tau + \delta}, z\right)$$

$$= i^{\frac{s}{2}\eta(\gamma)(\operatorname{sgn}\delta - 1)} \cdot i^{\frac{s}{4}(|\delta| - 1)^2} \left(\frac{\Delta}{|\delta|}\right)\left(\frac{\beta \operatorname{sgn}\delta}{|\delta|}\right)^s (\gamma\tau + \delta)^{\frac{s}{2}} |\gamma\tau + \delta|^z \Psi_s(\tau, z),$$

where $\eta(y) = 1$ *for* $y \geq 0$ *and* $\eta(y) = -1$ *for* $y < 0$.

Proof. In (3.1) take $(\alpha\tau' + \beta)/(\gamma\tau' + \delta)$ instead of τ, and assume that

$$\alpha q + \gamma H = q_0, \qquad \delta q_0 - \gamma H_0 = q,$$

i.e.

$$\beta q + \delta H = H_0, \qquad -\beta q_0 + \alpha H_0 = H;$$

then, omitting the prime in τ', we obtain

$$\Psi_s\left(\frac{\alpha\tau+\beta}{\gamma\tau+\delta}, z\right) = 1 + \frac{\left(\frac{i}{2}\right)^{\frac{s}{2}}}{2\Delta^{\frac{1}{2}}} \sum_{\substack{q,H=-\infty \\ (H,q)=1}}^{\infty}{}' i^{\frac{s}{2}(\operatorname{sgn} q - 1)} \frac{\prod_{k=1}^{s} S(-a_k H \operatorname{sgn} q, |q|)}{|q|^{\frac{s}{2}} \left(\frac{q_0\tau + H_0}{\gamma\tau+\delta}\right)^{\frac{s}{2}}} \left|\frac{\gamma\tau+\delta}{q_0\tau + H_0}\right|^z.$$

Then, proceeding as in the proof of Lemma 12 of [4], we obtain the required statement.

Consider now the function

$$\vartheta_{gh}(\tau; c, N) = \sum_{\substack{m=-\infty \\ m\equiv c \pmod N}}^{\infty} (-1)^{\frac{h(m-c)}{N}} e\left(\frac{\left(m+\frac{g}{2}\right)^2}{2N}\tau\right). \tag{3.8}$$

It is known (see for example [14], p. 318, formulas (1.2) and (1.4)), that

$$\vartheta_{g+2j,h}(\tau; c, N) = \vartheta_{gh}(\tau; c+j, N), \tag{3.9}$$

$$\vartheta_{gh}(\tau; c+Nj, N) = (-1)^{hj}\vartheta_{gh}(\tau; c, N). \tag{3.10}$$

Further, clearly

$$\vartheta_{-g,h}(\tau; 0, N) = \vartheta_{gh}(\tau; 0, N). \tag{3.11}$$

Lemma 19 ([4], Lemma 18). *Let* s *and* g_k $(k = 1, 2, \cdots, s)$ *be even, and let*

$$N_k \mid a, \quad 4 \,\Big|\, a\sum_{k=1}^{s} \frac{h_k^2}{N_k},$$

$$4 \,\Big|\, \sum_{l=1}^{m} \sum_{k=s_{l-1}+1}^{s_l} \frac{g_k^2}{4N_{s_l}} \; (s_0 = 0, s_m = s), \quad \sum_{k=s_{l-1}+1}^{s_l} \frac{g^2}{2N_{s_l}} \tag{3.12}$$

be integers. Further, for all α *and* δ *satisfying* $\alpha\delta \equiv 1 \pmod{4a}$ *let*

$$\prod_{k=1}^{s}\left\{\left(\frac{N_k}{|\delta|}\right)\vartheta_{\alpha g_k, h_k}(\tau; 0, 2N_k)\right\} = \left(\frac{\Delta}{|\delta|}\right)\prod_{k=1}^{s}\vartheta_{g_k h_k}(\tau; 0, 2N_k). \quad (3.13)$$

Then the function $\Pi_{k=1}^{s}\vartheta_{g_k h_k}(\tau; 0, 2N_k)$ *is an integral modular form of dimension* $-s/2$, *adjoined to the subgroup* $\Gamma_0(4a)$ *and of divisor* $4a$.

Remark. In particular, $\Pi_{k=1}^{s}\vartheta_{00}(\tau; 0, 2a_k)$ for any a_k is an integral modular form of dimension $-s/2$, adjoined to the subgroup $\Gamma_0(4a)$, and of divisor $4a$.

Lemma 20 ([4], Lemma 19). *Let* s *and* g_k $(k = 1, 2, \cdots, sq)$ *be even, let*

$$N_k \mid a, 4q \,\Big|\, a\sum_{k=1}^{sq}\frac{h_k^2}{N_k},$$

$$4q \,\Big|\, \sum_{k=1}^{m}\sum_{k=s_{l-1}+1}^{s_l}\frac{g_k^2}{4N_{s_l}} \quad (s_0 = 0, s_m = sq), \quad \sum_{k=s_{l-1}+1}^{s_l}\frac{g_k^2}{4N_{s_l}} \quad (3.14)$$

be integers, and for every k *let the following congruences be simultaneously satisfied:*

$$g_k \equiv 2N_k \pmod{4N_k}, \qquad h_k \equiv 1 \pmod 2. \quad (3.15)$$

Further, for all α *and* δ *satisfying* $\alpha\delta \equiv 1 \pmod{4a}$, *let the following condition hold:*

$$\prod_{k=1}^{sq}\left\{\left(\frac{N_k}{|\delta|}\right)\vartheta_{\alpha g_k, h_k}(\tau; 0, 2N_k)\right\} = \left(\frac{\Delta}{|\delta|}\right)^q\prod_{k=1}^{sq}\vartheta_{g_k h_k}(\tau; 0, 2N_k). \quad (3.16)$$

Then every branch of the function

$$X(\tau) = \left\{\prod_{k=1}^{sq}\vartheta_{g_k h_k}(\tau; 0, 2N_k)\right\}^{\frac{1}{q}}$$

is an integral modular form of dimension $-s/2$, *adjoined to the subgroup* $\Gamma_0(4a)$ *and of divisor* $4a$.

§4

In the case of a binary form we have $a = \Delta = a_1 a_2$, since, as was mentioned before, $(a_1, a_2) = 1$.

Lemma 21. *Let* $m/4\Delta = 2^\alpha m_1$ $(\alpha \geq 0,\ 2 \nmid m_1)$, $a_1 = 2^\gamma b$ $(\gamma \geq 0)$, $2 \nmid a_2$, $(a_1, a_2) = 1$. *Then the following assertions are valid:*

1) *If* $\lambda = 1$ *or* $\gamma + 1$, *then* $A_{2\lambda}(m/4\Delta) = 0$.

21) *If* $2 \le \lambda \le \gamma$, $2 \mid \lambda$, *then*

$$A_{2\lambda}\left(\frac{m}{4\Delta}\right) = \begin{cases} 2^{\frac{\lambda}{2}-1} & \text{for } \lambda < \alpha + 1, \\ -2^{\frac{1}{2}(\alpha-1)} & \text{for } \lambda = \alpha + 1, \\ (-1)^{\frac{1}{2}(m_1-a_2)} \cdot 2^{\frac{\alpha}{2}} & \text{for } \lambda = \alpha + 2, \\ 0 & \text{for } \lambda > \alpha + 2. \end{cases}$$

22) *If* $2 \le \lambda \le \gamma$, $2 \nmid \lambda$, *then*

$$A_{2\lambda}\left(\frac{m}{4\Delta}\right) = \begin{cases} (-1)^{\frac{1}{4}(m_1-a_2)} \cdot 2^{\frac{\alpha}{2}+1} & \text{for } \lambda = \alpha + 3,\ m_1 \equiv a_2,^{1)} \\ 0 & \text{otherwise.} \end{cases}$$

31) *If* $\lambda > \gamma + 1$, $2 \mid \lambda$, $2 \mid \gamma$, *then*

$$A_{2\lambda}\left(\frac{m}{4\Delta}\right) = \begin{cases} 2^{\frac{\gamma}{2}} & \text{for } \lambda < \alpha + 1,\ b \equiv -a_2, \\ -2^{\frac{\gamma}{2}} & \text{for } \lambda = \alpha + 1,\ b \equiv -a_2, \\ (-1)^{\frac{1}{2}(m_1-b)} \cdot 2^{\frac{\gamma}{2}} & \text{for } \lambda = \alpha + 2,\ b \equiv a_2, \\ 0 & \text{otherwise.} \end{cases}$$

32) *If* $\lambda > \gamma + 1$, $2 \nmid \lambda$, $2 \mid \gamma$, *then*

$$A_{2\lambda}\left(\frac{m}{4\Delta}\right) = \begin{cases} (-1)^{\frac{1}{4}(b+a_2)} \cdot 2^{\frac{\gamma}{2}} & \text{for } \lambda < \alpha + 1,\ b \equiv -a_2, \\ -(-1)^{\frac{1}{4}(b+a_2)} \cdot 2^{\frac{\gamma}{2}} & \text{for } \lambda = \alpha + 1,\ b \equiv -a_2, \\ (-1)^{\frac{1}{4}(b-a_2)+\frac{1}{2}(m_1-b)}\ 2^{\frac{\gamma}{2}} & \text{for } \lambda = \alpha + 2,\ b \equiv a_2, \\ 0 & \text{otherwise.} \end{cases}$$

33) *If* $\lambda > \gamma + 1$, $2 \mid \lambda$, $2 \nmid \gamma$, *then*

$$A_{2\lambda}\left(\frac{m}{4\Delta}\right) = \begin{cases} (-1)^{\frac{1}{4}(m_1-b)} \cdot 2^{\frac{1}{2}(\gamma-1)} & \text{for } \lambda = \alpha + 3,\ m_1 \equiv b, \\ (-1)^{\frac{1}{4}(m_1+b)+\frac{1}{2}(m_1-a_2)} \cdot 2^{\frac{1}{2}(\gamma-1)} & \text{for } \lambda = \alpha + 3,\ m_1 \equiv -b, \\ 0 & \text{otherwise.} \end{cases}$$

---

1) The congruences here are taken modulo 4.

34) *If* $\lambda > \gamma + 1$, $2 \nmid \lambda$, $2 \nmid \gamma$ *then*

$$A_{2^\lambda}\left(\frac{m}{4\Delta}\right) = \begin{cases} (-1)^{\frac{1}{4}(m_1 - a_2)} \cdot 2^{\frac{1}{2}(\gamma - 1)} & \text{for } \lambda = \alpha + 3,\ m_1 \equiv a_2, \\ (-1)^{\frac{1}{4}(m_1 + a_2) + \frac{1}{2}(m_1 - b)} \cdot 2^{\frac{1}{2}(\gamma - 1)} & \text{for } \lambda = \alpha + 3,\ m_1 \equiv -a_2, \\ 0 & \text{otherwise.} \end{cases}$$

Proof. If in (3.2) we take $s = 2$ and $q = 2^\lambda$ and then introduce instead of h a new variable of summability y defined by means of the congruence $h \equiv ba_2 y \pmod{2^\lambda}$, we obtain

$$A_{2^\lambda}\left(\frac{m}{4\Delta}\right) = 2^{-2\lambda} \sum_{y \bmod 2^\lambda}{}' e(-2^{\alpha-\lambda} m_1 b a_2 y)\, S(2^\gamma a_2 y,\ 2^\lambda)\, S(by,\ 2^\lambda). \quad (4.1)$$

Again, by reasoning similar to that used in the proof of Lemmas 22 and 31 of [2] (pp. 118–120, 127–129), we obtain the required statement from (4.1) by means of Lemmas 2, 4 and 6.

Lemma 22. *Let* $p > 2$, $p^\beta \| m/4\Delta$ $(\beta \geq 0)$, $p^l \| \Delta$, $(a_1, a_2) = 1$. *Further, let* $\bar{a}$ *and* $\underline{a}$ *be those of* a_1 *and* a_2 *for which* $p^l \mid \bar{a}$ *and* $p^l \nmid \underline{a}$ *respectively. Then the following assertions are valid.*

1) *If* $\lambda \leq l$, $2 \mid \lambda$, *then*

$$A_{p^\lambda}\left(\frac{m}{4\Delta}\right) = \begin{cases} p^{\frac{\lambda}{2}}(1 - p^{-1}) & \text{for } \lambda < \beta + 1, \\ -p^{\frac{1}{2}(\beta - 1)} & \text{for } \lambda = \beta + 1, \\ 0 & \text{for } \lambda > \beta + 1. \end{cases}$$

2) *If* $\lambda \leq l$, $2 \nmid \lambda$, *then*

$$A_{p^\lambda}\left(\frac{m}{4\Delta}\right) = \begin{cases} p^{\frac{\beta}{2}}\left(\dfrac{p^{-\beta}\frac{m}{4\Delta}\underline{a}}{p}\right) & \text{for } \lambda = \beta + 1, \\ 0 & \text{for } \lambda \neq \beta + 1. \end{cases}$$

3) *If* $\lambda > l$, $2 \mid l$, *then*

$$A_{p^\lambda}\left(\frac{m}{4\Delta}\right) = \begin{cases} \left(\dfrac{-p^{-l}\Delta}{p}\right)^{\lambda} p^{\frac{l}{2}}(1 - p^{-1}) & \text{for } \lambda < \beta + 1, \\ -\left(\dfrac{-p^{-l}\Delta}{p}\right)^{\beta+1} p^{\frac{l}{2} - 1} & \text{for } \lambda = \beta + 1, \\ 0 & \text{for } \lambda > \beta + 1. \end{cases}$$

4) *If* $\lambda > l$, $2 \nmid l$, *then*

$$A_{p^\lambda}\left(\frac{m}{4\Delta}\right) = \begin{cases} \left(\frac{p^{-l}\Delta}{p}\right)^{\beta+1} \left(\frac{p^{-(\beta+l)}\frac{m}{4\Delta}\bar{a}}{p}\right) p^{\frac{1}{2}(l-1)} & \text{for } \lambda = \beta+1, \\ 0 & \text{for } \lambda \neq \beta+1. \end{cases}$$

Proof. Let q be odd and $q = (q, a_k)q_k$, $a_k = (q, a_k)a'_k$ $(k = 1, 2)$. From (3.2), by Lemmas 2 and 5 for $s = 2$, it follows that

$$A_q\left(\frac{m}{4\Delta}\right) = \left(\frac{a'_1}{q_1}\right)\left(\frac{a'_2}{q_2}\right) I(q_1) I(q_2) (q, a_1)^{\frac{1}{2}} (q, a_2)^{\frac{1}{2}} q^{-1} \sum_{h \bmod q}{}' e\left(-\frac{mh}{4\Delta q}\right)\left(\frac{h}{q_1 q_2}\right),$$

from which, taking $q = p^\lambda$ and taking into account the fact that $(a_1, a_2) = 1$, we obtain

$$A_{p^\lambda}\left(\frac{m}{4\Delta}\right) = \left(\frac{-1}{p}\right)^{\min(\lambda, l)} \left(\frac{a}{p}\right)^\lambda \left(\frac{p^{-\min(\lambda, l)}\bar{a}}{p}\right)^{\lambda - \min(\lambda, l)} I(p^\lambda)$$

$$\times I(p^{\lambda - \min(\lambda, l)}) p^{\frac{1}{2}\min(\lambda, l)} \cdot p^{-\lambda} \sum_{h \bmod p^\lambda}{}' e\left(\frac{mh}{4\Delta p^\lambda}\right)\left(\frac{h}{p}\right)^{\min(\lambda, l)}. \quad (4.2)$$

From (4.2), according to Lemmas 6 and 7, the required statement follows.

Lemma 23. *Let* $4\Delta \mid m$ *and* $m \neq 0$. *Then there exists a sufficiently small* $\sigma > 0$ *such that the function* $V(m/4\Delta, z) = V_2(m/4\Delta, z)$, *defined for* $\operatorname{Re} z > 1$ *by* (3.5) *and* (3.2), *can be analytically continued into the domain* $\mathfrak{M} = \{\operatorname{Re} z \geq 0\} \cup \{|z| \leq \sigma\}$ *and is regular in this domain. Moreover, in an arbitrarily fixed bounded domain which is interior to* $\mathfrak{M}$, *we have*

$$\left|V\left(\frac{m}{4\Delta}, z\right)\right| < Km^2, \quad (4.3)$$

where K *is independent of* m *and* z.

Proof. I. If $2^\alpha \| m/4\Delta$ and $m \neq 0$ it follows from Lemma 21 that

$$A_{2^\lambda}\left(\frac{m}{4\Delta}\right) = 0 \quad \text{for } \lambda > \alpha + 3. \quad (4.4)$$

If $p > 2$, $p^\beta \| m/4\Delta$ and $m \neq 0$ it follows from Lemma 22 that

$$A_{p^\lambda}\left(\frac{m}{4\Delta}\right) = 0 \quad \text{for } \lambda > \beta + 1; \quad (4.5)$$

if, in addition, $p \nmid m$, i.e. $\beta = l = 0$, then we find by Lemma 22,3) that

$$V\left(\frac{m}{4\Delta}, z\right) = X_2\left(\frac{m}{4\Delta}, z\right) \prod_{\substack{p\mid m \\ p>2}} X_p\left(\frac{m}{4\Delta}, z\right) \prod_{p\nmid 2m} X_p\left(\frac{m}{4\Delta}, z\right). \tag{4.6}$$

From (3.7) for $\mathrm{Re}\, z > 1$ it follows that

$$A_{p\lambda}\left(\frac{m}{4\Delta}\right) = -\left(\frac{-\Delta}{p}\right) p^{-1} \text{ for } \lambda = 1. \tag{4.7}$$

From (3.6) by (4.4) and (4.5) it follows that the product

$$X_2\left(\frac{m}{4\Delta}, z\right) \prod_{p\mid m, p>2} X_p\left(\frac{m}{4\Delta}, z\right)$$

is an integral function.

Further, from (3.6), (4.5), and (4.6) according to Lemma 12, for $\mathrm{Re}\, z > 0$ it follows that

$$\prod_{p\nmid 2m} X_p\left(\frac{m}{4\Delta}, z\right) = \prod_{p\nmid 2m}\left(1 - \left(\frac{-\Delta}{p}\right) p^{-1-z}\right)$$
$$= \prod_{\substack{p\mid m \\ p>2}}\left(1 - \left(\frac{-\Delta}{p}\right) p^{-1-z}\right)^{-1} \frac{1}{L(1+z, \chi)}. \tag{4.8}$$

Here $\chi = (-4\Delta/n)$ is the Kronecker symbol, i.e. a nonprincipal character modulo 4Δ; thus the function $L(1+z, \chi)$ is an integral function and does not become zero for $\mathrm{Re}\, z \geq 0$. Consequently there exists a sufficiently small $\sigma > 0$ such that $L(1+z, \chi) \neq 0$ even in the domain $\mathfrak{M} = \{\mathrm{Re}\, z \geq 0\} \cup \{|z| \leq \sigma\}$. In this domain the first factor in the right-hand side of (4.8) is a regular function.

Therefore, according to (4.7) and (4.8),

$$X_2\left(\frac{m}{4\Delta}, z\right) \prod_{\substack{p\mid m \\ p>2}} X_p\left(\frac{m}{4\Delta}, z\right) \prod_{\substack{p\mid m \\ p>2}}\left(1 - \left(\frac{-\Delta}{p}\right) p^{-1-z}\right)^{-1} \frac{1}{L(1+z, \chi)} \tag{4.9}$$

is the required analytical continuation of $V(m/4\Delta, z)$ into $\mathfrak{M}$.

II. From (3.6) and Lemma 21 we find that

$$\left| X_2\left(\frac{m}{4\Delta}, z\right) \right| < 1 + 4 \cdot 2^{\alpha+\gamma} \sum_{\lambda=1}^{\infty} 2^{-\lambda(1+\mathrm{Re} z)};$$

consequently for $\mathrm{Re}\, z > -\sigma$ we have

$$\left| X_2\left(\frac{m}{4\Delta}, z\right) \right| < K \cdot 2^{\alpha+\gamma}. \tag{4.10}$$

For $p > 2$, from (3.6) and Lemma 22 we find that

$$\left| X_p\left(\frac{m}{4\Delta}, z\right) \right| < 1 + p^{\beta+l} \sum_{\lambda=1}^{\infty} p^{-\lambda(1+\mathrm{Re}\, z)},$$

consequently for $\mathrm{Re}\, z > -\sigma$ we have

$$\prod_{\substack{p|m \\ p>2}} \left| X_p\left(\frac{m}{4\Delta}, z\right) \right| < K \prod_{\substack{p|m \\ p>2}} p^{\beta+l}. \tag{4.11}$$

Further, for $\mathrm{Re}\, z > -\sigma$ we have

$$\prod_{\substack{p|m \\ p>2}} \frac{1}{\left|1 - \left(\frac{-\Delta}{p}\right) p^{-1-z}\right|} < \prod_{\substack{p|m \\ p>2}} \frac{p^{1-\sigma}}{p^{1-\sigma} - 1} < \prod_{\substack{p|m \\ p>2}} p < |m|. \tag{4.12}$$

Let us note that function $1/L(1+z, \chi)$, $\chi = (-4\Delta/n)$, is independent of m and is regular in the domain $\mathfrak{M}$, i.e. it is bounded in an arbitrary fixed bounded domain interior to $\mathfrak{M}$. Consequently the statement of the lemma follows from (4.9)–(4.12).

Lemma 24 *The function $V(0, z)$ may be analytically continued into the domain $\mathfrak{M}$ and is regular in this domain.*

Proof. From (3.7) for $\mathrm{Re}\, z > 1$ it follows that

$$V(0, z) = \prod_p X_p(0, z). \tag{4.13}$$

Let $m = 0$ and $p \nmid 2\Delta$. If in Lemma 22,3) we assume $\beta = \infty$ and $l = 0$, then from (3.6) we obtain

$$\prod_{p \nmid 2\Delta} X_p(0, z) = \prod_{p \nmid 2\Delta} \left\{ 1 + \sum_{\lambda=1}^{\infty} \left(\frac{-\Delta}{p}\right)^{\lambda} (1 - p^{-1}) p^{-\lambda z} \right\}$$

$$= \prod_{p \nmid 2\Delta} \frac{1 - \left(\frac{-\Delta}{p}\right) p^{-1-z}}{1 - \left(\frac{-\Delta}{p}\right) p^{-z}} = \prod_{p} \frac{1 - \left(\frac{-4\Delta}{p}\right) p^{-1-z}}{1 - \left(\frac{-4\Delta}{p}\right) p^{-z}}. \tag{4.14}$$

From (4.13) and (4.14) for $\mathrm{Re}\, z > 1$ it follows that

$$V(0, z) = X_2(0, z) \prod_{\substack{p \mid \Delta \\ p > 2}} X_p(0, z) \frac{L(z, \chi)}{L(1+z, \chi)}, \tag{4.15}$$

here $\chi = (-4\Delta/n)$ is the Kronecker symbol.

If in Lemma 21 we take $m = 0$, i.e. $\alpha = \infty$, then from (3.6) we obtain

$$X_2(0, z) = 1 + \sum_{\substack{\lambda=2 \\ 2 \mid \lambda}}^{\gamma} 2^{\frac{\lambda}{2}-1} \cdot 2^{-\lambda z} \quad \text{for } 2 \nmid \gamma \text{ or } 2 \mid \gamma,\ b \equiv a_2 \pmod 4; \quad (4.16)$$

$$X_2(0, z) = 1 + \sum_{\substack{\lambda=2 \\ 2 \mid \lambda}}^{\gamma} 2^{\frac{\lambda}{2}-1} \cdot 2^{-\lambda z} + \frac{2^{\frac{\gamma}{2}-(\gamma+2)z}}{1+2^{-z}} \quad \text{for } 2 \mid \gamma,\ b \equiv -a_2 \pmod 4,$$

$$\text{but } b \not\equiv -a_2 \pmod 8; \quad (4.17)$$

$$X_2(0, z) = 1 + \sum_{\substack{\lambda=2 \\ 2 \mid \lambda}}^{\gamma} 2^{\frac{\lambda}{2}-1} \cdot 2^{-\lambda z} + \frac{2^{\frac{\gamma}{2}-(\gamma+2)z}}{1-2^{-z}} \quad \text{for } 2 \mid \gamma,\ b \equiv -a_2 \pmod 8. \quad (4.18)$$

Thus $X_2(0, z)$ is regular for $\operatorname{Re} z > -\sigma$ apart from the case when $2 \mid \gamma$, $b \equiv -a_2 \pmod 8$. In this case it has a simple pole at $z = 0$.

Now let $m = 0$, $p > 2$, $p^l \| \Delta$. If in Lemma 22 we assume that $\beta = \infty$, then from (3.6) we obtain

$$X_p(0, z) = \begin{cases} Y_p(z) + p^{\frac{l}{2}}(1-p^{-1}) \dfrac{\left(\dfrac{-p^{-l}\Delta}{p}\right) p^{-(l+1)z}}{1-\left(\dfrac{-p^{-l}\Delta}{p}\right) p^{-z}} & \text{for } 2 \mid l, \\ Y_p(z) & \text{for } 2 \nmid l, \end{cases} \quad (4.19)$$

where

$$Y_p(z) = 1 + (1-p^{-1}) \sum_{\substack{\lambda=2 \\ 2 \mid \lambda}}^{l} p^{\frac{\lambda}{2}-\lambda z}. \quad (4.20)$$

Thus the functions $X_p(0, z)$, $p > 2$, are regular for $\operatorname{Re} z > -\sigma$ if $2 \nmid l$; they may have a simple pole at the point $z = 0$ if $2 \mid l$.

We now show that the function $V(0, z)$ may be analytically continued into the domain $\mathfrak{M}$ where it is regular. For this, consider two separate cases depending on whether -4Δ is or is not a fundamental discriminant.

a) Let -4Δ be a fundamental discriminant. Then by Lemma 10 the Kronecker symbol $(-4\Delta/c)$ for all natural c coincides with some primitive character $\chi^* = \chi^*(c)$ modulo 4Δ, where $\chi^*(-1) = -1$. Consequently, taking $z = 1$ in Lemma 13, we obtain

$$\frac{L(0, \chi^*)}{L(1, \chi^*)} = \frac{(4\Delta)^{\frac{1}{2}}}{\pi}. \quad (4.21)$$

Since -4Δ is a fundamental discriminant, it is not divisible by a square of an odd prime, i.e. if $p > 2$ and $p^l \| \Delta$ then $l = 1$. Further $-4\Delta \equiv 8$ or 12 (mod 16), i.e. $\Delta \equiv 2$ or 1 (mod 4); consequently, either Δ is an odd number multiplied by 2, i.e. $\gamma = 1$, or Δ is an odd number, i.e. $\gamma \equiv 0$ and $ba_2 \equiv 1$ (mod 4).

Thus in this case from (4.15), (4.16) and (4.19) we obtain

$$V(0, z) = \frac{L(z, \chi^*)}{L(1+z, \chi^*)}. \tag{4.22}$$

Consequently $V(0, z)$ may be analytically continued into $\mathfrak{M}$, where it is regular, and, by (4.21),

$$V(0, z)\big|_{z=0} = \frac{2\Delta^{\frac{1}{2}}}{\pi}. \tag{4.23}$$

b) Suppose -4Δ is not a fundamental discriminant. Then, according to Lemma 8, there exists the unique representation

$$-4\Delta = -fm^2, \tag{4.24}$$

where $m > 0$ and $-f\,(f > 0)$ is a fundamental discriminant. Consequently, according to Lemma 10, the Kronecker symbol $(-f/c)$ for all natural c coincides with some primitive character $\chi^* = \chi^*(c)$ modulo f, where $\chi^*(-1) = -1$.

The Kronecker symbol $(-4\Delta/c)$ for all natural c coincides with some nonprimitive character $\chi = \chi(c)$ modulo 4Δ, whose leading modulus, according to Lemma 9, is f, and the corresponding primitive character is χ^*. Consequently, by Lemma 14, we have

$$L(z, \chi) = \prod_{p \mid 4\Delta,\ p \nmid f} (1 - \chi^*(p)\, p^{-z})\, L(z, \chi^*),$$

$$L(1+z, \chi) = \prod_{p \mid 4\Delta,\ p \nmid f} (1 - \chi^*(p)\, p^{-1-z})\, L(1+z, \chi^*). \tag{4.25}$$

Here

$$\chi^*(p) = \begin{cases} 0 & \text{for } p > 2,\ p^l \| \Delta,\ 2 \nmid l, \qquad (4.26) \\ \left(\dfrac{-p^{-l}\Delta}{p}\right) & \text{for } p > 2,\ p^l \| \Delta,\ 2 \mid l. \qquad (4.27) \end{cases}$$

In fact, for prime p in the first line $p \mid f$ according to (4.24), i.e. $\chi^*(p) = (-f/p) = 0$. Further, since $-f$ is a fundamental discriminant, it is not divisible by a square of an odd p. Consequently, for prime p in the second line $p^l \mid m^2$, $p \nmid f$ according to (4.24). Thus

$$\left(\frac{-p^{-l}\Delta}{p}\right)=\left(\frac{-4p^{-l}\Delta}{p}\right)=\left(\frac{-f}{p}\right)=\chi^*(p).$$

Now, consider separately the following three subcases:

1) Let either $2\nmid\gamma$ or $2\mid\gamma$, $b\equiv a_2 \pmod 4$. Then always $2\mid f$. In the first case this follows directly from (4.24), from which it also follows that $8\parallel f$. If in the second case we had $2\nmid f$, then from $-ba_2\equiv 3 \pmod 4$ and (4.24) it would follow that $-f\equiv 3 \pmod 4$, which is impossible since $-f$ is a fundamental discriminant. In this case we even have $-f\equiv 12 \pmod{16}$, since the congruence $-f\equiv 48\equiv 0 \pmod{16}$ is impossible; hence $4\parallel f$.

Thus from (4.15), (4.19), (4.25) and (4.27) we obtain

$$V(0,z)$$

$$=\mathrm{X}_2(0,z)\prod_{\substack{p^l\parallel\Delta\\ p>2,\,2\nmid l}}Y_p(z)\prod_{\substack{p^l\parallel\Delta\\ p>2,\,2\mid l,\,p\nmid f}}\left\{Y_p(z)+\frac{p^{\frac{l}{2}}(1-p^{-1})\left(\frac{-p^{-l}\Delta}{p}\right)p^{-(l+1)z}}{1-\left(\frac{-p^{-l}\Delta}{p}\right)p^{-z}}\right\}$$

$$\times\prod_{\substack{p^l\parallel\Delta\\ p>2,2\mid l,\,p\nmid f}}\frac{1-\left(\frac{-p^{-l}\Delta}{p}\right)p^{-z}}{1-\left(\frac{-p^{-l}\Delta}{p}\right)p^{-1-z}}\frac{L(z,\chi^*)}{L(1+z,\chi^*)}=\mathrm{X}_2(0,z)\,Z(z),\tag{4.28}$$

where for brevity $Z(z)$ denotes the entire middle part without the factor $\mathrm{X}_2(0,z)$.

Consequently, in the case we have just considered, the function $V(0,z)$, according to (4.16) and (4.20), may be analytically continued into $\mathfrak{M}$ and is regular in this domain.

Since $4\Delta=fm^2$ and, as we have noted before, $8\parallel f$ for $2\nmid\gamma$ and $4\parallel f$ for $2\mid\gamma$, $b\equiv a_2 \pmod 4$, we have

$$m^2=\begin{cases}2^{\gamma-1}\prod\limits_{\substack{p^l\parallel\Delta\\ p>2,\,2\nmid l}}p^{l-1}\cdot\prod\limits_{\substack{p^l\parallel\Delta\\ p>2,\,2\mid l}}p^l & \text{for } 2\nmid\gamma,\\ 2^{\gamma}\prod\limits_{\substack{p^l\parallel\Delta\\ p>2,\,2\nmid l}}p^{l-1}\cdot\prod\limits_{\substack{p^l\parallel\Delta\\ p>2,\,2\mid l}}p^l & \text{for } 2\mid\gamma,\ b\equiv a_2 \pmod 4.\end{cases}\tag{4.29}$$

From (4.28), by (4.16), (4.19), (4.20), Lemma 13 and (4.29), we obtain

$$V(0,z)\big|_{z=0}=\frac{2\Delta^{\frac{1}{2}}}{\pi}.\tag{4.30}$$

2) Let $2 \mid \gamma$, $b \equiv -a_2 \pmod 4$ but $b \not\equiv -a_2 \pmod 8$. Then $2 \nmid f$ and $\chi^*(2) = (-f/2) = -1$. In fact, if $2 \mid f$ then from $-ba_2 \equiv 5 \pmod 8$ and (4.24) it would follow that $-f \equiv 0$ or $4 \pmod{16}$ which is impossible since $-f$ is a fundamental discriminant. Consequently, $2 \nmid f$ and $-f \equiv 5 \pmod 8$, i.e. $(-f/2) = -1$.

Thus from (4.15), (4.19), (4.25) and (4.27) we obtain

$$V(0, z) = \frac{1+2^{-z}}{1+2^{-1-z}} X_2(0, z) Z(z). \tag{4.31}$$

Consequently, in the present case, according to (4.17) and (4.20) the function $V(0, z)$ is analytically continuable into the domain $\mathfrak{M}$ and is regular in this domain.

Further, reasoning as in the first case, we obtain from (4.31) that equality (4.30) holds in the present case.

3) Let $2 \mid \gamma$, $b \equiv -a_2 \pmod 8$. In this case, similarly to the preceding one, $2 \nmid f$ and $-f \equiv 1 \pmod 8$, i.e. $\chi^*(2) = (-f/2) = 1$.

Thus from (4.15), (4.18)–(4.20), (4.25) and (4.27) we obtain

$$V(0, z) = \left\{ 1 + \sum_{\substack{\lambda=2 \\ 2 \mid \lambda}}^{\gamma} 2^{\frac{\lambda}{2} - 1 - \lambda z} + \frac{2^{\frac{\gamma}{2} - (\gamma+2)z}}{1 - 2^{-z}} \right\} \frac{1-2^{-z}}{1-2^{-1-z}} Z(z). \tag{4.32}$$

Reasoning in a way similar to the previous cases, we deduce that $V(0, z)$ may be analytically continued into $\mathfrak{M}$, is regular in this domain, and satisfies (4.30).

Lemma 25. *The function* $\Psi(\tau, z) = \Psi_2(\tau, z)$ *defined for* $\operatorname{Re} z > 1$ *by formula* (3.1) *may be analytically continued into the domain* $\mathfrak{M}$ *and is regular in this domain.*

Proof. Let D be an arbitrary fixed bounded domain interior to $\mathfrak{M}$, containing the point $z = 0$. The function $T_2(m, z) = U(z; 1, \tau/4\Delta, m)$, according to (3.4) and Lemma 11, is regular in D for $m \neq 0$, and in this domain

$$|T_2(m, z)| < Ke^{-K'|m|}, \tag{4.33}$$

where K and K' are independent of m and z.

Consider the function

$$T_2(0, z) = U\left(z; 1, \frac{\tau}{4\Delta}, 0\right) = \int_{-\infty}^{\infty} \frac{dt}{\left(t + \frac{\tau}{4\Delta}\right) \left| t + \frac{\tau}{4\Delta} \right|^z}. \tag{4.34}$$

Taking here $t + \operatorname{Re} \tau/4\Delta = \nu w$ and $\nu = \operatorname{Im} \tau/4\Delta > 0$, we obtain

$$T_2(0, z) = \frac{1}{\nu^z} \int_{-\infty}^{\infty} \frac{dw}{(w+i)\,|w+i|^z} = \frac{\Gamma\left(\frac{1}{2}\right)\Gamma\left(\frac{z}{2}+\frac{1}{2}\right)}{i\nu^z\,\Gamma\left(\frac{z}{2}+1\right)}. \tag{4.35}$$

Consequently $T_2(0, z)$ is regular in D, and in this domain

$$|T_2(0, z)| < K, \tag{4.36}$$

where K does not depend on z.

It follows from (4.35) that

$$T_2(0, z)\big|_{z=0} = \frac{\pi}{i}. \tag{4.37}$$

According to Lemmas 23 and 24 the function $V(m/4\Delta, z)$ for $4\Delta \mid m$ is also regular in D, and in this domain

$$|V(0, z)| < K, \quad \left|V\left(\frac{m}{4\Delta}, z\right)\right| < Km^2 \text{ for } m \neq 0, \tag{4.38}$$

where K is independent of m and z.

From (4.33), (4.36) and (4.38) it follows that in D we have

$$|T_2(0, z)\,V(0, z)| < K, \quad \left|T_2(m, z)\,V\left(\frac{m}{4\Delta}, z\right)\right| < Km^2 e^{-K'|m|} \tag{4.39}$$

for $4\Delta \mid m$, $m \neq 0$.

Thus the series $\sum_{m=-\infty}^{\infty} T_2(m, z)\,V(m/4\Delta, z)$ $(4\Delta \mid m)$ is absolutely and uniformly convergent in D and is a regular function of z in this domain. Consequently, according to (3.3)

$$1 + \frac{i}{2^{1+2z}\,\Delta^{\frac{1}{2}+z}} \sum_{\substack{m=-\infty \\ 4\Delta \mid m}}^{\infty} T_2(m, z)\,V\left(\frac{m}{4\Delta}, z\right) \tag{4.40}$$

is the required analytical continuation of $\Psi(\tau, z)$ into $\mathfrak{M}$.

Since $\Psi(\tau, z)$ is regular at $z = 0$, using (4.40), let us take

$$\theta(\tau) = \Psi(\tau, z)\big|_{z=0} = 1 + \frac{i}{2\Delta^{\frac{1}{2}}} \sum_{\substack{m=-\infty \\ 4\Delta \mid m}}^{\infty} T_2(m, z)\,V\left(\frac{m}{4\Delta}, z\right)\Big|_{z=0}. \tag{4.41}$$

Lemma 26. *The function $\theta(\tau)$ is an integral modular form of dimension -1 adjoined to the subgroup $\Gamma_0(4\Delta)$ and of divisor 4Δ.*

Proof. It follows from (4.41), (4.37), (4.30) and Lemma 11 that

$$\theta(\tau) = 2 + \frac{\pi}{\Delta^{\frac{1}{2}}} \sum_{\substack{m=1 \\ 4\Delta \mid m}}^{\infty} V\left(\frac{m}{4\Delta}, z\right)\Big|_{z=0} \cdot e\left(\frac{\tau m}{4\Delta}\right), \tag{4.42}$$

i.e. according to (4.38) $\theta(\tau)$ satisfies conditions 1) and 3) of Definition 2, and also Definition 3.

According to Lemma 18, for $\operatorname{Re} z > 1$, for every substitution of the group $\Gamma_0(4\Delta)$ we have

$$\Psi\left(\frac{\alpha\tau+\beta}{\gamma\tau+\delta}, z\right) = (-1)^{\frac{1}{2}(\operatorname{sgn}\delta - 1)} \left(\frac{-\Delta}{|\delta|}\right)(\gamma\tau+\delta)\,|\gamma\tau+\delta|^{z}\,\Psi(\tau, z). \tag{4.43}$$

According to Lemma 25 and the principle of analytical continuation the identity (4.43) holds in D; in particular, it holds for $z = 0$. Hence according to (4.41),

$$\theta\left(\frac{\alpha\tau+\beta}{\gamma\tau+\delta}\right) = (-1)^{\frac{1}{2}(\operatorname{sgn}\delta - 1)} \left(\frac{-\Delta}{|\delta|}\right)(\gamma\tau+\delta)\,\theta(\tau). \tag{4.44}$$

Thus condition 2) of Definition 2 is satisfied.

As in Lemma 25 of [17] (p. 144), we can show that in the neighborhood of $\tau = -\delta/\gamma$ ($\gamma \neq 0$, $(\gamma, \delta) = 1$, $\alpha\delta - \beta\gamma = 1$) the following expansion holds:

$$(\gamma\tau+\delta)\,\theta(\tau) = \sum_{m=0}^{\infty} A_m\, e\left(\frac{m}{4\Delta}\,\frac{\alpha\tau+\beta}{\gamma\tau+\delta}\right).$$

Consequently condition 4) of Definition 2 is also satisfied.

§5

In this and in the following sections α, β and γ will always denote nonnegative integers, m will denote odd natural numbers and ω will denote square-free numbers.

Let $r(n; a_1, a_2)$ denote the number of representations of n by a form $F = a_1x_1^2 + a_2x_2^2$, i.e. the number of solutions of the equation

$$n = a_1x_1^2 + a_2x_2^2. \tag{5.1}$$

Without loss of generality we assume that $(a_1, a_2) = 1$ and $2 \nmid a_2$. As above $\Delta = a_1a_2$. Further, if we assume $M = 4\Delta n$ then equation (5.1) takes the form

$$M = a_1y_1^2 + a_2y_2^2, \quad y_1 \equiv 0 \pmod{2a_2}, \quad y_2 \equiv 0 \pmod{2a_1}. \tag{5.2}$$

Denote by $R(M; a_1, a_2)$ the number of solutions of equation (5.2). Clearly,

$$r(n; a_1, a_2) = R(M; a_1, a_2). \tag{5.3}$$

It follows from (3.8), (5.2) and (5.3) that

$$\vartheta_{00}(\tau; 0, 2a_1)\,\vartheta_{00}(\tau; 0, 2a_2) = 1 + \sum_M R(M; a_1, a_2)\, e\left(\frac{M\tau}{4\Delta}\right) \tag{5.4}$$

$$= 1 + \sum_{n=1}^{\infty} r(n; a_1, a_2)\, e(n\tau). \tag{5.5}$$

Further, from (4.42) it follows that

$$\theta(\tau; a_1, a_2) = \theta(\tau) = 2 + \sum_M \mathrm{P}(M; a_1, a_2)\, e\left(\frac{M\tau}{4\Delta}\right), \tag{5.6}$$

where

$$\mathrm{P}(M; a_1, a_2) = \frac{\pi}{\Delta^{\frac{1}{2}}} V\left(\frac{M}{4\Delta}, z\right)\Big|_{z=0} = \frac{\pi}{\Delta^{\frac{1}{2}}} V(n, z)\,|_{z=0} = \rho(n; a_1, a_2). \tag{5.7}$$

It follows from (5.6) and (5.7) that

$$\theta(\tau; a_1, a_2) = 2 + \sum_{n=1}^{\infty} \rho(n; a_1, a_2)\, e(n\tau). \tag{5.8}$$

In the present section we deduce the formula for calculation of values of the functions $\rho(n; a_1, a_2)$.

Taking (3.6), (4.4) and (4.5) into account, assume that

$$\mathrm{X}_p = \mathrm{X}_p\left(\frac{M}{4\Delta}, 0\right) = \mathrm{X}_p(n, 0). \tag{5.9}$$

Lemma 27.

$$V\left(\frac{M}{4\Delta}, z\right)\Big|_{z=0} = \prod_p \mathrm{X}_p. \tag{5.10}$$

Proof. In the domain $\mathfrak{M}$ we have, by (4.9),

$$V\left(\frac{M}{4\Delta}, z\right)$$

$$= \mathrm{X}_2\left(\frac{M}{4\Delta}, z\right) \prod_{\substack{p \mid M \\ p>2}} \mathrm{X}_p\left(\frac{M}{4\Delta}, z\right) \prod_{\substack{p \mid M \\ p>2}} \left(1 - \left(\frac{-\Delta}{p}\right) p^{-1-z}\right)^{-1} \frac{1}{L(1+z, \chi)},$$

where $\chi = (-4\Delta/n)$ is the Kronecker symbol. Consequently, by (5.9) and Lemma 12

$$V\left(\frac{M}{4\Delta}, z\right)\Big|_{z=0} = \mathrm{X}_2 \prod_{\substack{p \mid M \\ p>2}} \mathrm{X}_p \prod_{\substack{p \mid M \\ p>2}} \left(1 - \left(\frac{-\Delta}{p}\right)\frac{1}{p}\right)^{-1} \prod_{p} \left(1 - \left(\frac{-4\Delta}{p}\right)\frac{1}{p}\right); \tag{5.11}$$

but it follows from (5.9), (3.6), (4.5) and (4.6) that

$$\prod_{p \nmid 2M} \mathrm{X}_p = \prod_{p \nmid 2M} \left(1 - \left(\frac{-\Delta}{p}\right)\frac{1}{p}\right)$$

$$= \prod_{\substack{p \mid M \\ p>2}} \left(1 - \left(\frac{-\Delta}{p}\right)\frac{1}{p}\right)^{-1} \prod_{p} \left(1 - \left(\frac{-4\Delta}{p}\right)\frac{1}{p}\right). \tag{5.12}$$

The required statement follows from (5.11) and (5.12).

Lemma 28. *Let* $n = 2^{\alpha} m$, $a_1 = 2^{\gamma} b$, $2 \nmid a_2$. *Then*

1) *for* $2 \mid \gamma$

$$\mathrm{X}_2 = 2^{\frac{\alpha}{2}+2}, \quad \text{if} \quad 0 \leqslant \alpha \leqslant \gamma - 3, \quad 2 \mid \alpha, \ m \equiv a_2 \pmod 8;$$

$$= 0, \ \text{if } 0 \leqslant \alpha \leqslant \gamma - 3, \ 2 \mid \alpha, \ m \not\equiv a_2 \pmod 8 \ \text{or} \ 0 \leqslant \alpha \leqslant \gamma - 1, 2 \nmid \alpha;$$

$$= \left(1 + (-1)^{\frac{1}{2}(m-a_2)}\right) 2^{\frac{\alpha}{2}}, \quad \text{if} \quad \alpha = \gamma - 2;$$

$$= \left(1 + (-1)^{\frac{1}{2}(m-b)}\right) 2^{\frac{\gamma}{2}}, \quad \text{if} \quad \alpha \geqslant \gamma, \ 2 \mid \alpha, \ b \equiv a_2 \pmod 4;$$

$$= 2^{\frac{\gamma}{2}}, \quad \text{if} \quad \alpha = \gamma, \ b \equiv -a_2 \pmod 4;$$

$$= \left(2 - (-1)^{\frac{1}{4}(b+a_2)}\right) 2^{\frac{\gamma}{2}} + \left(1 + (-1)^{\frac{1}{4}(b+a_2)}\right)(\alpha - \gamma - 2)\, 2^{\frac{\gamma}{2}-1}, \ \text{if} \ \alpha > \gamma, \ 2 \mid \alpha, \ b \equiv -a_2 \pmod 4;$$

$$= \left(1 + (-1)^{\frac{1}{4}(b-a_2)+\frac{1}{2}(m-b)}\right) 2^{\frac{\gamma}{2}}, \ \text{if } \alpha \geqslant \gamma + 1, \ 2 \nmid \alpha, \ b \equiv a_2 \pmod 4;$$

$$= \left(1 + (-1)^{\frac{1}{4}(b+a_2)}\right)(\alpha - \gamma - 1)\, 2^{\frac{\gamma}{2}-1}, \quad \text{if} \quad \alpha \geqslant \gamma + 1, \ 2 \nmid \alpha, \ b \equiv -a_2 \pmod 4;$$

2) *for* $2 \nmid \gamma$

$$\mathrm{X}_2 = 2^{\frac{\alpha}{2}+2}, \quad \text{if} \quad 0 \leqslant \alpha \leqslant \gamma - 3, \ 2 \mid \alpha, \ m \equiv a_2 \pmod 8;$$

$$= 0, \ \text{if } 0 \leqslant \alpha \leqslant \gamma - 3, \ 2 \mid \alpha, \ m \not\equiv a_2 \pmod 8 \ \text{or} \ 0 \leqslant \alpha \leqslant \gamma - 2, 2 \nmid \alpha;$$

$$\mathrm{X}_2 = \left(1 + (-1)^{\frac{1}{4}(m-a_2)}\right) 2^{\frac{1}{2}(\gamma-1)}, \quad \text{if} \quad \alpha \geqslant \gamma - 1, \ 2 \mid \alpha, \ m \equiv a_2 \pmod 4;$$

$$= \left(1 + (-1)^{\frac{1}{4}(m+a_2)+\frac{1}{2}(m-b)}\right) 2^{\frac{1}{2}(\gamma-1)}, \quad \textit{if} \quad \alpha \geqslant \gamma - 1,\ 2 \mid \alpha,\ m \equiv -a_2 \pmod 4;$$

$$= \left(1 + (-1)^{\frac{1}{4}(m-b)}\right) 2^{\frac{1}{4}(\gamma-1)}, \quad \textit{if} \ \iota \ \alpha \geqslant \gamma,\ 2 \nmid \alpha,\ m \equiv b \pmod 4);$$

$$= \left(1 + (-1)^{\frac{1}{4}(m+b)+\frac{1}{2}(m-a_2)}\right) 2^{\frac{1}{2}(\gamma-1)}, \quad \textit{if} \quad \alpha \geqslant \gamma,\ 2 \nmid \alpha,\ m \equiv -b \pmod 4).$$

Proof. In Lemma 21 replace m and m_1 by M and m and consider the following cases:

1) Let $2 \mid \gamma$. Then from (5.9), (3.6) and Lemma 21, 1)–32) we obtain:

11) for $0 \leqslant \alpha \leqslant \gamma - 3,\ \ 2 \mid \alpha,\ m \equiv a_2 \pmod 4$

$$X_2 = 1 + \sum_{\substack{\lambda=2 \\ 2 \mid \lambda}}^{\alpha} 2^{\frac{\lambda}{2}-1} + 2^{\frac{\alpha}{2}} + (-1)^{\frac{1}{4}(m-a_2)} \cdot 2^{\frac{\alpha}{2}+1};$$

12) for $0 \leqslant \alpha \leqslant \gamma - 3,\ 2 \mid \alpha,\ m \equiv -a_2 \pmod 4$

$$X_2 = 1 + \sum_{\substack{\lambda=2 \\ 2 \mid \lambda}}^{\alpha} 2^{\frac{\lambda}{2}-1} - 2^{\frac{\alpha}{2}};$$

13) for $0 \leqslant \alpha \leqslant \gamma - 1,\ \ 2 \nmid \alpha$

$$X_2 = 1 + \sum_{\substack{\lambda=2 \\ 2 \mid \lambda}}^{\alpha-1} 2^{\frac{\lambda}{2}-1} - 2^{\frac{1}{2}(\alpha-1)};$$

14) for $\alpha = \gamma - 2$

$$X_2 = 1 + \sum_{\substack{\lambda=2 \\ 2 \mid \lambda}}^{\alpha} 2^{\frac{\lambda}{2}-1} + (-1)^{\frac{1}{2}(m-a_2)} \cdot 2^{\frac{\alpha}{2}};$$

15) for $\alpha \geqslant \gamma,\ 2 \mid \alpha,\ b \equiv a_2 \pmod 4)$

$$X_2 = 1 + \sum_{\substack{\lambda=2 \\ 2 \mid \lambda}}^{\gamma} 2^{\frac{\lambda}{2}-1} + (-1)^{\frac{1}{2}(m-b)} \cdot 2^{\frac{\gamma}{2}};$$

16) for $\alpha = \gamma,\ b \equiv -a_2 \pmod 4)$

$$X_2 = 1 + \sum_{\substack{\lambda=2 \\ 2 \mid \lambda}}^{\gamma} 2^{\frac{\lambda}{2}-1};$$

17) for $\alpha > \gamma,\ 2 \mid \alpha,\ b \equiv -a_2 \pmod 4)$

$$X_2 = 1 + \sum_{\substack{\lambda=2 \\ 2 \mid \lambda}}^{\gamma} 2^{\frac{\lambda}{2}-1} + \sum_{\substack{\lambda=\gamma+2 \\ 2 \mid \lambda}}^{\alpha} 2^{\frac{\gamma}{2}} +$$

$$+\sum_{\substack{\lambda=\gamma+3\\ 2\nmid\lambda}}^{\alpha-1}(-1)^{\frac{1}{4}(b+a_2)}\cdot 2^{\frac{\gamma}{2}}-(-1)^{\frac{1}{4}(b+a_2)}\cdot 2^{\frac{\gamma}{2}};$$

18) for $\alpha\geqslant\gamma+1,\ 2\nmid\alpha,\ b\equiv a_2 \pmod 4$

$$X_2=1+\sum_{\substack{\lambda=2\\ 2\mid\lambda}}^{\gamma}2^{\frac{\lambda}{2}-1}+(-1)^{\frac{1}{4}(b-a_2)+\frac{1}{2}(m-b)}\cdot 2^{\frac{\gamma}{2}};$$

19) for $\alpha=\gamma+1,\ b\equiv -a_2 \pmod 4$

$$X_2=1+\sum_{\substack{\lambda=2\\ 2\mid\lambda}}^{\gamma}2^{\frac{\lambda}{2}-1}-2^{\frac{\gamma}{2}};$$

110) for $\alpha>\gamma+1,\ 2\nmid\alpha,\ b\equiv -a_2 \pmod 4$

$$X_2=1+\sum_{\substack{\lambda=2\\ 2\mid\lambda}}^{\gamma}2^{\frac{\lambda}{2}-1}+\sum_{\substack{\lambda=\gamma+2\\ 2\mid\lambda}}^{\alpha-1}2^{\frac{\gamma}{2}}-2^{\frac{\gamma}{2}}+\sum_{\substack{\lambda=\gamma+3\\ 2\nmid\lambda}}^{\alpha}(-1)^{\frac{1}{4}(b+a_2)}\cdot 2^{\frac{\gamma}{2}}.$$

2) Let $2\nmid\nu$. Then, from (5.9), (3.6) and Lemma 21, 1)–22) and 33)–34), we obtain:

21) for $0\leqslant\alpha\leqslant\gamma-3,\ 2\mid\alpha,\ m\equiv a_2 \pmod 4$

$$X_2=1+\sum_{\substack{\lambda=2\\ 2\mid\lambda}}^{\alpha}2^{\frac{\lambda}{2}-1}+2^{\frac{\alpha}{2}}+(-1)^{\frac{1}{4}(m-a_2)}\cdot 2^{\frac{\alpha}{2}+1};$$

22) for $0\leqslant\alpha\leqslant\gamma-3,\ 2\mid\alpha,\ m\equiv -a_2 \pmod 4$

$$X_2=1+\sum_{\substack{\lambda=2\\ 2\mid\lambda}}^{\alpha}2^{\frac{\lambda}{2}-1}-2^{\frac{\alpha}{2}};$$

23) for $0\leqslant\alpha\leqslant\gamma-2,\ 2\nmid\alpha$

$$X_2=1+\sum_{\substack{\lambda=2\\ 2\mid\lambda}}^{\alpha-1}2^{\frac{\lambda}{2}-1}-2^{\frac{1}{2}(\alpha-1)};$$

24) for $\alpha\geqslant\gamma-1,\ 2\mid\alpha,\ m\equiv a_2 \pmod 4$

$$X_2=1+\sum_{\substack{\lambda=2\\ 2\mid\lambda}}^{\gamma-1}2^{\frac{\lambda}{2}-1}+(-1)^{\frac{1}{4}(m-a_2)}\cdot 2^{\frac{1}{2}(\gamma-1)};$$

25) for $\alpha \geqslant \gamma - 1,\ 2 \mid \alpha,\ m \equiv -a_2 \pmod 4$

$$X_2 = 1 + \sum_{\substack{\lambda=2 \\ 2 \mid \lambda}}^{\gamma-1} 2^{\frac{\lambda}{2}-1} + (-1)^{\frac{1}{4}(m+a_2)+\frac{1}{2}(m-b)} \cdot 2^{\frac{1}{2}(\gamma-1)};$$

26) for $\alpha \geqslant \gamma,\ 2 \nmid \alpha,\ m \equiv b \pmod 4$

$$X_2 = 1 + \sum_{\substack{\lambda=2 \\ 2 \mid \lambda}}^{\gamma-1} 2^{\frac{\lambda}{2}-1} + (-1)^{\frac{1}{4}(m-b)} \cdot 2^{\frac{1}{2}(\gamma-1)};$$

27) for $\alpha \geqslant \gamma,\ 2 \nmid \alpha,\ m \equiv -b \pmod 4$

$$X_2 = 1 + \sum_{\substack{\lambda=2 \\ 2 \mid \lambda}}^{\gamma-1} 2^{\frac{\lambda}{2}-1} + (-1)^{\frac{1}{4}(m+b)+\frac{1}{2}(m-a_2)} \cdot 2^{\frac{1}{2}(\gamma-1)}.$$

On evaluating the sums on the right-hand sides of the above equalities we obtain the required statement.

Lemma 29. *Let* $p > 2$, $p^l \| \Delta$, $p^\beta \| n$. *Further, let* $\bar{a}$ *and* $\underline{a}$ *be those of* a_1 *and* a_2 *for which* $p^l \mid \bar{a}$ *and* $p^l \nmid \underline{a}$ *respectively. Then*

$$\begin{aligned} X_p &= \left(1 + \left(\frac{p^{-\beta} n \underline{a}}{p}\right)\right) p^{\frac{\beta}{2}}, && \text{if } l \geqslant \beta + 1,\ 2 \mid \beta; \\ &= 0, && \text{if } l \geqslant \beta + 1,\ 2 \nmid \beta; \\ &= \left(1 - \left(\frac{-p^{-l}\Delta}{p}\right)\frac{1}{p}\right)\left\{1 + \left(1 + \left(\frac{-p^{-l}\Delta}{p}\right)\right)\frac{\beta - l}{2}\right\} p^{\frac{l}{2}}, && \text{if } l \leqslant \beta,\ 2 \mid l,\ 2 \mid \beta; \\ &= \left(1 - \left(\frac{-p^{-l}\Delta}{p}\right)\frac{1}{p}\right)\left(1 + \left(\frac{-p^{-l}\Delta}{p}\right)\right)\frac{\beta - l + 1}{2} p^{\frac{l}{2}}, && \text{if } l \leqslant \beta,\ 2 \mid l,\ 2 \nmid \beta; \\ &= \left\{1 + \left(\frac{p^{-l}\Delta}{p}\right)^{\beta+1}\left(\frac{p^{-(\beta+l)} n \bar{a}}{p}\right)\right\} p^{\frac{1}{2}(l-1)}, && \text{if } l \leqslant \beta,\ 2 \nmid l. \end{aligned}$$

Proof. In Lemma 22 put m instead of $M = 4\Delta n$. Then, according to (5.9) and (3.6), we obtain:

1) for $l \geqslant \beta + 1,\ 2 \mid \beta$

$$X_p = 1 + \sum_{\substack{\lambda=2 \\ 2 \mid \lambda}}^{\beta} (1 - p^{-1}) p^{\frac{\lambda}{2}} + \left(\frac{p^{-\beta} n \underline{a}}{p}\right) p^{\frac{\beta}{2}};$$

2) for $l \geq \beta + 1,\ 2 \nmid \beta$

$$X_p = 1 + \sum_{\substack{\lambda=2 \\ 2 \mid \lambda}}^{\beta-1} (1 - p^{-1}) p^{\frac{\lambda}{2}} - p^{\frac{1}{2}(\beta-1)};$$

3) for $l \leq \beta,\ 2 \mid l$

$$X_p = 1 + \sum_{\substack{\lambda=2 \\ 2 \mid \lambda}}^{l} (1 - p^{-1}) p^{\frac{\lambda}{2}} + \sum_{\lambda=l+1}^{\beta} \left(\frac{-p^{-l}\Delta}{p}\right)^{\lambda} (1 - p^{-1}) p^{\frac{l}{2}} \tag{5.13}$$

$$- \left(\frac{-p^{-l}\Delta}{p}\right)^{\beta+1} p^{\frac{1}{2}l-1} = p^{\frac{l}{2}} \left(1 - \left(\frac{-p^{-l}\Delta}{p}\right) p^{-1}\right) \sum_{\lambda=0}^{\beta-l} \left(\frac{-p^{-l}\Delta}{p}\right)^{\lambda};$$

4) for $l \leq \beta,\ 2 \nmid l$

$$X_p = 1 + \sum_{\substack{\lambda=2 \\ 2 \mid \lambda}}^{l-1} (1 - p^{-1}) p^{\frac{\lambda}{2}} + \left(\frac{p^{-l}\Delta}{p}\right)^{\beta+1} \left(\frac{p^{-(\beta+l)} n\overline{a}}{p}\right) p^{\frac{1}{2}(l-1)}.$$

On evaluating the sums on the right-hand sides of the above equalities we find that the lemma is true.

Lemma 30. *Let* $\Delta = r^2\omega,\ u = \prod_{p \mid n,\ p \nmid 2\Delta} p^{\beta}$. *Then*

$$\rho(n; a_1, a_2) = \frac{\pi X_2 \prod\limits_{p \mid \Delta,\ p>2} X_p \sum\limits_{d \mid u} \left(\frac{-\Delta}{d}\right)}{\Delta^{\frac{1}{2}} \prod\limits_{p \mid r,\ p>2} \left(1 - \left(\frac{-\omega}{p}\right) \frac{1}{p}\right) L(1, -\omega)}, \tag{5.14}$$

where the values of X_2, X_p *and* $L(1, -\omega)$ *are given in Lemmas* 28, 29 *and* 15, *respectively.*

Proof. Let $p > 2,\ p^{\beta} \| n,\ p \nmid \Delta$. Then, assuming $l = 0$ in (5.13), we obtain

$$X_p = \left(1 - \left(\frac{-\Delta}{p}\right) p^{-1}\right) \sum_{d \mid p^{\beta}} \left(\frac{-\Delta}{d}\right). \tag{5.15}$$

If now $p \nmid 2\Delta n$, i.e. $l = \beta = 0$, then from (5.13) we obtain

$$X_p = 1 - \left(\frac{-\Delta}{p}\right) p^{-1}. \tag{5.16}$$

From (5.10), (5.15), (5.16) and Lemma 12 it follows that

$$V\left(\frac{M}{4\Delta}, z\right)\Bigg|_{z=0} = X_2 \prod_{\substack{p \mid \Delta \\ p>2}} X_p \prod_{p \nmid 2\Delta} \left(1 - \left(\frac{-\Delta}{p}\right) \frac{1}{p}\right) \sum_{d \mid u} \left(\frac{-\Delta}{d}\right)$$

$$= X_2 \prod_{\substack{p \mid \Delta \\ p>2}} X_p \prod_{\substack{p>2 \\ p \nmid r}} \left(1 - \left(\frac{-\omega}{p}\right) \frac{1}{p}\right) \sum_{d \mid u} \left(\frac{-\Delta}{d}\right) =$$

$$= X_2 \prod_{\substack{p \mid \Delta \\ p>2}} X_p \prod_{\substack{p \mid r \\ p>2}} \left(1 - \left(\frac{-\omega}{p}\right)\frac{1}{p}\right)^{-1} \prod_{p>2} \left(1 - \left(\frac{-\omega}{p}\right)\frac{1}{p}\right) \sum_{d \mid u} \left(\frac{-\Delta}{d}\right)$$

$$= X_2 \prod_{\substack{p \mid \Delta \\ p>2}} X_p \prod_{\substack{p \mid r \\ p>2}} \left(1 - \left(\frac{-\omega}{p}\right)\frac{1}{p}\right)^{-1} \frac{1}{L(1, \chi)} \sum_{d \mid u} \left(\frac{-\Delta}{d}\right), \tag{5.17}$$

where $\chi = (-4\omega/q)$ is the Kronecker symbol.

From (5.7), (5.17) and (2.5) we obtain the required statement.

In the following sections we assume throughout that $Q_\Delta = e(\tau/4\Delta)$, $Q = e(\tau)$.

§6

The present section is concerned with representations of numbers by the form $F = x_1^2 + 11x_2^2$.

Theorem 1. *The following identity holds*:

$$\vartheta_{00}(\tau; 0,2)\, \vartheta_{00}(\tau; 0,22) = \frac{1}{2}\, \theta(\tau; 1,11) + \frac{2}{3} X(\tau), \tag{6.1}$$

where we take the branch of the function

$$X(\tau)$$

$$= \{2\vartheta_{00}(\tau; 0,2)\, \vartheta_{01}(\tau; 0,2)\, \vartheta_{20}(\tau; 0,2) \vartheta_{00}(\tau; 0,22) \vartheta_{01}(\tau; 0,22)\, \vartheta_{22,0}(\tau; 0,22)\}^{\frac{1}{3}},$$

for which $\lim_{\operatorname{Im} \tau \to \infty} e(-\tau) X(\tau) = 2$.

Proof. The function

$$\psi(\tau; 1,11) = \vartheta_{00}(\tau; 0,2)\, \vartheta_{00}(\tau; 0,22) - \frac{1}{2}\, \theta(\tau; 1,11) - AX(\tau) \tag{6.2}$$

for a constant A is an integral modular form of dimension -1, adjoined to the subgroup $\Gamma_0(44)$ and of divisor 44.

In fact, by Lemma 26 and the remark to Lemma 19, the first two terms on the right-hand side of (6.2) are such forms. It is directly evident that the third term on the right-hand side of (6.2) satisfies conditions (3.14) and (3.15) of Lemma 20. Consequently, according to Lemma 20, it remains to prove that for all α and δ satisfying the condition $\alpha\Delta \equiv 1 \pmod{44}$ equality (3.16) is satisfied.

It follows from $\alpha\delta \equiv 1 \pmod{44}$ that $\alpha\delta \equiv 1 \pmod 2$, i.e.

$$\alpha \equiv 1 \pmod 2. \tag{6.3}$$

And from (3.9), (3.10) and (6.3) we obtain

$$\vartheta_{2\alpha,0}(\tau; 0,2) = \vartheta_{20}(\tau; \alpha - 1,2) = \vartheta_{20}(\tau; 0,2),$$

$$\vartheta_{22\alpha,0}(\tau; 0,2) = \vartheta_{22,0}(\tau; 11(\alpha - 1),\ 22) = \vartheta_{22,0}(\tau; 0,22),$$

from which (3.16) follows.

Thus, according to Lemma 1, function $\psi(\tau;\ 1,11)$ is identically zero if we can select a constant A in such a way that the coefficients of Q^n for all $n \leq (1/12)\, 44\, \Pi_{p|44}\,(1+p^{-1}) = 6$ in the expansion of $\psi(\tau;\ 1,11)$ in powers of Q are equal to zero.

Assuming in Lemmas 28–30 that

$$a_1 = b = 1,\ a_2 = 11,\ \gamma = 0,\ n = 2^{\alpha} 11^{\beta} u,\ (u,\ 22) = 1,\ \Delta = \omega = 11,\ r = 1,$$

we obtain

$$\rho(n;\ 1,11) = \frac{\pi X_2 X_{11}}{11^{\frac{1}{2}} \cdot L(1, -11)} \sum_{d \mid u} \left(\frac{-11}{d}\right),$$

where $X_2 = 1$ for $\alpha = 0$, $X_2 = 3$ for $\alpha > 0$, $2 \mid \alpha$, $X_2 = 0$ for $2 \nmid \underline{\alpha}$; $X_{11} = 1 + (-1)^{\alpha}(u/11)$, since $l = 1$, $\bar{a} = 11$, $\underline{a} = 1$; and $L(1, -11) = 3\pi/2 \cdot 11^{1/2}$ according to Lemma 15.

Consequently,

$$\rho(n;\ 1,11) = \begin{cases} \frac{2}{3}\left(1 + \left(\frac{u}{11}\right)\right) \sum\limits_{d \mid u} \left(\frac{d}{11}\right) & \text{for } \alpha = 0, \\ 2\left(1 + \left(\frac{u}{11}\right)\right) \sum\limits_{d \mid u} \left(\frac{d}{11}\right) & \text{for } 2 \mid \alpha,\ \alpha > 0, \\ 0 & \text{for } 2 \nmid \alpha. \end{cases} \tag{6.4}$$

From (3.8) we have

$$\vartheta_{00}(\tau;\ 0,2)\, \vartheta_{00}(\tau;\ 0,22) = \sum_{h=-\infty}^{\infty} Q^{h^2} \cdot \sum_{h=-\infty}^{\infty} Q^{11h^2} = 1 + 2Q + 2Q^4 + 2Q^9 + \ldots, \tag{6.5}$$

$$X^3(\tau) = 2 \sum_{h=-\infty}^{\infty} Q_{11}^{44h^2} \cdot \sum_{h=-\infty}^{\infty} (-1)^h Q_{11}^{44h^2} \cdot \sum_{h=-\infty}^{\infty} Q_{11}^{11(2h+1)^2} \cdot \sum_{h=-\infty}^{\infty} Q_{11}^{484h^2}$$
$$\times \sum_{h=-\infty}^{\infty} (-1)^h Q_{11}^{484h^2} \cdot \sum_{h=-\infty}^{\infty} Q_{11}^{(22h+11)^2} = 8Q^3(1 - 3Q^2 + 5Q^6 + \ldots), \tag{6.6}$$

and hence

$$X(\tau) = 2Q - 2Q^3 - 2Q^5 + 2Q^{11} + \ldots\,. \tag{6.7}$$

Evaluating $\rho(n;\ 1,11)$ for all $n \leq 6$ by (6.4) and substituting them in (5.8) we obtain

$$\frac{1}{2}\theta(\tau;\,1,11) = 1 + \frac{2}{3}Q + \frac{4}{3}Q^3 + 2Q^4 + \frac{4}{3}Q^5 + 2Q^9 + \dots \tag{6.8}$$

Now choose A in such a way that the coefficient of Q in the expansion of $\psi(\tau;\,1,11)$ in powers of Q is equal to zero, i.e. by (6.2), (6.5), (6.7) and (6.8) in such a way that $2 - 2/3 - 2A = 0$. Taking $A = 2/3$, we easily verify that all coefficients of Q^n $(n \le 6)$ in the expansion of $\psi(\tau;\,1,11)$ in powers of Q are equal to zero. Thus identity (6.1) is proved.

Theorem 1a. *Let* $n = 2^{\alpha}11^{\beta}u$, $(u, 22) = 1$. *Then*

$$r(n;\,1,11) = \begin{cases} \frac{1}{3}\left(1 + \left(\frac{u}{11}\right)\right)\sum\limits_{d\mid u}\left(\frac{d}{11}\right) + \frac{2}{3}\nu(n) & \text{for } \alpha = 0, \\ \left(1 + \left(\frac{u}{11}\right)\right)\sum\limits_{d\mid u}\left(\frac{d}{11}\right) & \text{for } 2\mid\alpha,\ \alpha > 0, \\ 0 & \text{for } 2\nmid\alpha, \end{cases}$$

where $\nu(n)$ *denotes the coefficient of* Q^n *in the expansion of* $X(\tau)$ *in powers of* Q.

Proof. It follows from (6.6) that

$$X^3(\tau) = 2Q^3 \sum_{\substack{h_1,\,h_2=-\infty \\ h_1\equiv h_2 \pmod 2}}^{\infty} (-1)^{h_2} Q^{h_1^2+h_2^2} \cdot \sum_{h=-\infty}^{\infty} Q^{h(h+1)}$$

$$\times \sum_{\substack{h_1,\,h_2=-\infty \\ h_1\equiv h_2 \pmod 2}}^{\infty} (-1)^{h_2} Q^{11\left(h_1^2+h_2^2\right)} \cdot \sum_{h=-\infty}^{\infty} Q^{11h(h+1)}.$$

Thus $X^3(\tau)$ is an odd function of Q. Therefore its expansion, and thus the expansion of $X(\tau)$ in powers of Q, does not contain an even power of Q, i.e.

$$\nu(n) = 0 \text{ for even } n. \tag{6.9}$$

Equating the coefficients of equal powers of Q on both sides of (6.1) and taking (5.5), (5.8), (6.4) and (6.9) into account, we obtain the required statement.

§7

The present section is concerned with representations of numbers by the form $F = x_1^2 + 17x_2^2$.

Theorem 2. *The following identity holds*:

$$\vartheta_{00}(\tau;\,0,2)\,\vartheta_{00}(\tau;\,0,34)=\frac{1}{2}\,\theta(\tau;\,1,17)+\vartheta_{40}(\tau;\,0,34)\,\vartheta_{16,0}(\tau;\,0,34)$$
$$+\vartheta_{80}(\tau;\,0,34)\,\vartheta_{32,0}(\tau;\,0,34)-\vartheta_{12,0}(\tau;\,0,34)\,\vartheta_{20,0}(\tau;\,0,34)$$
$$-\vartheta_{24,0}(\tau;\,0,34)\,\vartheta_{28,0}(\tau;\,0,34). \qquad (7.1)$$

Proof. The function

$$\psi(\tau;\,1,17)=\vartheta_{00}(\tau;\,0,2)\,\vartheta_{00}(\tau;\,0,34)-\frac{1}{2}\,\theta(\tau;\,1,17)$$
$$-\{\vartheta_{40}(\tau;\,0,34)\,\vartheta_{16,0}(\tau;\,0,34)+\vartheta_{80}(\tau;\,0,34)\,\vartheta_{32,0}(\tau;\,0,34)$$
$$-\vartheta_{12,0}(\tau;\,0,34)\,\vartheta_{20,0}(\tau;\,0,34)-\vartheta_{24,0}(\tau;\,0,34)\,\vartheta_{28,0}(\tau;\,0,34)\} \qquad (7.2)$$

is an integral modular form of dimension -1, adjoint to the subgroup $\Gamma_0(68)$, of divisor 68.

In fact, according to Lemma 26 and the remark to Lemma 19, the first two terms on the right-hand side of (7.2) are such functions. It is directly evident that the remaining terms on the right-hand side of (7.2) satisfy conditions (3.12) of Lemma 19.

Further, from $\alpha\delta \equiv 1 \pmod{68}$ we have

$$\alpha \equiv 1,\,3,\,5,\,7,\,9,\,11,\,13,\,15,\,19,\,21,\,23,\,25,\,27,\,29,\,31,\,33 \pmod{34},$$
$$\delta \equiv 1,\,6,\,7,\,5,\,2,\,14,\,4,\,8,\,9,\,13,\,3,\,15,\,12,\,10,\,11,\,16 \pmod{17}. \qquad (7.3)$$

Consider, in turn, the cases[1] $\alpha \equiv 1,33$; $\alpha \equiv 9,25$; $\alpha \equiv 13,21$; $\alpha \equiv 15,19$; $\alpha \equiv 3,31$; $\alpha \equiv 5,29$; $\alpha \equiv 7,27$ and $\alpha \equiv 11,23$ (according to (7.3) the first four cases have $(17/|\delta|)=1$, and in the last four $(17/|\delta|)=-1$). It is easy to verify by (3.9)–(3.11) and (7.3) that for all α and δ satisfying $\alpha\delta \equiv 1 \pmod{68}$, the following equality holds:

$$\vartheta_{4\alpha,0}(\tau;\,0,34)\,\vartheta_{16\alpha,0}(\tau;\,0,34)+\vartheta_{8\alpha,0}(\tau;\,0,34)\,\vartheta_{32\alpha,0}(\tau;\,0,34)$$
$$-\vartheta_{12\alpha,0}(\tau;\,0,34)\,\vartheta_{20\alpha,0}(\tau;\,0,34)-\vartheta_{24\alpha,0}(\tau;\,0,34)\,\vartheta_{28\alpha,0}(\tau;\,0,34)$$
$$=\vartheta_{40}(\tau;\,0,34)\,\vartheta_{16,0}(\tau;\,0,34)+\vartheta_{80}(\tau;\,0,34)\,\vartheta_{32,0}(\tau;\,0,34)$$
$$-\vartheta_{12,0}(\tau;\,0,34)\,\vartheta_{20,0}(\tau;\,0,34)-\vartheta_{24,0}(\tau;\,0,34)\,\vartheta_{28,0}(\tau;\,0,34),$$

since $\Pi_{k=1}^{2}(N_k/|\delta|)=(17/|\delta|)^2=1$.

Thus, according to Lemma 19, the function in curly brackets on the right-hand side of (7.2) is also an integral modular form of dimension -1 adjoined to the subgroup $\Gamma_0(68)$ and of divisor 68

Thus, by Lemma 1, the function $\psi(\tau;\,1,17)$ is identically zero if in its expansion in powers of Q the coefficients of Q^n for all

---

1) All the congruences here are taken modulo 34.

$$n \leqslant \frac{1}{12} 68 \prod_{p|68} (1+p^{-1}) = 9$$

are equal to zero.

Further, reasoning as in the proof of Theorem 1, we obtain

$$\rho(n;\, 1{,}17) = \begin{cases} \left(1+\left(\frac{u}{17}\right)\right)\sum\limits_{d\,|\,u}\left(\frac{-17}{d}\right) & \text{for } u \equiv 1 \pmod 4, \\ 0 & \text{for } u \equiv 3 \pmod 4. \end{cases} \tag{7.4}$$

From (3.8) we obtain

$$\vartheta_{00}(\tau;\, 0{,}2)\,\vartheta_{00}(\tau;\, 0{,}34) = \sum_{h=-\infty}^{\infty} Q^{h^2} \cdot \sum_{h=-\infty}^{\infty} Q^{17h^2} = 1+2Q+2Q^4+2Q^9+\ \dots, \tag{7.5}$$

$$\vartheta_{40}(\tau;\, 0{,}34)\,\vartheta_{16,0}(\tau;\, 0{,}34) = \sum_{h=-\infty}^{\infty} Q_{17}^{(34h+2)^2} \sum_{h=-\infty}^{\infty} Q_{17}^{(34h+8)^2} \tag{7.6}$$

$$= Q+Q^{10}+\ \dots, \tag{7.7}$$

$$\vartheta_{80}(\tau;\, 0{,}34)\,\vartheta_{32,0}(\tau;\, 0{,}34) = \sum_{h=-\infty}^{\infty} Q_{17}^{(34h+4)^2} \cdot \sum_{h=-\infty}^{\infty} Q_{17}^{(34h+16)^2} \tag{7.8}$$

$$= Q^4+Q^5+Q^{17}+\ \dots, \tag{7.9}$$

$$\vartheta_{12,0}(\tau;\, 0{,}34)\,\vartheta_{20,0}(\tau;\, 0{,}34) = \sum_{h=-\infty}^{\infty} Q_{17}^{(34h+6)^2} \cdot \sum_{h=-\infty}^{\infty} Q_{17}^{(34h+10)^2} \tag{7.10}$$

$$= Q^2+Q^9+\ \dots, \tag{7.11}$$

$$\vartheta_{24,0}(\tau;\, 0{,}34)\,\vartheta_{28,0}(\tau;\, 0{,}34) = \sum_{h=-\infty}^{\infty} Q_{17}^{(34h+12)^2} \cdot \sum_{h=-\infty}^{\infty} Q_{17}^{(34h+14)^2} \tag{7.12}$$

$$= Q^5+Q^8+Q^{10}+\ \dots . \tag{7.13}$$

Evaluating $\rho(n;\, 1{,}17)$ for all $n \leq 9$ by (7.4), we obtain

$$\frac{1}{2}\theta(\tau;\, 1{,}17) = 1+Q+Q^2+Q^4+Q^8+3Q^9+\ \dots . \tag{7.14}$$

It is not difficult to verify, taking (7.2), (7.5), (7.7), (7.9), (7.11), (7.13) and (7.14) into account, that all coefficients of Q^n $(n \leq 9)$ in the expansion of $\psi(\tau;\, 1{,}17)$ in powers of Q are equal to zero. Thus (7.1) is proved.

Theorem 2a. *Let* $n = 2^{\alpha} 17^{\beta} u$, $(u, 34) = 1$. *Then*

$$r(n;\, 1{,}17) = \begin{cases} \frac{1}{2}\left(1+\left(\frac{u}{17}\right)\right)\sum\limits_{d\,|\,u}\left(\frac{-17}{d}\right) + \nu_1(n)+\nu_2(n)-\nu_3(n)-\nu_4(n) & \\ \qquad\qquad \textit{for } u \equiv 1 \pmod 4, & \\ 0 \qquad\qquad \textit{for } u \equiv 3 \pmod 4, & \end{cases}$$

where $\nu_1(n)$, $\nu_2(n)$, $\nu_3(n)$ *and* $\nu_4(n)$ *denote, respectively, the coefficients of* Q^n *in the expansion of the functions* $\vartheta_{40}(\tau; 0,34)\vartheta_{16,0}(\tau; 0,34)$, $\vartheta_{80}(\tau; 0,34)\vartheta_{32,0}(\tau; 0,34)$, $\vartheta_{12,0}(\tau; 0,34)\vartheta_{20,0}(\tau; 0,34)$ *and* $\vartheta_{24,0}(\tau; 0,34)\cdot\vartheta_{28,0}(\tau; 0,34)$ *in powers of* Q.

Proof. From (7.6), (7.8), (7.10) and (7.12) we have

$$\vartheta_{40}(\tau; 0,34)\vartheta_{16,0}(\tau; 0,34) = \sum_{h_1,h_2=-\infty}^{\infty} Q_{17}^{\frac{1}{4}\{(17h_1+1)^2+(17h_2+4)^2\}},$$

$$\vartheta_{80}(\tau; 0,34)\vartheta_{32,0}(\tau; 0,34) = \sum_{h_1,h_2=-\infty}^{\infty} Q_{17}^{\frac{1}{4}\{(17h_1+2)^2+(17h_2+8)^2\}},$$

$$\vartheta_{12,0}(\tau; 0,34)\vartheta_{20,0}(\tau; 0,34) = \sum_{h_1,h_2=-\infty}^{\infty} Q_{17}^{\frac{1}{4}\{(17h_1+3)^2+(17h_2+5)^2\}},$$

$$\vartheta_{24,0}(\tau; 0,34)\vartheta_{28,0}(\tau; 0,34) = \sum_{h_1,h_2=-\infty}^{\infty} Q_{17}^{\frac{1}{4}\{(17h_1+6)^2+(17h_2+7)^2\}}.$$

The exponents of powers in the decomposition of these functions in powers of Q_{17} are of the form

$$M = 68n = 4\cdot 17n = 2^{\alpha+2}\cdot 17m = 2^{\alpha+2}\cdot 17(4\varkappa+1).$$

In fact, after taking powers of 2 common to both squares outside the curly brackets, there remain inside the curly brackets either two odd squares whose sum $\equiv 2 \pmod 8$ or one even and one odd square whose sum is $\equiv 1 \pmod 4$.

Consequently the exponents of powers in the expansion of the same functions in powers of Q are of the form $n = 2^\alpha(4\kappa+1) = 2^\alpha\cdot 17^\beta(4l+1)$, i.e.

$$\nu_1(n) = \nu_2(n) = \nu_3(n) = \nu_4(n) = 0 \quad \text{for} \quad u \equiv 3 \pmod 4. \tag{7.15}$$

Equating the coefficients of equal powers on both sides of (7.1), according to (7.4) and (7.15), we obtain the required statement.

§8

This section is concerned with representation of numbers by forms $F_1 = 4x_1^2 + 5x_2^2$ and $F = x_1^2 + 20x_2^2$.

Theorem 3. *The following identity holds:*

$$\vartheta_{00}(\tau; 0,8)\vartheta_{00}(\tau; 0,10) = \frac{1}{2}\theta(\tau; 4,5) - \vartheta_{41}(\tau; 0,20)\vartheta_{12,1}(\tau; 0,20). \tag{8.1}$$

Proof. Consider the function

$$\psi(\tau;\,4,5)$$
$$=\vartheta_{00}(\tau;\,0,8)\,\vartheta_{00}(\tau;\,0,10)-\frac{1}{2}\,\theta(\tau;\,4,5)-A\vartheta_{41}(\tau;\,0,20)\,\vartheta_{12,1}(\tau;\,0,20). \quad (8.2)$$

From $\alpha\delta \equiv 1 \pmod{80}$ we obtain

$$\alpha \equiv 1,\,3,\,7,\,9,\,11,\,13,\,17,\,19 \pmod{20},$$
$$\delta \equiv 1,\,2,\,3,\,4,\,1,\,2,\,3,\,4 \pmod 5). \quad (8.3)$$

Considering in turn the cases [1] $\alpha \equiv 1,19$; $\alpha \equiv 9,11$; $\alpha \equiv 3,17$ and $\alpha \equiv 7,13$ (according to (8.3) in the first two cases $(5/|\delta|) = 1$ and in the last two cases $(5/|\delta|) = -1$ it is not difficult to verify by means of (3.9)–(3.11) and (8.3) that for all α and δ satisfying the condition $\alpha\delta \equiv 1 \pmod{80}$ equality (3.13) holds, since $\Pi_{k=1}^{2}\,(N_k/|\delta|) = (10/|\delta|)^2 = 1$.

Consequently, by Lemmas 19 and 26, the function $\psi(\tau;\,4,5)$ is an integral modular form of dimension -1, adjoined to the subgroup $\Gamma_0(80)$ and of divisor 80.

Thus, according to Lemma 1, the function $\psi(\tau;\,4,5)$ is identically equal to zero if we can select the constant A in such a way that the coefficients of Q^n for all $n \leq (1/12)\,80\,\Pi_{p|80}\,(1+p^{-1}) = 12$ in the expansion of $\psi(\tau;4,5)$ in powers of Q are equal to zero.

Further, reasoning as in the proof of Theorem 1, we obtain

$$\rho(n;\,4,5) = \begin{cases} \left(1+\left(\frac{u}{5}\right)\right)\sum\limits_{d\mid u}\left(\frac{-5}{d}\right) & \text{for } \alpha = 0,\ u \equiv 1 \pmod 4, \\ 2\left(1+(-1)^{\alpha}\left(\frac{u}{5}\right)\right)\sum\limits_{d\mid u}\left(\frac{-5}{d}\right) & \text{for } 2\mid\alpha,\ \alpha>0, \\ & u\equiv 1 \pmod 4 \text{ and for } 2\nmid\alpha,\ \alpha>1,\ u\equiv 3 \pmod 4, \\ 0 \ \text{ for } 2\mid\alpha,\ u\equiv 3 \pmod 4, & \text{for } \alpha = 1 \text{ and for } 2\nmid\alpha,\ \alpha>1, \\ & u\equiv 1 \pmod 4. \end{cases} \quad (8.4)$$

From (3.8) we have

$$\vartheta_{00}(\tau;\,0,8)\,\vartheta_{00}(\tau;\,0,10) = \sum_{h=-\infty}^{\infty} Q^{4h^2}\cdot\sum_{h=-\infty}^{\infty} Q^{5h^2}$$
$$= 1+2Q^4+2Q^5+4Q^9+2Q^{16}+\ldots, \quad (8.5)$$

---

1) All congruences here are taken modulo 20.

$$\vartheta_{41}(\tau;\,0{,}20)\,\vartheta_{12,1}(\tau;\,0{,}20) = \sum_{h=-\infty}^{\infty}(-1)^h Q_{20}^{2(20'\iota +} \cdot \sum_{h=-\infty}^{\infty}(-1)^h Q_{20}^{2(20h+6)^2} \tag{8.6}$$

$$= Q - Q^5 - Q^9 + Q^{25} + \ \ldots\ . \tag{8.7}$$

Evaluating $\rho(n;\,4{,}5)$ for all $n \leq 12$ by (8.4), we obtain

$$\frac{1}{2}\,\theta(\tau;\,4{,}5) = 1 + Q + 2Q^4 + Q^5 + 3Q^9 + 2Q^{16} + \ \ldots\ . \tag{8.8}$$

It is not difficult to verify that for $A = -1$ the coefficients of Q^n $(n \leq 12)$ in the expansion of $\psi(\tau;\,4{,}5)$ in powers of Q are equal to zero. Thus (8.1) is proved.

Theorem 3a. *Let* $n = 2^{\alpha}5^{\beta}u,\ (u,\,10) = 1$. *Then*

$$r(n;\,4{,}5) = \begin{cases} \frac{1}{2}\left(1+\left(\frac{u}{5}\right)\right)\sum\limits_{d\mid u}\left(\frac{-5}{d}\right) - \nu(n) & \text{for } \alpha = 0,\ u \equiv 1 \pmod 4, \\ \left(1+\left(\frac{u}{5}\right)\right)\sum\limits_{d\mid u}\left(\frac{-5}{d}\right) & \text{for } 2\mid\alpha,\ \alpha>0,\ u\equiv 1 \pmod 4, \\ \left(1-\left(\frac{u}{5}\right)\right)\sum\limits_{d\mid u}\left(\frac{-5}{d}\right) & \text{for } 2\nmid\alpha,\ \alpha>1,\ u\equiv 3 \pmod 4, \\ 0 \ \text{for } 2\mid\alpha,\ u\equiv 3 \pmod 4, & \text{for } \alpha = 1 \text{ and for } 2\nmid\alpha,\ \alpha>1,\ u \equiv 1 \pmod 4, \end{cases}$$

where $\nu(n)$ *denotes the coefficient of* Q^n *in the expansion of* $\vartheta_{41}(\tau;\,0{,}20)\,\vartheta_{12,1}(\tau;\,0{,}20)$ *in powers of* Q.

Proof. It follows from (8.6) that

$$\vartheta_{41}(\tau;\,0{,}20)\,\vartheta_{12,1}(\tau;\,0{,}20) = \sum_{h_1,\,h_2=-\infty}^{\infty}(-1)^{h_1+h_2}\,Q_{20}^{8\,\{(10h_1+1)^2+(10h_2+3)^2\}}.$$

The exponents of powers of Q_{20} in this expansion are of the form $M = 80n = 16\,(4\kappa + 1)$. In fact, the curly brackets contain two odd squares whose sum $\equiv 2 \pmod 8$. Consequently, in the same expansion in powers of Q the exponents of powers are of the form $n = 4l + 1$, i.e.

$$\nu(n) = 0 \ \text{ for } \ n \not\equiv 1 \pmod 4. \tag{8.9}$$

Equating the coefficients of equal powers of Q on both sides of (8.1), according to (8.4) and (8.9) we obtain the required assertion.

Theorem 4. *The following identity holds:*

$$\vartheta_{00}(\tau;\,0{,}2)\,\vartheta_{00}(\tau;\,0{,}40) = \frac{1}{2}\,\theta(\tau;\,1{,}20) + \vartheta_{41}(\tau;\,0{,}20)\,\vartheta_{12,1}(\tau;\,0{,}20). \tag{8.10}$$

Proof. Assuming in Lemmas 28–30 that

$$a_1 = 20,\ a_2 = 1,\ b = 5,\ \gamma = 2,$$

$$n = 2^\alpha 5^\beta u,\ (u,\ 10) = 1,\ \Delta = 20,\ r = 2,\ \omega = 5,$$

we obtain

$$\rho(n;\ 1{,}20) = \rho(n;\ 4{,}5),\ \text{ i.e. }\ \theta(\tau;\ 1{,}20) = \theta(\tau;\ 4{,}5).$$

From (3.8) we have

$$\vartheta_{00}(\tau;\ 0{,}2)\,\vartheta_{00}(\tau;\ 0{,}40) = \sum_{h=-\infty}^{\infty} Q^{h^2}\cdot \sum_{h=-\infty}^{\infty} Q^{20h^2}$$
$$= 1 + 2Q + 2Q^4 + 2Q^9 + 2Q^{16} + \ldots .$$

Further, reasoning in the same way as in the proof of Theorem 3, we find that the function

$$\psi(\tau;\ 1{,}20)$$
$$= \vartheta_{00}(\tau;\ 0{,}2)\,\vartheta_{00}(\tau;\ 0{,}40) - \frac{1}{2}\,\theta(\tau;\ 1{,}20) - \vartheta_{41}(\tau;\ 0{,}20)\,\vartheta_{12,1}(\tau;\ 0{,}20)$$

is identically equal to zero. Thus identity (8.10) is proved.

Similarly to the proof of Theorem 3a, the following theorem may be deduced from (8.10).

Theorem 4a. *Let* $n = 2^\alpha 5^\beta u$, $(u,\ 10) = 1$. *Then*

$$r(n;\ 1{,}20) = \begin{cases} \frac{1}{2}\left(1 + \left(\frac{u}{5}\right)\right) \sum_{d \mid u} \left(\frac{-5}{d}\right) + \nu(n) & \text{for } \alpha = 0,\ u \equiv 1 \pmod 4, \\ r(n;\ 4{,}5) & \text{otherwise,} \end{cases}$$

where $\nu(n)$ *is defined in Theorem 3a.*

§9 [1)]

In this section we discuss the representation of numbers by the forms $F_1 = x_1^2 + 19x_2^2$, $F_2 = x_1^2 + 23x_2^2$ and $F_3 = x_1^2 + 32x_2^2$.

Theorem 5. *The following identity holds:*

$$\vartheta_{00}(\tau;\ 0{,}2)\,\vartheta_{00}(\tau;\ 0{,}38) = \frac{1}{2}\,\theta(\tau;\ 1{,}19) - \frac{1}{3}\,\vartheta_{20}(\tau;\ 0{,}2)\,\vartheta_{38,0}(\tau;\ 0{,}38)$$
$$+ \frac{2}{3}\,\vartheta_{80}(\tau;\ 0{,}8)\,\vartheta_{00}(\tau;\ 0{,}152) + \frac{2}{3}\,\vartheta_{00}(\tau;\ 0{,}8)\,\vartheta_{152,0}(\tau;\ 0{,}152).$$

---

1) This section was written by the author especially for the translation.

Theorem 5*a*. *Let* $n = 2^{\alpha}19^{\beta}u$, $(u, 38) = 1$. *Then*

$$r(n;\,1,19)=\begin{cases}\frac{1}{3}\left\{1+\left(\frac{u}{19}\right)\right\}\sum_{d|u}\left(\frac{-19}{d}\right)-\frac{1}{3}\nu_1(n)+\frac{2}{3}\nu_2(n) & \text{for } \alpha=0,\ u\equiv 1 \pmod 4,\\ \frac{1}{3}\left\{1+\left(\frac{u}{19}\right)\right\}\sum_{d|u}\left(\frac{-19}{d}\right)-\frac{1}{3}\nu_1(n)+\frac{2}{3}\nu_3(n) & \text{for } \alpha=0,\ u\equiv 3 \pmod 4,\\ \left\{1+\left(\frac{u}{19}\right)\right\}\sum_{d|u}\left(\frac{-19}{d}\right) & \text{for } 2\mid\alpha,\ \alpha>0,\\ 0 & \text{for } 2\nmid\alpha,\end{cases}$$

where $\nu_1(n)$, $\nu_2(n)$ *and* $\nu_3(n)$ *respectively denote the coefficients of* Q^n *in the expansions of the functions* $\vartheta_{20}(\tau;\,0,2)\,\vartheta_{38,0}(\tau;\,0,38)$, $\vartheta_{80}(\tau;\,0,8)\,\vartheta_{00}(\tau;\,0,152)$ *and* $\vartheta_{00}(\tau;\,0,8)\,\vartheta_{152,0}(\tau;\,0,152)$ *in powers of* Q.

Theorem 6. *The following identity holds*:

$$\vartheta_{00}(\tau;\,0,2)\,\vartheta_{00}(\tau;\,0,46)=\frac{1}{2}\,\theta(\tau;\,1,23)+\frac{2}{3}\,\mathrm{X}_1(\tau)-\frac{2}{3}\,\mathrm{X}_2(\tau),$$

for those branches of the functions

$$\mathrm{X}_1(\tau)=\{16\vartheta_{00}^4(\tau;\,0,2)\vartheta_{01}(\tau;\,0,2)\vartheta_{20}(\tau;\,0,2)\vartheta_{00}^4(\tau;\,0,46)\vartheta_{01}(\tau;\,0,46)\vartheta_{46,0}(\tau;\,0,46)\}^{\frac{1}{6}}$$

and

$$\mathrm{X}_2(\tau)=\{2\vartheta_{00}(\tau;\,0,2)\,\vartheta_{01}(\tau;\,0,2)\,\vartheta_{20}(\tau;\,0,2)\partial_{00}(\tau;\,0,46)\vartheta_{01}(\tau;\,0,46)\,\vartheta_{46,0}(\tau;\,0,46)\}^{\frac{1}{3}},$$

for which $\lim_{\operatorname{Im}\tau\to\infty} e(-\tau)\,\mathrm{X}_k(\tau)=2$ $(k=1,\,2)$.

Theorem 6a. *Let* $n = 2^{\alpha}23^{\beta}u$, $(u, 46) = 1$. *Then*

$$r(n;\,1,23)=\begin{cases}\frac{1}{3}\left\{1+\left(\frac{u}{23}\right)\right\}\sum_{d|u}\left(\frac{-23}{d}\right)+\frac{2}{3}\,\nu_1(n) & \text{for } \alpha=0,\\ 0 & \text{for } \alpha=1,\\ \frac{1}{3}(\alpha-1)\left\{1+\left(\frac{u}{23}\right)\right\}\sum_{d|u}\left(\frac{-23}{d}\right)+\frac{2}{3}(\nu_1(n)-\nu_2(n)) & \text{for } \alpha>1,\end{cases}$$

where $\nu_1(n)$ *and* $\nu_2(n)$ *respectively denote the coefficients of* Q^n *in the expansions of the functions* $\mathrm{X}_1(\tau)$ *and* $\mathrm{X}_2(\tau)$ *in powers of* Q.

Theorem 7. *The following identity holds*:

$$\vartheta_{00}(\tau;\,0,2)\,\vartheta_{00}(\tau;\,0,64)=\frac{1}{2}\,\theta(\tau;\,1,32)+\frac{1}{2}\,\vartheta_{80}(\tau;\,0,8)\,\vartheta_{01}(\tau;\,0,16).$$

Theorem 7a. *Let* $n = 2^{\alpha}u$. *Then*

$$r(n;\ 1,32) = \begin{cases} \sum\limits_{d|n} \left(\frac{-2}{d}\right) + \frac{1}{2}\nu(n) & \text{for } n \equiv 1 \pmod 8, \\ 2\sum\limits_{d|u} \left(\frac{-2}{d}\right) & \text{for } \alpha = 2,\ u \equiv 1 \pmod 8 \text{ and for} \\ & \quad \alpha > 3,\ u \equiv 1 \text{ or } 3 \pmod 8, \\ 0 & \text{otherwise,} \end{cases}$$

where $\nu(n)$ *denotes the coefficient of* Q^n *in the expansion of the function* $\vartheta_{80}(\tau;\ 0,8)\vartheta_{01}(\tau;\ 0,16)$ *in powers of* Q.

The proofs of these theorems are similar to those of the theorems in the preceding three sections.

BIBLIOGRAPHY

[1] A. Val'fiš, *Gitterpunkte in mehrdimensionalen Kugeln*, Monografie Matematyczne, vol. 33, PWN, Warsaw, 1957; Russian transl., Akad. Nauk Gruzin. SSR Mat. Inst. A. M. Razmadze, Izdat. Akad. Nauk Gruzin. SSR, Tbilisi, 1959. MR 20 #3826; MR 22 #10982.

[2] G. A. Lomadze, *On the representation of numbers by certain quadratic forms in four variables*, Tbiliss. Gos. Univ. Trudy Ser. Meh.-Mat. Nauk 76 (1959), 107–159. (Russian)

[3] ———, *On the representation of numbers by binary quadratic forms*, Tbiliss. Gos. Univ. Trudy Ser. Meh.-Mat. Nauk 84 (1962), 285–290. (Russian) MR 27 #2486.

[4] ———, *Representation of numbers by certain quadratic forms in six variables*. I, Thbilis. Sahelmç. Univ. Šrom. Mekh.-Math. Mecn. Ser. 117 (1966), 7–43. (Russian) MR 34 #7460.

[5] H. Hasse, *Vorlesungen über Zahlentheorie*, Die Grundlehren der Math. Wissenschaften, Band 59, Springer-Verlag, Berlin, 1950; Russian transl., IL, Moscow, 1953. MR 14, 534; MR 16, 569.

[6] N. G. Čudakov, *Introduction to the theory of Dirichlet's L-functions*, OGIZ, Moscow, 1947. (Russian) MR 11, 234.

[7] F. van der Blij, *Binary quadratic forms of discriminant* -23, Nederl. Akad. Wetensch. Proc. Ser. A 55 = Indag. Math. 14 (1952), 498–503. MR 14, 623.

[8] P. Bronkhorst, *On the number of solutions of the system of Diophantine equations*:

$$\left.\begin{aligned} x_1^2 + x_2^2 + \ldots + x_s^2 &= n \\ x_1 + x_2 + \ldots + x_s &= m \end{aligned}\right\} \text{for } s = 6 \text{ and } s = 8,$$

Thesis, University of Groningen, 1943, North-Holland, Amsterdam, 1943. (Dutch) MR 14, 1063.

[9] P. Dirichlet, "Recherches sur diverses applications de l'analyse infinitésimale à la théorie des nombres" in *Werke*. I, Berlin, 1899, pp. 411–496.

[10] E. Hecke, "Analytische Funktionen und algebraische Zahlen," in *Mathematische Werke*, Vandenhoeck & Ruprecht, Göttingen, 1959, pp. 381–404. MR 21 #3303.

[11] ———, "Über Dirichlet-Reihen mit Funktionalgleichung und ihre Nullstellen auf den Mittelgeraden," in *Mathematische Werke*, Vandenhoeck & Ruprecht, Göttingen, 1959, pp. 708–730. MR 21 #3303.

[12] ———, "Analytische Arithmetik der positiven quadratischen Formen," in *Mathematische Werke*, Vandenhoeck & Ruprecht, Göttingen, 1959, pp. 789–918. MR 21 #3303.

[13] H. D. Kloosterman, *On the representation of numbers in the form* $ax^2 + by^2 + cz^2 + dt^2$, Proc. London Math. Soc. 25 (1926), 143–173.

[14] ———, *The behaviour of general theta functions under the modular group and the characters of binary modular congruence groups*. I, Ann. of Math. (2) 47 (1946), 317–375. MR 9, 12.

[15] E. Landau, *Handbuch der Lehre von der Verteilung der Primzahlen*. I, Leipzig, 1909.

[16] ———, *Vorlesungen über Zahlentheorie*. I, Leipzig, 1927.

[17] G. A. Lomadze, *Über die Darstellung der Zahlen durch einige quaternäre quadratische Formen*, Acta Arith. 5 (1959), 125–170. MR 21 #7193.

[18] H. Streefkerk, *On the number of solutions of the Diophantine equation*, $U = \Sigma_{i=1}^{s} (Ax_i^2 + Bx_i + C)$, Thesis, Free University of Amsterdam, 1943. (Dutch) MR 7, 414.

Translated by:
Helen Alderson

ASYMPTOTIC BEHAVIOR AND ERGODIC PROPERTIES OF SOLUTIONS OF THE GENERALIZED HARDY-LITTLEWOOD EQUATION

UDC 511.5

B. M. BREDIHIN AND Ju. V. LINNIK

Introduction

In [1] and [2] we indicated a method of studying the asymptotic behavior of the number of solutions of certain diophantine equations on the basis of ergodic considerations connected with the "trajectory" of these solutions. Such an approach may be applied to equations of the type

$$a_k + \varphi_\rho(\xi, \eta) = n. \tag{0.1}$$

Here $\{a_k\}$ is a numerical sequence sufficiently well distributed in arithmetical progressions (allowing the application of Eratosthenes' sieve) and $\phi_\rho(\xi, \eta)$ runs through a sequence of numbers representable by a given genus of positive binary quadratic forms.

When the forms are primitive it is possible to utilize the correspondence between quadratic forms and ideals. As a result equation (0.1) reduces to an equation which, in terms of the theory of quadratic fields, can be written in the following manner which is suitable for further study:

$$a_k + N_\rho(\mathfrak{a}) = n. \tag{0.2}$$

In equation (0.2) $N_\rho(\mathfrak{a})$ runs through the sequence of numbers which are norms of integral ideals $\mathfrak{a}$ of a given genus R.

Consequently, with every solution of (0.2) we can associate an integral ideal $\mathfrak{a} \in R$, satisfying that equation for a corresponding value a_k. In the canonical decomposition of the ideal $\mathfrak{a}$ we exhibit the prime ideals $\mathfrak{p}_i$ which appear in the first power (if there are any). Then this ideal will be represented as

$$\mathfrak{a} = \mathfrak{p}_1 \mathfrak{p}_2 \cdots \mathfrak{p}_l \mathfrak{q}. \tag{0.3}$$

The canonical decomposition of the overwhelming majority of the ideals of R contains sufficiently many prime ideals raised to the first power; more precisely, the decomposition (0.3) of these ideals contains among the $\mathfrak{p}_i$ sufficiently many representatives of every class.

With any given ideal $\mathfrak{a}$ we associate its "trajectory". We perform a norm-preserving mapping of the ideal $\mathfrak{a}$: $\mathfrak{p}_i \to \tilde{\mathfrak{p}}_i$, where $\tilde{\mathfrak{p}}_i = \mathfrak{p}_i$ or $\tilde{\mathfrak{p}}_i = \mathfrak{p}_i'$ $(i = 1, 2, \cdots, l)$ ($\mathfrak{p}_i'$ is the ideal conjugate to $\mathfrak{p}_i$). Multiplying the resulting images yields 2^l ideals of the form $\tilde{\mathfrak{a}} = \tilde{\mathfrak{p}}_1\tilde{\mathfrak{p}}_2 \cdots \tilde{\mathfrak{p}}_l\mathfrak{q}$, satisfying the equation (0.2) and belonging to certain classes of genus R. The ideals $\mathfrak{a}$ will be called the points of the trajectory. We can view the motion of a point along the trajectory as having constant velocity. Adopting elementary concepts of ergodic theory, we shall consider a trajectory "good" if the time its points stay in every class of ideals is conversely proportional to the number of classes of ideals in the given genus R. Other trajectories, for instance those that have only few prime factors $\mathfrak{p}_i$, will be considered "bad".

Such a division of the trajectories into two classes allows the posing and solution of the following problem: find among all the solutions of equation (0.2) those for which the corresponding integral ideals $\mathfrak{a}$ belong to a given class of ideals of a given genus, and find the asymptotic behavior of the number of such solutions. For this purpose we first pick out, through the sieve of Eratosthenes, those solutions of (0.2) for which the corresponding ideals $\mathfrak{a}$ have bad trajectories. We shall call them solutions with bad trajectories. The remaining solutions, which constitute the overwheming majority, will have good trajectories (this is in accord with the notion of ergodicity). Essential use of this property will be made in order to find the required asymptotic behavior. We must, of course, also know the asymptotic behavior of the number of solutions of the basic equation (0.2).

We regret that at present we can find the asymptotic behavior only in a few cases. We mention some of them:

I. $a_k = ck^2$ (a problem on ternary quadratic forms);

II. $a_k = \psi(\xi', \eta')$, where $\psi(\xi', \eta')$ is an arbitrary binary quadratic form (a problem on quaternary quadratic forms);

III. $a_k = p$, where p is a prime number (the generalized Hardy-Littlewood problem).

In the present paper we consider only the last problem. We present a method for finding the asymptotic behavior of the number of solutions of the equation

$$p + \varphi(\xi, \eta) = n, \tag{0.4}$$

where p runs through the primes and $\phi(\xi, \eta) = a\xi^2 + b\xi\eta + c\eta^2$ is a given positive quadratic form whose discriminant is not a square. The equation (0.4) is a natural generalization of the Hardy-Littlewood equation, which is (0.4) with $\phi(\xi, \eta) = \xi^2 + \eta^2$ and is studied in [3] and [4].

In [5] the solvability of the equation (0.4) for sufficiently large n was proved and a lower bound was obtained for the number of solutions of (0.4).

However, the question of the existence of an asymptotic formula for the number of solutions of (0.4) remained open.

The basic difficulty consisted in the fact that the form $\phi(\xi, \eta)$ is in general "multiclass", and, therefore, we can solve the equation $\phi(\xi, \eta) = m$ only for the case when the values $\phi(\xi, \eta)$ run through the whole genus of quadratic forms. The transition to an individual form involves a loss of a certain portion of the asymptotics.

In [2], a short outline is given of the deduction of the asymptotic formula for the number of solutions of the equation (0.4). This outline is a specialization, applicable to the equation (0.4), of the above-stated considerations on the ergodic properties of the solutions of equations of type (0.1). The asymptotic behavior in note [2] is obtained for even n $(n \longrightarrow \infty)$ and with a remainder of order $O((\ln \ln n)^{-\eta})$ (relative to the main term), where η, $0 < \eta < 1$, is a constant.

In the present paper an asymptotic formula for the number of solutions of the equation (0.4) is deduced for any integer n $(n \longrightarrow \infty)$ with a remainder term of (relative) order $O((\ln n)^{-\eta})$, where η is a constant depending on the number of classes of quadratic forms having the given discriminant. For the sake of clearer presentation of the method we confine our considerations to primitive forms with a negative discriminant (the discriminant of the quadratic field).

We use the following notation and terminology. The sign O has its usual meaning. The letters $c_1, c_2, \cdots$ and the constants under the sign O will denote positive constants depending only on the given quadratic form; C_0 is a sufficiently large positive number; n is an arbitrary large positive integer (the basic parameter of the present study): p is a prime; $\phi(m)$ is the Euler function; $\mu(m)$ is the Möbius function; $\tau(m)$ is the number of divisors of m; (m_1, m_2) is the largest common divisor of m_1 and m_2; $K(\sqrt{d})$ is the quadratic field, where K is the field of rational numbers and d a negative square-free integer; D is the discriminant of the field $K(\sqrt{d})$; $P = p_1 p_2 \cdots \cdots p_t$ is the product of all the distinct prime divisors of the discriminant D; $\chi(x)$ is the quadratic character (mod $|D|$) of the field $K(\sqrt{d})$; $\mathfrak{a}$, $\mathfrak{b}$, $\mathfrak{c}$, $\mathfrak{q}$ are integral ideals of the field $K(\sqrt{d})$; $\mathfrak{p}$, π are prime ideals of this field; $N(\mathfrak{a})$ is the norm of the ideal $\mathfrak{a}$; $\sim$ denotes equivalence of ideals; $\mathfrak{a}'$ denotes the conjugate ideal; A, B, C are classes of ideals; G is the group of classes of ideals of the quadratic field $K(\sqrt{d})$; $R = R(A)$ is the genus of the ideals defined by the class $A \in R$; R_σ is a genus of ideals depending on σ; G_0 is the principal genus of ideals; h is the number of classes in G, t_0 is the the number of classes in G_0 and g is the number of genera; $(a, b/p)$, where a

and b are rational numbers, is the Hilbert symbol.

In our paper we make essential use of some results of the arithmetical theory of quadratic forms, studied in the monographs [6], [7] and [8].

§1. Formulation of the basic results

We consider the equation (0.4) in the case when $\phi(\xi, \eta)$ is a given positive primitive form with discriminant $D = b^2 - 4ac$, coinciding with the discriminant of the field $K(\sqrt{d})$; ξ and η run independently through the integers under the condition $0 < \phi(\xi, \eta) < n$, which is automatically satisfied in view of the positivity of the form $\phi(\xi, \eta)$. We introduce the notation $\phi(\xi, \eta) = \phi_K(\xi, \eta)$.

Let $Q(n) = \sum_{p+\phi(\xi,\eta)=n} 1$ be the number of solutions of the equation (0.4) under the indicated conditions and let ϵ_0 be the smallest positive root of the equation

$$\frac{1}{h} - 2\varepsilon \ln 2 - \varepsilon + \varepsilon \ln \varepsilon = 0. \tag{1.1}$$

Then we have

Theorem A. *As $n \to \infty$, we have*

$$Q(n) = C_\varphi(n) A_D(n) \frac{n}{\ln n} + O(n(\ln n)^{-1-\gamma}), \tag{1.2}$$

where $C_\phi(n) \geq c_0 > 0$, the constant c_0 depending only on the discriminant of the form $\phi(\xi, \eta)$,

$$A_D(n) = \frac{2\pi}{\sqrt{|D|}} \prod_p \left(1 + \frac{\chi(p)}{p(p-1)}\right) \prod_{p/D_n} \frac{(p-1)(p-\chi(p))}{p^2 - p + \chi(p)}$$

and $\gamma = \min\{\epsilon_1; 0.042\}$, $0 < \epsilon_1 < \epsilon_0 \ln 2$.

Here $C_\phi(n)$ is the sum of a singular series, namely

$$C_\varphi(n) = \frac{2g}{\varphi(|D|)} \sum_{\substack{\sigma/P \\ (\sigma, n)=1}} \frac{1}{\sigma} \sum_{L\in(L)_\sigma} \sum_{\substack{r=1,\ (r,n)=1 \\ (n-r^2\sigma L,\ D)=1}}^{\infty} \frac{1}{r^2}, \tag{1.3}$$

where $(L)_\sigma$ is a set of $\phi(|D|)/2g$ numbers L, where $0 < L < |D|$, $(L, D) = 1$, $\chi(L) = 1$. The set $\{L\}_\sigma$ depends on the form $\phi(\xi, \eta)$ and on the number σ. In view of its cumbersomeness we refrain from expressing this dependence explicitly (it is implicitly contained in the considerations of §2). We set $r = \Pi_{p/P} P^{\alpha_p}$, $\alpha_p \geq 0$. It is possible to take $c_0 = 1/P^2$ and $\epsilon_1 = \ln 2/4h \ln(h+1)$. We also have the inequalities

$$c_1 (\ln\ln n)^{-1} < A_D(n) < c_2 \ln\ln n. \tag{1.4}$$

From (1.2) and (1.4) it follows that Theorem A yields the asymptotic behavior of the number of solutions of the generalized Hardy-Littlewood equation (0.4) in the case under consideration. We note an interesting corollary of this theorem.

Corollary of Theorem A. *Every sufficiently large integer n can be represented in the form $n = p + \phi(\xi, \eta)$, where $\phi(\xi, \eta) = \phi_K(\xi, \eta)$.*

This result strengthens the corollary of Theorem 5.1.1 of [5] for the forms $\phi_K(\xi, \eta)$, since we do not now require that the number n have no common divisor with a fixed number r determined by the form $\phi(\xi, \eta)$.

There are cases when the singular series (1.3) can be computed without difficulty. For instance, let $n \longrightarrow \infty$ through a sequence of values $n = |D|m$, where m runs through the positive integers. The condition $(|D|m - r^2\sigma L, D) = 1$ is satisfied for such n if and only if $r = 1$ and $\sigma = 1$. Hence

$$C_\varphi(n) = \frac{2g}{\varphi(|D|)} \sum_{L \in (L)_1} 1 = 1.$$

Thus when $n = |D|m \longrightarrow \infty$ we have

$$Q(n) = \frac{2\pi}{\sqrt{|D|}} \prod_p \left(1 + \frac{\chi(p)}{p(p-1)}\right) \prod_{p/Dn} \frac{(p-1)(p-\chi(p))}{p^2 - p + \chi(p)} \frac{n}{\ln n} + O(n(\ln n)^{-1-\gamma}).$$

The problem of finding the asymptotic behavior of $Q(n)$ can be conveniently formulated in terms of the theory of quadratic fields. Establishing a correspondence between forms of the type $\phi_K(\xi, \eta)$ and ideals of the field $K(\sqrt{d})$, we make correspond to a given form $\phi(\xi, \eta)$ a class of ideals A_ϕ. Then the problem of representing the number m by the form $\phi(\xi, \eta)$ reduces to the problem of existence of integral ideals $\mathfrak{a}$ of the field $K(\sqrt{d})$ having norm m and belonging to the class $A = A_\phi^{-1}$; more precisely,

$$\sum_{\varphi(\xi, \eta) = m} 1 = w \sum_{N(\mathfrak{a}) = m,\ \mathfrak{a} \in A} 1, \tag{1.5}$$

where

$$w = \begin{cases} 6 & \text{for } D = -3, \\ 4 & \text{for } D = -4, \\ 2 & \text{for } D < -4. \end{cases}$$

Thus

$$Q(n) = w \sum_{p+N(\mathfrak{a})=n,\ \mathfrak{a}\in A} 1. \tag{1.6}$$

We consider the genus $R = R(A)$. Put

$$\widetilde{Q}(n) = w \sum_{p+N(\mathfrak{a})=n,\ \mathfrak{a}\in R} 1, \tag{1.7}$$

where $\mathfrak{a}$ runs through the integral ideals of all classes of genus R.

Theorem B. *As $n \to \infty$, we have*

$$\widetilde{Q}(n) = t_0 C_\varphi(n) A_D(n) \frac{n}{\ln n} + O(n(\ln n)^{-1,042}); \tag{1.8}$$

where $C_\phi(n)$ and $A_D(n)$ are as in (1.2) *and* (1.3), *and t_0 is the number of classes of ideals in the genus.*

The proof of Theorem B is based on a dispersion method developed in [9] and [5]. It can also be obtained with the aid of very recent results of E. Bombieri [10], who substantially strengthened and simplified the "large sieve" method. In the case when $t_0 = 1$, i.e. in the case of one-class genera of forms, Theorem B already yields the asymptotic behavior in the generalized Hardy-Littlewood problem. In the general case the transition from the solution of the equation (0.4) for the genus to the solution of this equation for the individual form $\phi(\xi, \eta)$ is accomplished with the aid of the following theorem.

Theorem C. *As $n \to \infty$, we have*

$$Q(n) = \frac{1}{t_0}\widetilde{Q}(n) + O(n(\ln n)^{-1-\gamma}), \tag{1.9}$$

where $\gamma = \min\{\epsilon_1;\ 0.042\}$, $0 < \epsilon_1 < \epsilon_0 \ln 2$, with ϵ_0 defined by (1.1).

Theorem A immediately follows from (1.8) and (1.9). The validity of Theorem A will therefore be established as soon as Theorems B and C are proved.

§2. Proof of Theorem B

Every integral ideal $\mathfrak{a} \in R(A)$ can be uniquely represented in the form

$$\mathfrak{a} = \mathfrak{b}_1^2 \mathfrak{b}_2 \mathfrak{c} \tag{2.1}$$

with $N(\mathfrak{b}_1^2 \mathfrak{b}_2)$ consisting of prime divisors of the determinant D, $(N(\mathfrak{c}), D) = 1$ and $\mathfrak{b}_2$ a square-free ideal.

Let A, B and C be the classes containing respectively the ideals $\mathfrak{a}$, $\mathfrak{b}_2$ and $\mathfrak{c}$. Since $\mathfrak{b}_1^2 \mathfrak{b}_2 \sim \mathfrak{b}_2$ we have $A = BC$, and hence

$$R(A) = R(B)\, R(C) \tag{2.2}$$

(here group notation is used for the multiplication of classes and genera).

Since all ideals with a given norm belong to the same genus, for a fixed norm $\sigma = N(\mathfrak{b}_2)$ and a given genus $R(A)$ the genus $R_\sigma = R(C)$ will be uniquely determined by the equation (2.2). It is known [6] that

$$f(m) = \sum_{N(\mathfrak{a}),\ \mathfrak{a} \in R} 1 = \sum_{x/m} \chi(x), \tag{2.3}$$

where x runs through all the divisors of the integer m. Hence for numbers m containing only prime divisors of the discriminant D we have

$$\sum_{N(\mathfrak{b}) = m,\ \mathfrak{b} \in R(B)} 1 = 1. \tag{2.4}$$

From (1.7), (2.1), (2.2) and (2.4) we find that

$$\widetilde{Q}(n) = w \sum_{\sigma/P} \widetilde{Q}_\sigma(n), \tag{2.5}$$

where

$$\widetilde{Q}_\sigma(n) = \sum_{\substack{p + r^2 \sigma N(\mathfrak{c}) = n \\ \mathfrak{c} \in R_\sigma,\ (N(\mathfrak{c}),\ D) = 1}} 1 \tag{2.6}$$

In (2.6) r runs through the positive integers containing, in the canonical decomposition, only powers of prime divisors of P; σ is a fixed divisor of P.

We use the Hilbert symbol in order to express the conditions under which ideals belong to a given genus. From (2.3) and (2.6) we obtain

$$\widetilde{Q}_\sigma(n) = \sum_{p + r^2 \sigma m = n} \sum_{x/m} \chi(x), \tag{2.7}$$

where m runs through the positive integers satisfying the condition

$$(m, D)=1, \quad \chi(m)=1, \quad \left(\frac{m, D}{p}\right)=e_p(R_\sigma) \quad \text{for} \quad p/D. \tag{2.8}$$

The numbers $e_p(R_\sigma)$ in (2.8) are invariants of the genus R_σ.

With the aid of the properties of Hilbert's symbol conditions (2.8) may be given in the following form: $m \equiv L(\text{mod}\,|D|)$, where L runs through some set $(L)_\sigma$ of size $\phi(|D|)/2g$; here $0 < L < |D|$, $(L, D) = 1$, $\chi(L) = 1$.

It follows from (2.7) that

$$\widetilde{Q}_\sigma(n)=\sum_{L\in(L)_\sigma} \widetilde{Q}_L(n), \tag{2.9}$$

where

$$\widetilde{Q}_L(n)=\sum_{\substack{p+r^2\sigma xy=n \\ xy\equiv L(\text{mod}\,|D|)}} \chi(x) \tag{2.10}$$

with fixed σ and L.

Sums of the type (2.10) have been studied in [5] with the aid of a dispersion method and the method of C. Hooley [11], supplemented by estimates due to P. Erdös [12].

Lemma 1 (cf. [5], (5.4.23)). *Let $\chi(x)$ be a real nonprincipal character with a given positive period $\tilde{d}$;*

$$0<L<\tilde{d}, \quad (L, \tilde{d})=1, \quad \chi(L)=1.$$

Let K be an integer satisfying the conditions

$$1\leqslant K\leqslant(\ln n)^{C_0}, \quad (n-KL, \tilde{d}K)=1.$$

Put

$$Q_L(n)=\sum_{\substack{p+Kxy=n \\ xy\equiv L(\text{mod}\,\tilde{d})}} \chi(x).$$

Then

$$Q_L(n)=\frac{2}{\varphi(\tilde{d})K} L(1, \chi) \prod_p \left(1+\frac{\chi(p)}{p(p-1)}\right) \prod_{p/\tilde{d}n} \frac{(p-1)(p-\chi(p))}{p^2-p+\chi(p)}$$
$$\times \prod_{p/K,\, p\nmid\tilde{d}n} \frac{p^2}{p^2-p+\chi(p)} \frac{n}{\ln n}+O(n(\ln n)^{-1{,}0425}). \tag{2.11}$$

We evaluate the sum (2.10) with the aid of Lemma 1. For this purpose we consider the sum (2.10) with the additional conditions

$$(r, n) = 1, \quad (\sigma, n) = 1, \qquad r^2\sigma \leqslant (\ln n)^{C_0},$$

which introduces into (2.10) an error

$$O(n(\ln n)^{-\frac{C_0}{2}}). \tag{2.12}$$

For the solvability of (2.10) it is necessary to add another condition:

$$(n - r^2\sigma L,\ r^2\sigma D) = (n - r^2\sigma L, D) = 1.$$

The conditions of Lemma 1 will be satisfied if we assume $K = r^2\sigma$ and $\tilde{d} = |D|$. Summing up these results for r satisfying $r^2\sigma \leq (\ln n)^{C_0}$, we note that the following additional error is introduced into the remainder term of the formula (2.11):

$$\sum_{r^2\sigma \leqslant (\ln n)^{C_0}} 1 \leqslant \sum_{r \leqslant (\ln n)^{C_0}} 1 = O((\ln \ln n)^t).$$

The extension of the summation on r in the main term to the interval $[1, +\infty)$ involves an error (2.12). Consequently Lemma 1 yields the following asymptotic formula for $\tilde{Q}_L(n)$:

$$\tilde{Q}_L(n) = \frac{2}{\varphi(|D|)\sigma} L(1, \chi) \prod_p \left(1 + \frac{\chi(p)}{p(p-1)}\right) \prod_{p/Dn} \frac{(p-1)(p-\chi(p))}{p^2 - p + \chi(p)}$$

$$\times \left(\sum_{r=1}^{\infty} \frac{1}{r^2}\right) \frac{n}{\ln n} + O(n(\ln n)^{-1,042}), \tag{2.13}$$

where $(\sigma, n) = 1$, $(r, n) = 1$, $(n - r^2\sigma L, D) = 1$; $r = \Pi_{p/P} P^{\alpha_p}$, $\alpha_0 \geq 0$.

It is known [13] that

$$h = \frac{w\sqrt{|D|}}{2\pi} L(1, \chi). \tag{2.14}$$

Moreover,

$$h = g t_0. \tag{2.15}$$

Now (1.8) can easily be deduced from (2.5), (2.9), (2.13), (2.14) and (2.15). To complete the proof of Theorem B we have only to show that $C_\phi(n) \geq c_0 > C$, where the constant c_0 depends only on the discriminant D. In order to prove this, it is sufficient to take $\sigma = 1$, $L \in (L)_1$, $r = r_0 = \Pi_{i=1}^t p_i^{s_i}$, where $p_1 p_2 \cdots p_t = P$ and

$$s_i = \begin{cases} 1, & \text{if} \quad p_i \nmid n, \\ 0, & \text{if} \quad p_i / n. \end{cases}$$

The condition $(n - r^2\sigma L, D) = 1$ is satisfied for the above values of σ and r. Consequently

$$C_\varphi(n) \geqslant \frac{2g}{\varphi(|D|)} \frac{1}{r_0^2} \sum_{L \in (L)_1} 1 = \frac{1}{r_0^2} \geqslant \frac{1}{P^2}.$$

Thus we may take $c_0 = 1/P^2$. This completes the proof of Theorem B.

§3. Proof of Theorem C

Put

$$\widetilde{S}(n) = \frac{1}{w} \widetilde{Q}(n) = \sum_{p + N(\mathfrak{a}) = n,\ \mathfrak{a} \in R} 1, \tag{3.1}$$

where R is a fixed genus determined by the given form $\phi(\xi, \eta)$. From (2.3) and (3.1) it follows that

$$\widetilde{S}(n) = \sum_{p+m=n} f(m), \tag{3.2}$$

where $f(m)$ is the number of solutions of the equation

$$N(\mathfrak{a}) = m \tag{3.3}$$

under the condition that the integral ideals $\mathfrak{a}$ run through the given genus R; for m we have the conditions

$$\left(\frac{m, D}{p}\right) = e_p(R) \text{ for all } p \text{ including } p = \infty.$$

We divide into two classes the set of numbers m satisfying (3.2). In class E we put those m whose canonical decomposition contains at least K_0 primes p_{ij} $(j = 1, 2, \cdots, K_0)$ raised exactly to the first power and for which $\chi(p_{ij}) = 1$. Consequently $p_{ij} = \mathfrak{p}_{ij}\mathfrak{p}'_{ij}$, $\mathfrak{p}_{ij} \neq \mathfrak{p}'_{ij}$, $N(\mathfrak{p}_{ij}) = N(\mathfrak{p}'_{ij}) = p_{ij}$. Moreover $\mathfrak{p}_{ij}$ (or $\mathfrak{p}'_{ij}$) $\in C_i$ for every $i = 1, 2, \cdots, h$. Here $K_0 = [\epsilon_0 \ln\ln n]$, where ϵ_0 is defined by (1.1). In other words, the decompositions of the numbers m of class E in the field $K(\sqrt{d})$ contain sufficiently many prime ideals raised to the first power in every class of that field. In class F we put the remaining m. Thus

$$\widetilde{S}(n) = \sum_{p+m=n,\ m \in E} f(m) + \sum_{p+m=n,\ m \in F} f(m) = \Sigma_E + \Sigma_F. \tag{3.4}$$

We estimate the sum Σ_F from above by sieve methods based on the following lemmas.

Lemma 2 (see [11]). *Let $S_r(n)$ be the number of solutions of the equation $p + p'r = n$, where p and p' run through the primes, r is a fixed positive integer; $r < n/2$. Then, for $n \to \infty$,*

$$S_r(n) = O\left(\frac{n \ln \ln n}{r \ln^2 \frac{n}{r}}\right). \tag{3.5}$$

The estimate (3.5) *is uniform in r.*

Lemma 3 (see [14]). *Let $F(x, z)$ be the number of positive integers $m \le x$ whose canonical decomposition contains only primes for which*

$$p \leqslant z; \quad \ln x \leqslant z \leqslant x^{0.01}, \quad \alpha = \frac{\ln z}{\ln x}.$$

Then

$$F(x, z) = O\left(x \exp\left(-\frac{1}{\alpha} \ln \frac{1}{\alpha}\right)\right). \tag{3.6}$$

Lemma 4 (see [15]). *Let K_n be a finite extension of degree $n \ge 2$ of the field of rational numbers and h the number of classes of ideals of this field. Then the set P of all prime ideals of the field K_n, belonging to at least one of r given classes of ideals of the field K_n, generates a semigroup g_r of integral ideals $\mathfrak{a}$ for which*

$$\nu_{g_r}(x) = \sum_{N(\mathfrak{a}) \leqslant x,\ \mathfrak{a} \in g_r} 1 = C_{g_r} x \ln^{\frac{r}{h}-1} x + O\left(\frac{x \ln^{\frac{r}{h}-1} x}{\ln \ln x}\right). \tag{3.7}$$

Let F_i denote the set of those m whose decompositions contains less than K_0 primes p_{ij} raised exactly to the first power and for which $\chi(p_{ij}) = 1$, $p_{ij} = \mathfrak{p}_{ij} \mathfrak{p}'_{ij}$, $\mathfrak{p}_{ij}$ or $\mathfrak{p}'_{ij} \in C_i$. It is clear that

$$\Sigma_F \leqslant h \max_i \Sigma_{F_i}, \tag{3.8}$$

where

$$\Sigma_{F_i} = \sum_{p+m=n,\ m \in F_i} f(m) \quad (i = 1, 2, \ldots, h).$$

The conditions on m in (3.3) are eliminated in (3.8). Put

$$\Sigma_{F_i} = \Sigma_{F_i'} + \Sigma_{F_i''}, \tag{3.9}$$

where F_i' contains the numbers m all of whose prime divisors satisfy $p \le n^{1/\ln \ln n}$ and F_i'' contains the numbers m which have at least one prime divisor satisfying $p > n^{1/\ln\ln n}$; $F_i = F_i' \cup F_i''$.

We have

$$\Sigma_{F_i'} \leqslant \sum_{m\leqslant n,\ m\in F_i'} \tau(m).$$

Hence

$$\Sigma_{F_i'} \leqslant \Bigl(\sum_{m\leqslant n,\ m\in F_i'} 1\Bigr)^{\frac{1}{2}} \Bigl(\sum_{m\leqslant n} \tau^2(m)\Bigr)^{\frac{1}{2}}. \tag{3.10}$$

Take $x = n$ and $z = n^{1/\ln\ln n}$ in Lemma 3. Then it follows from (3.6) that

$$\sum_{m\leqslant n,\ m\in F_i'} 1 = O(n(\ln n)^{-C_0}). \tag{3.11}$$

At the same time

$$\sum_{m\leqslant n} \tau^2(m) = O(n\ln^3 n). \tag{3.12}$$

Since C_0 is a sufficiently large positive number, we deduce from (3.10)–(3.12) the estimate

$$\Sigma_{F_i'} = O(n(\ln n)^{-\frac{C_0}{2}}). \tag{3.13}$$

We now estimate the sum $\Sigma_{F_i''}$. Taking (2.3) into consideration we find that

$$\Sigma_{F_i''} \leqslant \sum_{p\cdot p'^{\alpha}m=n} \sum_{x/p'^{\alpha}m} \chi(x) = \Sigma_1 + \Sigma_2, \tag{3.14}$$

where

$$p' > n^{\frac{1}{\ln\ln n}}, \quad (p', m) = 1, \quad m \in F_i; \quad \alpha = 1 \text{ in } \Sigma_1, \quad \alpha \geqslant 2 \text{ in } \Sigma_2.$$

For Σ_2 the following rough estimate will suffice:

$$\Sigma_2 \leqslant 2 \sum_{p'^{\alpha}m\leqslant n,\ \alpha\geqslant 2} \alpha\tau(m) = O(n(\ln n)^{-C_0}). \tag{3.15}$$

We estimate Σ_1 with the aid of Lemma 2, which is based on the Brunn sieve. First we have

$$\Sigma_1 \leqslant 2 \sum_{\substack{p+p'N(\mathfrak{a})=n \\ N(\mathfrak{a})\in F_i,\ p'>n^{\frac{1}{\ln\ln n}}}} 1 \tag{3.16}$$

Taking $r = N(\mathfrak{a})$ in Lemma 2 and noting that

$$r < n^{1-\frac{1}{\ln\ln n}} < \frac{1}{2} n$$

for sufficiently large n, we deduce from (3.5) the inequality

$$\Sigma_1 \leqslant c_3 \frac{n(\ln\ln n)^3}{\ln^2 n} \sum_{N(\mathfrak{a})\leqslant n,\ N(\mathfrak{a})\in F_i} \frac{1}{N(\mathfrak{a})}. \tag{3.17}$$

We note that the summation in (3.16) and (3.17) extends over all integral ideals of the field $K(\sqrt{d})$ whose norms belong to the set F_i.

From the definition of F_i it follows that every ideal $\mathfrak{a}$ of (3.17) can be represented by

$$\mathfrak{a} = \mathfrak{b}\mathfrak{c}_1\mathfrak{c}_2, \tag{3.18}$$

where the canonical decomposition of $\mathfrak{b}$ contains only prime ideals $\mathfrak{p}$ with $N(\mathfrak{p})/D$; $\mathfrak{c}_1$ has a canonical decomposition of the following form:

$$\mathfrak{c}_1 = \mathfrak{p}_1\mathfrak{p}_2\ldots\mathfrak{p}_s\mathfrak{p}_{s+1}^{\alpha_{s+1}}\ldots\mathfrak{p}_{s+k}^{\alpha_{s+k}}\mathfrak{q}_1^{\beta_1}\ldots\mathfrak{q}_l^{\beta_l},\quad \mathfrak{p}_j\in C_i,\quad \chi(N(\mathfrak{p}_j))=1,\quad s<K_0,$$
$$\alpha_{s+2}\geqslant 2,\quad \mathfrak{q}_m\in C_i,\quad \chi(N(\mathfrak{q}_m))=-1,\quad \beta_m\geqslant 1;$$

$\mathfrak{c}_2$ does not contain in its canonical decomposition prime ideals $\mathfrak{p}\in C_i$; $(N(\mathfrak{c}_1\mathfrak{c}_2), D) = 1$. From (3.17) and (3.18) we obtain

$$\Sigma_1 < c_3 \frac{n(\ln\ln n)^3}{\ln^2 n} \sum_{N(\mathfrak{b})\leqslant n} \frac{1}{N(\mathfrak{b})} \sum_{N(\mathfrak{c}_1)\leqslant n} \frac{1}{N(\mathfrak{c}_1)} \sum_{N(\mathfrak{c}_2)\leqslant n} \frac{1}{N(\mathfrak{c}_2)} \tag{3.19}$$

with the above-mentioned conditions on $\mathfrak{b}$, $\mathfrak{c}_1$ and $\mathfrak{c}_2$. Next we estimate the three sums in (3.19).

It is not difficult to see that

$$\sum_{N(\mathfrak{b})\leqslant n} \frac{1}{N(\mathfrak{b})} < c_4. \tag{3.20}$$

Further,

$$\sum_{N(\mathfrak{c}_1)\leqslant n}\frac{1}{N(\mathfrak{c}_1)}\leqslant\sum_{\substack{N(\mathfrak{p}_1\ldots\mathfrak{p}_s)\leqslant n\\ s<K_0}}\frac{1}{N(\mathfrak{p}_1\ldots\mathfrak{p}_s)}\sum_{\substack{N(\mathfrak{a})\leqslant n\\ \mathfrak{a}=\Pi\mathfrak{p}^{\alpha_p},\\ \alpha_\mathfrak{p}\geqslant 2}}\frac{1}{N(\mathfrak{a})}\sum_{\substack{N(\mathfrak{a})\leqslant n\\ \mathfrak{a}=\Pi\mathfrak{p}^{\beta_p}\\ \chi(N(\mathfrak{p}))=-1}}\frac{1}{N(\mathfrak{a})},\tag{3.21}$$

where we omitted the condition that the prime ideals belong to the class C_i.

We denote the sums appearing on the right side in (3.21) by S_1, S_2 and S_3 respectively.

$$S_1\leqslant\sum_{s=1}^{K_0}\sum_{p_1p_2\ldots p_s\leqslant n}\frac{2^s}{p_1p_2\ldots p_s}<\sum_{s=1}^{K_0}\frac{1}{s!}\left(\sum_{p\leqslant n}\frac{2}{p}\right)^s.\tag{3.22}$$

We apply the estimate

$$\sum_{p\leqslant n}\frac{1}{p}=\ln\ln n+O(1).\tag{3.23}$$

From (3.22) and (3.23) we deduce the inequality

$$S_1<c_5\sum_{s=1}^{K_0}\frac{1}{s!}(2\ln\ln n)^s.$$

Hence by Stirling's formula we obtain

$$S_1<c_6\sum_{s=1}^{K_0}\exp\left(s\ln 2e+s\ln\frac{\ln\ln n}{s}\right).\tag{3.24}$$

The function $y=s\ln 2e+s\ln((\ln\ln n)/s)$ is monotone increasing on the segment $[0,\,2\ln\ln n]$. Since $K_0\leq\epsilon_0\ln\ln n$, with $\epsilon_0<1$, we find from (3.24) that

$$S_1<c_6(\ln n)^{\varepsilon_0\ln 2e-\varepsilon_0\ln\varepsilon_0}\ln\ln n.\tag{3.25}$$

We also have

$$S_2<\sum_{m\leqslant n}\frac{\tau^2(m)}{m^2}<c_7,\tag{3.26}$$

$$S_3\leqslant\sum_{m^2\leqslant n}\frac{1}{m^2}<c_8.\tag{3.27}$$

Thus we deduce from (3.21) and (3.25)–(3.27) the inequality

$$\sum_{N(\mathfrak{c}_1)\leqslant n} \frac{1}{N(\mathfrak{c}_1)} < c_9 (\ln n)^{\varepsilon_0 \ln 2e - \varepsilon_0 \ln \varepsilon_0} \ln \ln n. \tag{3.28}$$

To estimate the last sum in (3.19) we make use of the fact that the ideals $\mathfrak{c}_2$ do not contain prime ideals of the given class C_i. Consequently the ideals $\mathfrak{c}_2$ can be generated by prime ideals belonging at most to $h-1$ classes of the field $K(\sqrt{d})$. Taking $n=2$, $K_n = K(\sqrt{d})$ and $r = h-1$ in Lemma 4, we obtain from (3.7) the estimate $\sum_{N(\mathfrak{c}_2)\leq x} 1 < c_{10} x(\ln x)^{-1/h}$. Therefore we have

$$\sum_{N(\mathfrak{c}_2)\leqslant n} \frac{1}{N(\mathfrak{c}_2)} < c_{11} (\ln n)^{1-\frac{1}{h}}, \tag{3.29}$$

as is easily seen by applying Abel's summation formula.

The estimate for Σ_F is obtained from (3.8) and (3.9) through the estimates (3.19), (3.20), (3.28) and (3.29), which yield the estimate for Σ_1, and the estimates (3.13)–(3.15). We thus obtain

$$\Sigma_F = O\left(\frac{n (\ln \ln n)^4 (\ln n)^{\varepsilon_0 \ln 2e - \varepsilon_0 \ln \varepsilon_0}}{(\ln n)^{1+\frac{1}{h}}}\right). \tag{3.30}$$

We now pass to the estimate of the sum Σ_E. For $m \in E$ we have the asymptotic uniformity of distribution of the integral ideals $\mathfrak{a}$, satisfying the equation (3.3), over all the classes of ideals of a given genus. More precisely, we have the following assertion.

Lemma 5. *Let $f_A(m)$ be the number of solutions of the equation* (3.3) *under the condition that the integral ideals $\mathfrak{a}$ run through the given class A of genus R. Then, if $m \in E$, we have the asymptotic equality*

$$f_A(m) = \frac{f(m)}{t_0}\left(1 + O\left(\frac{1}{(\ln n)^{\varepsilon_0 \ln 2}}\right)\right). \tag{3.31}$$

Lemma 5 is of interest by itself. It yields, in particular, the asymptotic behavior of the number of representations $m \in E$ of the individual quadratic form $\phi(\xi, \eta) = \phi_K(\xi, \eta)$. For such a form we have the asymptotic formula

$$\sum_{\varphi(\xi,\eta)=m} 1 = \frac{1}{t_0}\left(w \sum_{x/m} \chi(x)\right)\left(1 + O\left(\frac{1}{(\ln n)^{\varepsilon_0 \ln 2}}\right)\right), \tag{3.32}$$

where $m \in E$ ($m \to \infty$ as $n \to \infty$).

Formula (3.32) is an obvious corollary of (1.5) and (3.31).

We prove Lemma 5 following the general scheme about the "good" trajectories, stated in the Introduction. We consider the equation (3.3) in the genus R. Since $m \in E$, we can find a solution of (3.3) having the form

$$\mathfrak{a} = \left(\prod_{\substack{i=1,2,\dots,h \\ j=1,2,\dots,K_0}} \mathfrak{p}_{ij} \right) \mathfrak{q}, \tag{3.33}$$

where $\mathfrak{p}_{ij} \in C_i$ for $i = 1, 2, \cdots, h$, with distinct $\mathfrak{p}_{ij}$, $N(\mathfrak{p}_{ij}) = N(\mathfrak{p}'_{ij}) = p_{ij}$, $p_{ij} = \mathfrak{p}_{ij}\mathfrak{p}'_{ij}$, $\mathfrak{p}_{ij} \neq \mathfrak{p}'_{ij}$ and, finally, $(N(\Pi_{i,j}\, \mathfrak{p}_{ij}), N(\mathfrak{q})) = 1$.

We make use of the fact that the group G_0 of the classes of ideals of the principal Abel genus consists of classes each of which is a square of some class of ideals of the field $K(\sqrt{d})$. The group G_0 can be decomposed into a direct product of cyclical subgroups, i.e.

$$G_0 = \prod_{s=1}^{s_0} g_s, \tag{3.34}$$

where g_s has the order $h_s = p_s^{\alpha_s}$, and every class D_s generating the group g_s $(s = 1, 2, \cdots, s_0)$ can be represented as

$$D_s = C_t^2, \quad t = t(s), \tag{3.35}$$

where C_t is some class of the field $K(\sqrt{d})$. From (3.33) it follows that the canonical decomposition of the ideal $\mathfrak{a}$ contains prime ideals raised to the first power and satisfy the conditions

$$\pi_{s1} \in C_{t(s)}, \quad \pi_{s2} \in C_{t(s)}, \quad \pi_{s3} \in D_s, \quad \pi_{s4} \in D_s^2, \dots, \quad \pi_{sK_0} \in D_s^{2^{K_0-3}}, \tag{3.36}$$

where $s = 1, 2, \cdots, s_0$.

Hence we obtain the following representation of the ideal $\mathfrak{a}$:

$$\mathfrak{a} = \left(\prod_{\substack{s=1,2,\dots,s_0 \\ j=1,2,\dots,K_0}} \pi_{sj} \right) \mathfrak{q}_k, \tag{3.37}$$

where all the π_{sj} are distinct, $N(\pi_{sj}) = p_{sj}$, $N(\mathfrak{q}_k) = q_k$, $\chi(p_{sj}) = 1$ and $(\Pi_{s,j} p_{sj}, q_k) = 1$.

From (2.3), (3.3) and (3.37) we find that

$$f(m) = 2^{K_0 s_0} f(q). \tag{3.38}$$

It is not difficult to see that all $f(m)$ solutions of the equation (3.3) are obtained from the initial solution (3.37) through the mappings

$$\pi_{sj} \to \pi'_{sj}, \quad \pi_{sj_1} \to \pi'_{sj_1}, \quad \pi_{sj_1}\pi_{sj_2} \to \pi'_{sj_1}\pi'_{sj_2}, \ldots, \pi_{s1}\pi_{s2}\ldots\pi_{sK_0} \to \pi'_{s1}\pi'_{s2}\ldots\pi'_{sK_0}$$

for $s = 1, 2, \cdots, s_0$ with fixed q_k, and q_k ranging over $k = 1, 2, \cdots, f(q)$. We thus obtain a set of ideals $\tilde{\mathfrak{a}}$, which can be presented in the form

$$\tilde{\mathfrak{a}} = \Big(\prod_{s,j} \tilde{\pi}_{sj}\Big) \mathfrak{q}_k, \tag{3.39}$$

where $\tilde{\pi}_{sj} = \pi_{sj}$ or $\tilde{\pi}_{sj} = \pi'_{sj}$. Moreover $\pi'_{sj} \sim \pi_{sj}^{-1}$.

Therefore we see from (3.35), (3.36) and (3.39) that

$$\mathfrak{a} \in A_i = \Big(\prod_{s=1}^{s_0} C_{t(s)}^{\pm 1 \pm 1 \pm 2 \pm 2^2 \pm \ldots \pm 2^{K_0-2}}\Big) C(\mathfrak{q}_k), \tag{3.40}$$

where $A_i \in R$. Since the numbers of the form $\tilde{\alpha}_s = \pm 1 \pm 1 \pm 2 \pm 2^2 \pm \cdots \pm 2^{K_0-2}$ are even, the class $\prod_{s=1}^{s_0} C_{t(s)}^{\tilde{\alpha}_s}$ will belong to the principal genus. Consequently the class $C(q_k) \in R$ is fixed for fixed q_k.

We show that the number of factors in $C(q_k)$ is asymptotically uniformly distributed over all classes of the principal genus and, hence, in view of (3.40), we find that the classes A_i are asymptotically distributed over all the classes of the genus R.

Consider the set of numbers $\tilde{\alpha}_s$. Put $\tilde{\alpha}_s = \pm 1 + \tilde{\beta}_s$, where $\tilde{\beta}_s = \pm 1 \pm 2 \pm 2^2 \pm \cdots \pm 2^{K_0-2}$. All the numbers $\tilde{\beta}_s$ are distinct. Indeed, let 2^k be the highest power of the number 2 which appears with different signs in the representation $\tilde{\beta}_s$ of the numbers β_1 and β_2. Then

$$|\beta_1 - \beta_2| \geqslant 2\cdot 2^k - 2(1 + 2 + \cdots + 2^{k-1}), \text{ i.e. } |\beta_1 - \beta_2| \geqslant 2, \ \beta_1 \neq \beta_2.$$

Hence it easily follows that the numbers $\tilde{\beta}_s$ assume as values all the odd numbers from $-(2^{K_0-1} - 1)$ to $2^{K_0-1} - 1$, each exactly once. Accordingly the set of numbers $\tilde{\alpha}_s$ decomposes into four sets of consecutive numbers:

$$\begin{array}{c} -(2^{K_0-1} - 2), \ldots, -2, 0; \\ 2, 4, \ldots, 2^{K_0-1}; \\ -2^{K_0-1}, \ldots, -4, -2; \\ 0, 2, \ldots, 2^{K_0-1} - 2. \end{array} \tag{3.41}$$

Each of these sets contains the same number 2^{K_0-2}, of members. We can arrange the factors $C_{t(s)}^{\tilde{\alpha}_s}$ from a fixed set of values $\tilde{\alpha}_s$ according to increasing $\tilde{\alpha}_s$. We split each of the sets of numbers (3.41) from left to right into subsets of h_s numbers (the last subset may be incomplete). We thus finally have $4[2^{K_0^s-2}/h_s]$ full subsets of factors $C_{t(s)}^{\tilde{\alpha}_s}$ of the form

$$C_{t(s)}^{a_s},\ C_{t(s)}^{a_s+2},\ldots,\ C_{t(s)}^{a_s+2l_s},\ldots,\ C_{t(s)}^{a_s+2(h_s-1)}, \tag{3.42}$$

where a_s is even and $l_s = 0, 1, 2, \cdots, h_s - 1$ $(s = 1, 2, \cdots, s_0)$. In addition there may be M_s incomplete subclasses . Obviously $M_s < 4h_s$. The factors of (3.42) represent all classes of the subgroup g_s.

From (3.34) and (3.35) it follows that all the classes of the principal genus G_0 are obtained, each exactly once, if we take any fixed collection of sets (3.42) for $s = 1, 2, \cdots, s_0$ and from all possible products $\Pi_{s=1}^{s_0} C_{t(s)}^{a_s+2l_s}$ with fixed numbers a_s determining the sets taken. Examining all collections of such type and taking into consideration the existence of incomplete subsets, we find that, for a fixed q_k, every class A_i of a given genus R will be counted

$$N_k = \prod_{s=1}^{s_0} 4\left[\frac{2^{K_0-2}}{h_s}\right] + O\left(\prod_{s=1}^{s_0-1} 4\left[\frac{2^{K_0-2}}{h_s}\right]\right) \tag{3.43}$$

times.

Hence, taking into consideration that $\Pi_{s=1}^{s_0} h_s = t_0$ and that the estimate (3.43) is uniform in q_k, we deduce an estimate for $f_A(m)$. For this purpose it is sufficient to put $A_i = A$ and sum up (3.43) for $k = 1, 2, \cdots \cdots, f(q)$. We obtain

$$f_A(m) = \frac{2^{K_0 s_0}}{t_0} f(q)\left(1 + O\left(\frac{1}{2^{K_0}}\right)\right). \tag{3.44}$$

From (3.38) and (3.44) for $K_0 = [\epsilon_0 \ln \ln n]$ we deduce (3.31), which proves Lemma 5.

With the aid of Lemma 5, the sum Σ_E can be represented as

$$\Sigma_E = t_0 \sum_{p+m=n,\ m\in E} f_A(m)\left(1 + O\left(\frac{1}{(\ln n)^{\varepsilon_0 \ln 2}}\right)\right). \tag{3.45}$$

We now complete the proof of Theorem C. We have

$$Q(n) = w \sum_{p+m=n} f_A(m),$$

whence

$$Q(n) = w \sum_{p+m=n,\, m \in E} f_A(m) + O(\Sigma_F). \tag{3.46}$$

From (3.45) and (3.46) we find that

$$Q(n) = \frac{w}{t_0} \Sigma_E \left(1 + O\left(\frac{1}{(\ln n)^{\varepsilon_0 \ln 2}}\right)\right) + O(\Sigma_F).$$

From (3.1) and (3.4) we obtain $\Sigma_E + \Sigma_F = \widetilde{Q}(n)/w$. Thus

$$Q(n) = \frac{1}{t_0} \widetilde{Q}(n) \left(1 + O\left(\frac{1}{(\ln n)^{\varepsilon_0 \ln 2}}\right)\right) + O(\Sigma_F). \tag{3.47}$$

Let the fixed number ϵ_1 satisfy the condition

$$0 < \varepsilon_1 < \varepsilon_0 \ln 2, \tag{3.48}$$

where ϵ_0 is the smallest positive root of the equation (1.1). Then it follows from (1.8), (3.30) and (3.47) that

$$Q(n) = \frac{1}{t_0} \widetilde{Q}(n) + O(n(\ln n)^{-1-\gamma}), \tag{3.49}$$

where $\gamma = \min\{\epsilon_1;\ 0.042\}$.

It is not difficult to verify that the number $\epsilon_1 = \ln 2/4h \ln (h+1)$ satisfies the inequality (3.48).

Theorem C is thus proved.

BIBLIOGRAPHY

[1] B. M. Bredihin and Ju. V. Linnik, *Binary additive problems with ergodic properties of the solutions*, Dokl. Akad. Nauk SSSR 166 (1966), 1267–1269 = Soviet Math. Dokl. 7 (1966), 254–257. MR 34 #5785.

[2] ———, *Asymptotic behavior in the general Hardy-Littlewood problem*, Dokl. Akad. Nauk SSSR 168 (1966), 975–977 = Soviet Math. Dokl. 7 (1966), 740–743. MR 33 #4029.

[3] Ju. V. Linnik, *An asymptotic formula in an additive problem of Hardy-and Littlewood*, Izv. Akad. Nauk SSSR Ser. Mat. 24 (1960), 629–706; English transl., Amer. Math. Soc. Transl. (2) 46 (1965), 65–148. MR 23 #A130.

[4] B. M. Bredihin, *Sharpening of the bound for the remainder term in Hardy-Littlewood type problems*, Vestnik Leningrad. Univ. 17 (1962), no. 19, 133–137. (Russian) MR 26 #1293.

[5] B. M. Bredihin, *The dispersion method and binary additive problems of definite type*, Uspehi Mat. Nauk 20 (1965), no. 2 (122), 89–130 = Russian Math. Surveys 20 (1965), no. 2, 85–125. MR 32 #5618.

[6] Z. I. Borevič and I. R. Šafarevič, *Number theory*, "Nauka", Moscow, 1964; English transl., Pure and Appl Math., vol. 20, Academic Press, New York, 1966. MR 30 #1080; MR 33 #4001.

[7] B. A. Venkov, *Elementary number theory*, GITTL, Moscow, 1937. (Russian)

[8] B. W. Jones, *The arithmetic theory of quadratic forms*, Carus Monographs, no. 10, Wiley, New York, 1950. MR 12, 244.

[9] Ju. V. Linnik, *The dispersion method in binary additive problems*, Izdat. Leningrad. Univ., Leningrad 1961; English transl., Transl. Math. Monographs, vol. 4, Amer. Math. Soc., Providence, R. I., 1963. MR 25 #3920.

[10] E. Bombieri, *On the large sieve*, Collectanea Mathematica. Publ. Ist. di Mat. Univ. Milano, no. 281, Tamburini, Milan, 1965. MR 33 #5591.

[11] C. Hooley, *On the representation of a number as the sum of two squares and a prime*, Acta Math. 97 (1957), 189–210. MR 19, 532.

[12] P. Erdös, *An asymptotic inequality in the theory of numbers*, Vestnik Leningrad. Univ. 5 (1960), no. 13, 41–49. (Russian) MR 23 #A3720.

[13] E. Landau, *Vorlesungen über Zahlentheroie*. Vol. 1, part 1: *Elementaren Zahlentheorie*, Chelsea, New York, 1946/47.

[14] A. I. Vinogradov, *On numbers with small prime divisors*, Dokl. Akad. Nauk SSSR 109 (1956), 683–688. (Russian) MR 19, 16.

[15] B. M. Bredihin, *The remainder term in the asymptotic formula for* $\nu_G(x)$, Izv. Vysš. Učebn. Zaved. Matematika 1960, no. 6 (19), 40–49. (Russian) MR 26 #92.

Translated by:
A. Dvoretzky

SETS OF NONNEGATIVE INTEGERS NOT CONTAINING AN ARITHMETIC PROGRESSION OF LENGTH p

UDC 511.2

Ju. T. TKAČENKO

By the length of an arithmetic progression we mean the number of successive terms in it which are not equal to each other (for example, the numbers 5, 8, 11, 14, 17 make up an arithmetic progression of length 5); p denotes a prime number not less than three. We have [1]

Van der Waerden's Theorem. *If the set of natural numbers is divided into k classes in any way, then there is at least one class which contains a progression of any preassigned length.*

I became interested in a problem which is in a certain sense converse to this. Let M be a set of natural numbers not containing a progression of a given length $r \geq 3$. We call M a maximal set not containing a progression of length r, or simply a maximal r-set, if joining to it any natural number not in M gives a set containing a progression of length r.

The question arises of whether maximal r-sets exist. For $r = 3$ this problem was examined in a note I sent to the magazine "Matematika v Škole" in 1959 and which was not printed: a similar note by V. A. Gauhman appeared in 1962 in No. 6 of the same magazine, which, however, presented a different solution from mine. In the present note maximal r-sets are proved to exist and a method of constructing them is indicated for $r = p$. Our proof of the existence theorem was communicated by I. Il'jasov.

Theorem 1. *There are infinitely many maximal sets not containing an arithmetic progression of length r.*

Let N_1 be any finite set of nonnegative integers which does not contain an arithmetic progression of length $r \geq 3$ and let M be a set of nonnegative integers. We take from $M \setminus N_1$ the least nonnegative integer n_1 which does not belong to N_1 and which is such that $N_2 = N_1 \cup n_1$ does not contain an arithmetic progression of length r. We take from $M \setminus N_2$ the least natural number n_2 which does not belong to N_1 such that $N_3 = N_2 \cup n_2$ does not contain an arithmetic progression of length r. Continuing this process, we obtain a sequence of imbedded sets $N_1 \subset N_2 \subset N_3 \subset \cdots$.

We show that $N = \bigcup_{k=1}^{\infty} N_k$ is a maximal r-set.

a) N does not contain an arithmetic progression of length r. In fact, the assumption that N contains an arithmetic progression of length r implies that some N_i contains all the terms of this progression, while by construction N_i does not contain an arithmetic progression of length r.

b) Let the nonnegative integer a not belong to the set N. We take from the sequence $\{n_k\}$ $(k = 1, 2, \cdots)$ the first number n_{k_0} which exceeds a. Then $N_{k_0} \cup a$ contains an arithmetic progression of length r. Hence $N \cup a$ contains an arithmetic progression of length r.

We prove that there are infinitely many such maximal sets. We first note that if N is a maximal set not containing a progression of length r, then $M \setminus N$ is an infinite set (otherwise we would have all natural numbers from some point on belonging to N, which is not the case).

Let N_1 be one of the maximal sets not containing a progression of length r, and let N_1' be any finite subset of the set N_1. We take from $M_1 \setminus N_1$ a number n_1 such that the set $N_1' \cup n_1 = N_2'$ does not contain an arithmetic progression of length r, and we extend N_2' to a maximal set N_2 not containing a progression of length r. We take from $M \setminus N_2$ a number n_2 such that $N_2' \cup n_2 = N_3'$ does not contain an arithmetic progression of length r. We extend N_3' to a maximal set N_3 not containing an arithmetic progression of length r, and so on.

Continuing this process, we obtain an infinite sequence of maximal sets not containing an arithmetic progression of length r: $N_1, N_2, N_3, \cdots, N_k, \cdots$. For $i \neq j$ we have $N_i \neq N_j$; in fact $n_i \notin N_i$ and $n_i \in N_j$ $(i < j)$. This completes the proof.

For given p and $0 \leq l \leq p$ we let N_l designate the set of nonnegative integers whose p-adic representation does not contain the digit l.

Theorem 2. *The sets N_l are p-maximal.*

We consider the following arithmetic progression of length p: $a_i = a_1 + d(i-1)$ $(i = 1, 2, \cdots, p)$, and we let $d = kp^n$, where $(k, p) = 1$. Then a_i $(i = 1, 2, \cdots, p)$ are congruent $\bmod p^n$, so that all the a_i have n identical digits on the right (since they have the same residues r). We now have $(a_i - r)/p^n = (a_1 - r)/p^n + k(i-1)$. But $(i-1)$ runs through a complete system of residues $\bmod p$, so that, because $(k, p) = 1$, the numbers $(a_i - r)/p^n$ also run through a complete system of residues, i.e. the first digit on the right of the number $(a_i - r)/p^n$ also runs through the values $0, 1, \cdots, (p-1)$. For the number a_i this digit is $(n+1)$th from the right. Consequently, no arithmetic progression of length p can be contained in any of the sets N_l.

We now prove that the sets N_l are maximal. Let the nonnegative integer a_l not belong to N_l, i.e. the digit l occurs in certain places in the p-adic representation of the number a_l. Successively replacing this digit wherever it occurs by k $(k = 0, 1, \cdots, l-1, l+1, \cdots, p-1)$, we obtain $p-1$ numbers a_k in N_l. Obviously, a_l together with these a_k forms a progression of length p, as was to be proved.

Our sets can be ordered. This can be done especially simply if the last digit $p-1$ is absent. In fact, in the number system with base $p-1$ the set N_{p-1} contains all natural numbers together with zero, which can be taken as the set of indices. Then the index of a term in the sequence and the term itself have the same representation, the first in the $(p-1)$-adic system and the other in the p-adic system. For example, if $p = 3$ the digit 2 is absent. Then the index $(1011010)_2 = 64 + 16 + 8 + 2 = 90$ corresponds to the term of the sequence $(1011010)_3 = 729 + 81 + 27 + 3 = 840$, i.e. $a_{90} = 840$. The problem becomes complicated for composite lengths, and its solution has not been found.

BIBLIOGRAPHY

[1] A. Ja. Hinčin, *Three pearls of number theory*, OGIZ, Moscow-Leningrad, 1947; English transl., Graylock Press, Rochester, N. Y., 1952. MR 11, 83; MR 13, 724.

Translated by:
N. Koblitz

representation of the number q successively replacing the last digit wherever [illegible] [illegible] [illegible] [illegible] pattern bers [illegible] Conversely, together with [illegible] forms a progression of length [illegible] to be proved.

[illegible] can be [illegible]. This can be done especially simply if the [illegible] [illegible] is absent. In fact, in the number system with base [illegible]. If the set [illegible] contains all natural numbers together with zero which cannot be taken as the [illegible] indices. Then [illegible] in the [illegible] and [illegible] itself have the same [illegible] in the [illegible] system and the [illegible] progression. For example, suppose the digit 2 is absent. [illegible] the index [illegible] [illegible] [illegible] the sequence [illegible] [illegible] $= 840$. The problem becomes complicated [illegible] and [illegible] solution has not been found.

[illegible]

BIBLIOGRAPHY

[illegible]

AN ESTIMATE FOR A CERTAIN SUM EXTENDED OVER THE PRIMES OF AN ARITHMETICAL PROGRESSION

UDC 511

I. M. VINOGRADOV

In this paper we give an estimate for the sum $\sum_{p \leq N} \chi(p+k)$, where the summation is performed over the primes of an arithmetical progression.

Notation. We shall use the following notation:

q is a prime. We assume that $q \geq q_0$ where q_0 is a sufficiently large constant exceeding 3.

k is a number relatively prime to q.

χ is a nonprincipal character modulo q: $\overline{\chi}(h)$ is the character conjugate to $\chi(h)$.

When $(x, q) = 1$, the symbol x' denotes the number satisfying the condition $xx' \equiv 1 \pmod q$. When z is an integer the product zx' may be denoted by the symbolic fraction z/x.

p denotes a variable which runs through prime numbers.

ϵ is an arbitrarily small positive constant.

c is a positive constant.

When B is positive, $A \ll B$ means that the inequality $|A| \leq cB$ holds. For positive A and B the notation $A \approx B$ is equivalent to $B \ll A \ll B$.

$\psi(z)$ for $z \geq 1$ denotes a function satisfying a condition of the form $\psi(z) \ll z^{\epsilon}$.

Σ_z denotes the summation performed over all indicated values of z.

N is a positive number and a is a positive integer, satisfying the conditions

$$\max(a^3 q^{3/4}, a^5) \leqslant N \leqslant a^{5/4} q^{5/4};$$

in Lemmas 2, 3, 4a, 4, 5 and in the theorem they also satisfy the condition $|k|^{3/2} \ll N$.

l, f, g are numbers relatively prime to a, satisfying the conditions

$$0 \leqslant l < a, \quad 0 \leqslant f < a, \quad 0 \leqslant g < a.$$

The article presents the following estimate ($|k|^{3/2} \ll N$):

$$\sum_{\substack{p\leqslant N \\ p\equiv g\,(\mathrm{mod}\ a)}} \chi(p+k) \ll \frac{Nq^{\varepsilon}}{a}\left(\frac{q^{1/4}}{N^{1/3}}a + N^{-0,1}a^{0,5}\right),$$

which is nontrivial for $q^{\epsilon_1}\max(a^3q^{3/4}, a^5) \leq N \leq a^{5/4}q^{5/4}$. The case $a^{5/4}q^{5/4} < N$ is not investigated here but can be easily considered by means of the method in [1].

The method of the present article is a variation of the method of [2] and [4], which is an extension of the method of [1].

Lemma 1. *Let Y be a number satisfying the condition $q^{1/4}N^{1/3} < Y < N$. Then for the sum*

$$S = \sum_x \sum_y \psi(x)\chi(xy+k),$$

extended over the domain

$$Y < y \leqslant 2Y, \quad 0 < xy \leqslant N, \quad xy \equiv g\,(\mathrm{mod}\ a),$$

the following estimate holds:

$$S \ll \frac{Nq^{\varepsilon}}{a}\frac{q^{1/4}}{N^{1/3}}a.$$

Proof. Here $x \ll NY^{-1} \ll N^{2/3}q^{-1/4}$ and, for a given x, y runs through the values which satisfy the condition $xy \equiv g\,(\mathrm{mod}\ a)$ and lie in an interval of the form $y' < y \leq y''$. Applying a well-known estimate for the sum of values of characters, we obtain

$$S \ll q^{\varepsilon}N^{2/3}q^{-1/4}\sqrt{q}\ln q \ll \frac{Nq^{\varepsilon}}{a}\frac{q^{1/4}}{N^{1/3}}a.$$

Lemma 2. *Let Y_0, Y, X_0, X be numbers satisfying the conditions*

$$0 \leqslant Y_0 < Y, \quad N^{1/3}a < Y \leqslant q^{1/4}N^{1/3}, \quad 0 \leqslant X_0 < X \leqslant NY^{-1}.$$

Then the sum

$$\sum_x \sum_y \psi(x)\chi(xy+k),$$

extended over the domain

$$x \equiv l\,(\mathrm{mod}\ a), \quad y \equiv f\,(\mathrm{mod}\ a), \quad X_0 < x \leqslant X, \quad Y_0 < y \leqslant Y,$$

admits the following estimate:

$$S \ll \frac{Nq^{\varepsilon'}}{a^2}\frac{q^{1/4}}{N^{1/8}}a.$$

Proof. We consider only the case $X_0 = 0$, $X = NY^{-1}$. The general case also reduces to this case if for $0 < x \le X_0$ and for $X < x \le NY^{-1}$ we assume $\psi(x) = 0$. Taking $\eta_0 = [a^{-1/2} q^{1/2} X^{-1/2}]$, we find that

$$S = \frac{1}{\eta_0}\sum_x \psi(x)\sum_{\eta=1}^{\eta_0}\sum_y \chi(x(y+a\eta)+k) + O(Xa^{-1}\eta_0 q^{\varepsilon})$$

$$\ll \frac{q^{\varepsilon}}{\eta_0}\sum_x\sum_\eta\Big|\sum_y \chi(a\eta+kx'+y)\Big| + Xa^{-1}\eta_0 q^{\varepsilon}$$

$$\ll \frac{q^{\varepsilon}}{\eta_0}\sum_{u=0}^{q-1}\lambda(u)\Big|\sum_y \chi(u+y)\Big| + Xa^{-1}\eta_0 q^{\varepsilon},$$

where $\lambda(u)$ denotes the number of solutions of the congruence $a\eta + kx' \equiv u \pmod q$. Therefore

$$S^2 \ll \frac{q^{2\varepsilon}}{\eta_0^2}\sum_u(\lambda(u))^2\sum_{y_1}\sum_y\sum_u \chi(u+y_1)\bar{\chi}(u+y) + X^2a^{-2}\eta_0^2 q^{2\varepsilon},$$

where y_1 runs through the same set of values as y. The latter summation over u gives either $q-1$ or -1 depending on whether y_1 and y are or are not equal to each other. Thus,

$$S^2 \ll \frac{q^{2\varepsilon}}{\eta_0^2}qYa^{-1}\sum_u(\lambda(u))^2 + X^2a^{-2}\eta_0^2 q^{2\varepsilon}.$$

The sum $\Sigma_\mu(\lambda(u))^2$ is equal to the number of solutions of the congruence

$$a(\eta_1-\eta) + k(x_1'-x') \equiv 0 \pmod q,$$

where η_1 runs through the same values as η, and x_1 runs through the same values as x. Therefore, the given sum is $\ll \eta_0 Xa^{-1} + \eta_0 T$, where T is the number of solutions of the congruence

$$a\eta + k(x_1'-x') \equiv 0 \pmod q$$

and, consequently, is equal to the number of solutions of the diophantine equation

$$(a\eta x_1 + k)(a\eta x - k) + k^2 = a^2\eta z q,$$

which for given η and z has $\ll q^{\epsilon}$ solutions. Therefore, noting that η runs through η_0 values, and z through $\ll \eta_0 X^2 q^{-1} + 1$ values, we obtain

$$T \ll (\eta_0^2 X^2 q^{-1} + \eta_0)q^{\varepsilon}.$$

Consequently,

$$S^2 \ll \frac{q^{3\varepsilon}}{\eta_0^2} qYa^{-1}(\eta_0 Xa^{-1} + \eta_0^3 X^2 q^{-1} + \eta_0^2) + X^2 a^{-2}\eta_0^2 q^{2\varepsilon}$$
$$\ll q^{3\varepsilon}(XY\sqrt{qX}a^{-3/2} + qYa^{-1} + qXa^{-3}).$$

Hence

$$S \ll Nq^{\varepsilon'}\left(\left(\frac{q}{NY}\right)^{1/4} a^{-3/4} + \left(\frac{qY}{N^2}\right)^{1/2} a^{-1/2}\right) \ll \frac{Nq^{\varepsilon'}}{a^2}\frac{q^{1/4}}{N^{1/3}}a.$$

Lemma 3. *Let Y be a number satisfying the condition $N^{1/3}a < Y \le q^{1/4}N^{1/3}$ and let $X = NY^{-1}$. Let $\beta = 2\beta_0$ be an even number satisfying $\beta^{-1} \ge q^{1/4}a/N^{1/3}$, and let $y_0 = Y\beta^{-1}$, and for $s = 0, 1, \cdots, \beta$ let the notation $Y_s = 2Y - sy_0$, $X_s = NY_s^{-1}$ be introduced. (Consequently $Y_0 = 2Y$, $Y_\beta = Y$, $X_0 = N(2Y)^{-1}$ and $X_\beta = NY^{-1}$.) For $r = 0, \cdots, \beta_0 - 1$ consider the sum*

$$S_r = \sum_x \psi(x) \sum_y \chi(xy + k),$$

where the summation is extended over the domain

$$X_{2r} < x \le X_{2r+1}, \quad x \equiv l (\operatorname{mod} a),$$
$$Y_{2r+2} < y \le Y_{2r+1}, \quad y \equiv f (\operatorname{mod} a).$$

Then

$$\sum_{r=0}^{\beta_0 - 1} S_r \ll \frac{Nq^{\varepsilon_1}}{a^2}\frac{q^{1/4}}{N^{1/3}}a.$$

Proof. We have

$$\left(\sum_{r=0}^{\beta_0-1} S_r\right)^2 \ll \beta_0 \sum_{r=0}^{\beta_0-1} |S_r|^2.$$

Taking $\eta_0 = [a^{-1/2}q^{1/2}X^{-1/2}]$ and $x_0 = X\beta^{-1}$, we find that

$$S_r = \frac{1}{\eta_0}\sum_x \psi(x) \sum_{\eta=1}^{\eta_0}\sum_y \chi(x(y + a\eta) + k) + O(x_0 a^{-1}\eta_0 q^{\varepsilon})$$
$$\ll \frac{q^{\varepsilon}}{\eta_0}\sum_x\sum_\eta\left|\sum_y \chi(a\eta + kx' + y)\right| + x_0 a^{-1}\eta_0 q^{\varepsilon}$$
$$\ll \frac{q^{\varepsilon}}{\eta_0}\sum_{u=0}^{q-1}\lambda(u)\left|\sum_y \chi(u + y)\right| + x_0 a^{-1}\eta_0 q^{\varepsilon},$$

where $\lambda(u)$ denotes the number of solutions of the congruence $a\eta + kx' \equiv u (\operatorname{mod} q)$. Thus we have

$$S_r^2 \ll \frac{q^{2\varepsilon}}{\eta_0^2} \sum_u (\lambda(u))^2 \sum_{y_1} \sum_y \sum_u \chi(u+y_1)\bar{\chi}(u+y) + x_0^2 a^{-2} \eta_0^2 q^{2\varepsilon},$$

where y_1 runs through the same set of values as y. But the last summation over u gives $q-1$ or -1 depending on whether y and y_1 are or are not equal to each other. Therefore

$$S_r^2 \ll \frac{q^{2\varepsilon}}{\eta_0^2} q y_0 a^{-1} \sum_u (\lambda(u))^2 + x_0^2 a^{-2} \eta_0^2 q^{2\varepsilon}.$$

The sum in this formula is equal to the number of solutions of the congruence

$$a(\eta_1 - \eta) + k(x_1' - x') \equiv 0 \pmod{q},$$

where η_1 runs through the same set of values as η, and x_1 runs through the same values as x. Therefore the above sum is $\ll \eta_0 x_0 a^{-1} + \eta_0 T_r$, where T_r is the number of solutions of the congruence

$$a\eta + k(x_1' - x') \equiv 0 \pmod{q}$$

and, consequently, is also equal to the number of solutions of the diophantine equation

$$(a\eta x_1 + k)(a\eta x - k) + k^2 = a^2 \eta z q$$

equivalent to it. Let us estimate the sum $\sum_{r=0}^{\beta_0 - 1} T_r$. Its part corresponding to the given pair of values η and z is, clearly, $\ll q^{\epsilon}$. Also, since η runs through η_0 values, and z runs through $\ll \eta_0 X^2 q^{-1} + 1$ values, we have

$$\sum_r T_r \ll (\eta^2 X^2 q^{-1} + \eta_0) q^{\varepsilon}.$$

Consequently

$$\begin{aligned} \Big(\sum_r S_r\Big)^2 &\ll \frac{q^{3\varepsilon}}{\eta_0^2} q y_0 a^{-1} \beta_0 (\eta_0 x_0 a^{-1} \beta_0 + \eta_0^3 X^2 q^{-1} + \eta_0^2) + \beta_0^2 x_0^2 a^{-2} \eta_0^2 q^{2\varepsilon} \\ &\ll q^{3\varepsilon}(qXY\eta_0^{-1}a^{-2} + X^2 Y \eta_0 a^{-1} + qYa^{-1} + X^2 a^{-2} \eta_0^2) \\ &\ll q^{2\varepsilon}(XY\sqrt{qX} a^{-3/2} + qYa^{-1} + qXa^{-3}). \end{aligned}$$

From this result it is easy to obtain (see the proof of Lemma 2) the estimate of the lemma.

Lemma 4a. *Let* Y *be a number satisfying the condition* $N^{1/2} a < Y \leq q^{1/4} N^{1/3}$, *and let*

$$S = \sum_x \sum_y \psi(x)\chi(xy+k),$$

where the summation is extended over the domain

$$x \equiv l (\mathrm{mod}\, a), \quad y \equiv f (\mathrm{mod}\, a), \quad Y < y \leqslant 2Y, \quad xy \leqslant N.$$

Then

$$S \ll \frac{Nq^{\varepsilon_2}}{a^2} \frac{q^{1/4}}{N^{1/3}} a.$$

Proof. Denoting by h the smallest number satisfying $2^{-h} \geq q^{1/4}/N^{1/3} a$, we select in the summation domain the rectangles numbered 0, 1, 2, $\cdots$, h as in Figure 1. The height of the triangle with number t is equal to $Y2^{-t}$. The parts of the sum S corresponding to the rectangles with numbers 0 and 1 are estimated by Lemma 2. For $t > 1$ the part of the sum S corresponding to the set of all rectangles with number t is estimated by means of Lemma 3 (taking $\beta = 2^t$). The remaining part of the sum S is estimated trivially. Thus we obtain the estimate for S indicated in the lemma.

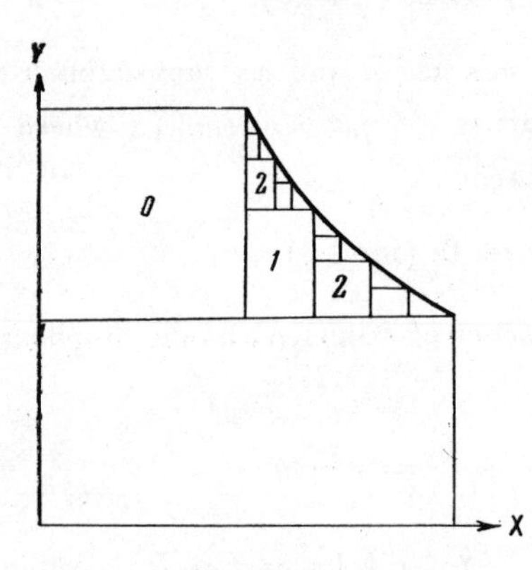

Figure 1

Lemma 4. *Let Y be a number satisfying $N^{1/3} a < Y \leq q^{1/4} N^{1/3}$, and let*

$$S = \sum_x \sum_y \psi(x)\chi(xy+k),$$

where the summation is extended over the domain

$$xy = g (\mathrm{mod}\, a), \quad Y < y \leqslant 2Y, \quad 0 < xy \leqslant N.$$

Then

$$S \ll \frac{Nq^{\varepsilon_2}}{a} \frac{q^{1/4}}{N^{1/3}} a.$$

Proof. Let S_l be the part of the sum S containing the terms under the condition $x \equiv l \pmod a$, $y \equiv f \pmod a$, where f is selected to satisfy the requirement $lf \equiv g \pmod a$. The sum S is a sum of $\phi(a) \le a$ sums of the form S_l, each of which is estimated by Lemma 4a.

Lemma 5. *Let x_1 and x_2 be numbers satisfying the conditions $0 \le x_1 < x_2 \le q\sqrt{q}$, where b and a are integers such that $b - a$ is not divisible by q. Then the sum*

$$S = \sum_{x_1 < x \leqslant x_2} \chi\left(\frac{x+a}{x+b}\right)$$

has the estimate $S \ll q \ln q$.

Proof. This lemma is a slight variation of a well-known lemma which we used earlier (Lemma 4 of [2]). The change of the condition $0 \le x_1 < x_2 \le q$ into a new one is possible since for $x_2 - x_1 = q$ the sum S is equal to -1.

Lemma 6. *Let X and Y be nonnegative numbers not exceeding N and let X_0 and Y_0 be numbers satisfying $a \le X_0 \le N$, $1 \le Y_0 \le N$. Then for the sum*

$$S = \sum_x \sum_y \psi(x)\psi_1(y)\chi(xy+k),$$

extended over the domain

$$xy \equiv g \pmod a, \quad X < x \leqslant X + X_0, \quad Y < y \leqslant Y + Y_0,$$

the following estimate holds:

$$S \ll \frac{X_0 Y_0 q^{\varepsilon_1}}{a} \sqrt{\frac{q^{0.5} a}{X_0} + \frac{a}{Y_0}}.$$

Proof. For $Y_0 < a$ the lemma follows from the fact that

$$S \ll \frac{X_0 Y_0 q^{\varepsilon_1}}{a} \ll \frac{X_0 Y_0 q^{\varepsilon_1}}{a} \sqrt{\frac{a}{Y_0}}.$$

Therefore only the case $Y_0 \ge a$ need be considered. Let S_l denote the part of sum S containing the values x and y subject to the condition $x \equiv l \pmod a$, $y \equiv f \pmod a$, where f is selected according to the requirement $lf \equiv g \pmod a$. We have

$$S_l^2 \ll X_0 a^{-1} q^{\varepsilon} \sum_{y_1} \sum_{y} \psi_1(y_1)\psi_1(y) \sum_x \chi\left(\frac{xy_1+k}{xy+k}\right),$$

where y_1 runs through the same values as y. The last sum over x is $\ll X_0 a^{-1}$, or (Lemma 5) $\ll \sqrt{q}\log q$, depending on whether y_1 and y are or are not equal to each other. Therefore

$$S_l^2 \ll X_0 a^{-1} q^{\varepsilon_1}(X_0 Y_0 a^{-2} + q^{0,5} Y_0^2 a^{-2}),$$

$$S_l \ll \frac{X_0 Y_0}{a^2}\sqrt{\frac{q^{0.5}a}{X_0} + \frac{a}{Y_0}}.$$

From this (see the proof of Lemma 4) the estimate of the lemma follows.

Lemma 7. *Let X and Y be numbers such that $a \le X$, $1 \le Y \le N$, $XY \le N$. Then the sum*

$$S = \sum_x \sum_y \psi(x)\psi_1(y)\chi(xy+k),$$

extended over a domain of the form

$$xy \equiv g(\mathrm{mod}\, a), \quad X < x < Xcq^{\varepsilon_0},$$
$$Y < y \leqslant Ycq^{\varepsilon_0}, \quad xy \leqslant N,$$

has the estimate

$$S \ll \frac{Nq^{\varepsilon_2}}{a}\sqrt{\frac{q^{0.5}Ya}{N} + \frac{Xa}{N}}.$$

Proof. In the case $c^2 q^{2\epsilon_0} XY \le N$, by the inequality of Lemma 6, which for $X_0 Y_0 \le N$ may be replaced by the less precise one

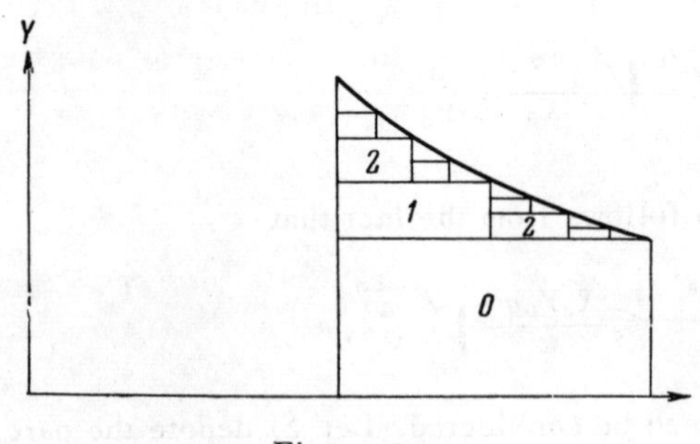

Figure 2

$$S \ll \frac{Nq^{\varepsilon_1}}{a}\sqrt{\frac{q^{0.5}Y_0 a}{N} + \frac{X_0 a}{N}},$$

we obtain directly the estimate of the present lemma with $\epsilon_2 = \epsilon_1 + \epsilon_0/2$. Therefore we consider only the case $c^2 q^{2\epsilon_0} XY > N$. Let S_1 denote the part of the sum S extended over the domain

$$xy \equiv g(\mathrm{mod}\, a), \quad X_1 < x \leqslant X_1 + X_0, \quad xy \leqslant N,$$

where the interval indicated for x is a part of the previous one (indicated in the lemma for the domain of summation of S), where $X_1/2 \le X_0 \le X_1$, and in the case where y lies outside the previous interval of summation we assume $\psi_1(y) = 0$.

Since the sum S may be split into $\ll \log N$ sums of the type S_1, the estimate for S given in the lemma will be obtained immediately when we obtain a similar estimate (with its own value of ϵ) for S_1. We shall do that now. Taking

$$\sqrt{\frac{q^{0.5}Ya}{N} + \frac{Xa}{N}} = \Delta,$$

denote by h the greatest integer satisfying the condition $2^{-h} \ge \Delta q^{\epsilon_0}$. Then we easily find that

$$XY > Nq^{-2\varepsilon_0}c^{-2},$$

$$1 \geqslant 2^h \Delta q^{\varepsilon_0} = 2^h q^{\varepsilon_0} \sqrt{\frac{q^{0.5}XYa}{NX} + \frac{XYa}{NY}} > \sqrt{\frac{q^{0.5}a2^{2h}}{Xc^2} + \frac{a2^{2h}}{Yc^2}},$$

$$X > q^{0.5}a2^{2h}c^{-2}, \quad Y > a2^{2h}c^{-2}.$$

We select rectangles from the summation domain of S_1 with numbers 0, $1, \cdots, h$ in accordance with Figure 2. Here the base of each rectangle with number t is equal to $X_0 2^{-t} \approx X_1 2^{-t}$. Consequently, the height of each such rectangle is $\approx NX_1^{-1}2^{-t}$. The parts of S_1 corresponding to rectangles with numbers 0 and 1 are considered in a similar manner and with the same result as for S in the case already considered.

In estimating the part $S^{(t)}$ of the sum S_1, which corresponds for $t > 1$ to the rectangle with number t, it should be noted that $S^{(t)}$ may be nonzero only in the case $N/2X_1 < Ycq^{\epsilon_0}$, and in this case, applying Lemma 6, we find that

$$S^{(t)} \ll \frac{Nq^{\varepsilon_1}2^{-2t}}{a} \sqrt{\frac{q^{0.5}a}{X_1 2^{-t}} + \frac{a}{NX_1^{-1}2^{-t}}} \ll \frac{Nq^{\varepsilon_1 + 0.5\varepsilon_0}}{a} \Delta 2^{-1.5t}.$$

Consequently, the sum of all $S^{(t)}$ correspond to a given t is

$$\ll \frac{Nq^{\varepsilon_1+0.5\varepsilon_0}}{a} \Delta 2^{-0.5t}.$$

Moreover, a trivial argument shows that the remaining part of S_1 is

$$\ll \frac{Nq^{\varepsilon_1}}{a}2^{-h} \ll \frac{Nq^{\varepsilon_1+\varepsilon_0}}{a}\Delta.$$

Therefore

$$S_1 \ll \frac{Nq^{\varepsilon_1+0.5\varepsilon_0}}{a}\Delta\sum_{t=0}^{h}2^{-0.5t} + \frac{Nq^{\varepsilon_1+\varepsilon_0}}{a}\Delta \ll \frac{Nq^{\varepsilon_1+\varepsilon_0}}{a}\Delta.$$

Lemma 8. *Let* X *and* Y *be numbers such that*

$$a < X \leqslant N^{4/5}, \quad 1 \leqslant Y \leqslant N^{1/3}a, \quad XY \leqslant N.$$

Then the sum

$$S = \sum_{x}\sum_{y} \psi(x)\psi_1(y)\chi(xy+k),$$

extended over the domain

$$xy \equiv g \pmod a, \quad X < x < Xcq^{\varepsilon_0},$$

$$Y < y \leqslant Ycq^{\varepsilon_0}, \quad xy \leqslant N,$$

can be estimated by

$$S \ll \frac{Nq^{\varepsilon_2}}{a}\left(\frac{q^{1/4}}{N^{1/3}}a + N^{-0,1}a^{0,5}\right).$$

Proof. This lemma is a simple corollary of Lemma 7.

Lemma 9. *Let* X *and* Y *be numbers satisfying the conditions*

$$a \leqslant X \leqslant N^{4/5}, \quad 1 \leqslant Y \leqslant N^{1/3}a, \quad XY \leqslant N,$$

and let each variable u *and* v *be a product of a finite number of factors which run through increasing sequences of natural numbers; let* x *and* y *each run through its increasing sequence of natural numbers. Moreover, let* u, x, y *and* v *satisfy the conditions* $X < ux < Xcq^{\epsilon_0}$ *and* $Y < yv < Ycq^{\epsilon_0}$. *Then the sum*

$$S = \sum_{u}\sum_{x}\sum_{y}\sum_{v} \chi(uxyv+k),$$

extended over the domain

$$uxyv \equiv g \pmod a, \quad (x, y) = 1, \quad uxyv \leqslant N,$$

can be estimated by

$$S \ll \frac{Nq^{\varepsilon_2}}{a}\left(\frac{q^{1/4}}{N^{1/3}}a + N^{-0.1}a^{0.5}\right).$$

Proof. We consider only the case when

$$X \geqslant N^{0.25}, \qquad Y \geqslant N^{0.1}a^{-0.5},$$

since otherwise $XY < N^{0.9}a^{-0.5}$ and the lemma is trivial. The sum S may be represented in the form $S = \Sigma_\sigma \mu(\sigma) S_\sigma$, where σ runs through natural numbers such that $(\sigma, a) = 1$ and

$$S_\sigma = \sum_u \sum_\xi \sum_\eta \sum_v \chi(\sigma^2 u\xi\eta v + k),$$

where ξ and η run through the quotients of division of σ by x and y, which are multiples of σ, and the variables u, ξ, η, v satisfy the conditions

$$X\sigma^{-1} < u\xi < X\sigma^{-1}cq^{\varepsilon_0}, \qquad Y\sigma^{-1} < \eta v < Y\sigma^{-1}cq^{\varepsilon_0}$$

the summation being extended over the domain

$$\sigma^2 u\xi\eta v \equiv g \pmod a, \quad u\xi\eta v \leqslant N\sigma^{-2}.$$

Let $X_0 = X\sigma^{-1}$, $Y_0 = Y\sigma^{-1}$, $N_0 = N\sigma^{-2}$. Further, let k_0 be defined by the conditions $0 < k_0 < q$, $k \equiv k_0\sigma^2 \pmod q$, and let g_0 be defined by the conditions $0 \le g_0 < a$, $g \equiv q_0\sigma^2 \pmod a$. Finally, let $\psi(x_0)$ denote the number of solutions of $u\xi = x_0$ and let $\psi_1(y_0)$ denote the number of solutions of $\eta v = y_0$. For some ϵ we have $\psi(x_0) << q^\epsilon$ and $\psi(y_0) << q^\epsilon$. The modulus of the sum S_σ is equal to the modulus of the sum

$$S'_\sigma = \sum_{x_0} \sum_{y_0} \psi(x_0)\psi_1(y_0)\chi(x_0y_0 + k_0),$$

extended over the domain

$$x_0y_0 \equiv g_0 \pmod a, \qquad X_0 < x_0 < X_0cq^{\varepsilon_0},$$

$$Y_0 < y_0 < Y_0cq^{\varepsilon_0}, \quad x_0y_0 \leqslant N_0.$$

Take

$$\frac{q^{1/4}}{N^{1/3}}a + N^{-0.1}a^{0.5} = \Delta$$

and denote by σ_0 the greatest integer such that $\sigma_0^{-1} \ge \Delta$.

First, let us estimate the sum S'_σ in the case $\sigma \le \sigma_0$. Here we can easily verify the inequalities

$$a < X_0 < N_0, \quad 1 < Y_0 < N_0, \quad q^{3/4}a^3 \leqslant N_0, \quad a^5 \leqslant N_0$$

and, consequently, Lemma 7 may be applied to the sum S'_σ (with X_0, Y_0, N_0, k_0, g_0 instead of X, Y, N, k, g). We obtain

$$S'_\sigma \ll \frac{N_0 q^{\varepsilon_2}}{a} \sqrt{\frac{q^{0,5} Y_0 a}{N_0} + \frac{X_0 a}{N_0}} \ll \frac{N q^{\varepsilon_2}}{a} \Delta \sigma^{-1.5}.$$

Further, let us estimate the sum S'_σ in the case $\sigma > \sigma_0$. Here we can easily verify that the sum of terms of S'_σ corresponding to the condition $x_0 y_0 = w$ for a given w is $\ll q^{3\epsilon}$, whilst the number of distinct solutions w not exceeding N_0 is $\leq N_0 a^{-1}$. Therefore

$$S'_\sigma \ll \frac{N_0 q^{3\varepsilon}}{a} \ll \frac{N q^{3\varepsilon}}{a\sigma^2}.$$

Combining all the above results, we have

$$S \ll \sum_\sigma |S'_\sigma| \ll \frac{N q^{\varepsilon_2}}{a} \Delta \sum_{\sigma \leqslant \sigma_0} \sigma^{-1.5} + \frac{N q^{3\varepsilon}}{a} \sum_{\sigma > \sigma_0} \frac{1}{\sigma^2} \ll \frac{N q^{\varepsilon_2}}{a} \Delta.$$

Lemma 10. *Let $H = N^{0.1}$, let P be a product of primes satisfying $p \leq H$, and let Q be a product of primes satisfying $H < p \leq N$. Further, let $\theta(x)$ be a function such that $|\theta(x)| \leq 1$, and let*

$$S = \sum_{\substack{p \leqslant N \\ p \equiv g \pmod a}} \theta(p),$$

$$W_s = \sum_{\substack{d_1 \backslash P \\ \\ d_1 m_1 \ldots d_s m_s \equiv g \pmod a \\ d_1 m_1 \ldots d_s m_s \leqslant N}} \sum_{m_1 > 0} \cdots \sum_{d_s \backslash P} \sum_{m_s > 0} \mu(d_1) \ldots \mu(d_s)\, \theta(d_1 m_1 \ldots d_s m_s);$$

$$s = 1, \ldots, 10.$$

Then for some constants $\alpha_1, \cdots, \alpha_{10}$ we have

$$S = \alpha_1 W_1 + \ldots + \alpha_{10} W_{10} + O(NH^{-1} q^\varepsilon).$$

Proof. It is easy to verify that

$$W_s = \sum_{\substack{y_1 \backslash Q \\ y_1 \ldots y_s \equiv g \pmod a \\ y_1 \ldots y_s \leqslant N}} \cdots \sum_{y_s \backslash Q} \theta(y_1 \ldots y_s).$$

In fact, the right-hand side of the last equality can be represented as the sum

$$\sum_{\eta_1>0} \cdots \sum_{\eta_s>0} \theta(\eta_1 \ldots \eta_s) \sum_{\substack{d_1 \setminus (\eta_1, P) \\ \eta_1 \ldots \eta_s \equiv g \pmod a \\ \eta_1 \ldots \eta_s \leq N}} \mu(d_1) \ldots \sum_{d_s \setminus (\eta_s, P)} \mu(d_s),$$

which is identical with the sum expressing W_s in the lemma.

Further, it is easy to see that among the products $y_1 \cdots y_s$ entering into the indicated expression for W_s, there are n, $1 \leq n \ll NH^{-1}q^{\epsilon}$ products divisible by squares of prime divisors of Q. Among the remaining products $y_1 \cdots y_s$ a given number z_h which is a product of h distinct prime divisors of Q occurs s^h times, since its every prime factor may enter into $y_1 \cdots y_s$. Therefore, taking $S^{(h)} = \Sigma_{z_h} \theta(z_h)$, we have

$$sS^{(1)} + s^2S^{(2)} + \ldots + s^sS^{(s)} = W_s + O(NH^{-1}q^{\varepsilon}).$$

Writing the latter equality for all $S = 1, \cdots, 10$ and determining $S^{(1)}$ from the system of equations thus obtained, we find that the lemma is true, if we note that $S^{(1)} = S + O(H)$.

Lemma 11. *Let $\epsilon_0 < 0.05$ and let P be a product of primes satisfying $p \leq N^{0.1}$. Then, assuming that*

$$D = r^{\frac{\ln r}{\ln(1+\varepsilon_0)}}, \quad r = \ln N,$$

the divisors d of number P, which do not exceed N, may be distributed among $< D$ sets with the following properties.

a) *Numbers d belonging to the same set have the same number of prime divisors and, consequently, they have the same $\mu(d) = (-1)^{\beta}$.*

b) *One of the sets, which we shall call the simplest, consists of a single number $d = 1$. For this set we assume $\phi = 1$ and thus obtain $d = \phi = 1$. To each of the remaining sets corresponds its own ϕ such that all numbers d of this set satisfy the condition $\phi < d \leq \phi^{1+\epsilon_0}$.*

c) *For every set distinct from the simplest set, and for every U satisfying $0 < U \leq \phi$ there exist two sets (the second of which may be the simplest set) of numbers d: numbers d' and d'' with corresponding values ϕ' and ϕ'' of ϕ, which satisfy the condition $U < \phi' \leq UN^{0.1}$, $\phi'\phi'' = \phi$, such that for some positive integer B we obtain every number d of the given set B times, if we select from all products $d'd''$ only those for which $(d', d'') = 1$.*

Proof. All prime divisors p of the number P (in other words, all primes not exceeding $N^{0.1}$) we distribute inside $\tau + 1$ intervals of the form

$$\left(\frac{3}{2}\right)^{(1+\varepsilon_0)^t} < p \leqslant \left(\frac{3}{2}\right)^{(1+\varepsilon_0)^{t+1}},$$

which are obtained by making the number t ("the number of the interval") run through values $t = 0, 1, \cdots \tau$, where τ denotes the greatest integer satisfying the condition

$$\left(\frac{3}{2}\right)^{(1+\varepsilon_0)^\tau} < N^{0.1}.$$

From this condition which defines τ we easily find that the number $\tau + 1$ of all indicated intervals satisfies

$$\tau + 1 < \frac{\ln r}{\ln(1+\varepsilon_0)} - 1.$$

We associate every divisor d of P, $1 < d \leq N$, with a sequence $l_0, l_1, \cdots$ $\cdots, l_\tau$ where l_t denotes the number of prime factors of d lying in the interval with number t. The set of values d associated with the same sequence is, in fact, the required set of values of d described in the lemma. Since every such d is a product of less than $\log N - 1$ prime factors $(N \geq a^3 q^{3/4})$, we have $l_t < r - 1$ for every $t = 0, 1, \cdots, \tau$, and consequently the number of distinct sets is $< r^{\tau \pm 1} + 1 < D$.

It follows from the above definition of the set of numbers d that numbers d belonging to it do, in fact, have the same number β of prime factors, and consequently they have the same value of $\mu(d) = (-1)^\beta$.

Further, consider an arbitrary (but not the simplest) set of numbers d. Let $d = p_1 \cdots p_\beta$ be the factorization of one of its members into prime factors, where the factors are written in nondecreasing order. Denote the number of the interval containing p_s by t_s and the number $\left(\frac{3}{2}\right)^{(1+\varepsilon_0)^{t_s}}$ by ϕ_s. We have $\phi_s < p_s < \phi_s^{1+\epsilon_0}$. Assigning to this selected set the number $\phi = \phi_1 \cdots \phi_\beta$, we obtain $\phi < d \leq \phi^{1+\epsilon_0}$.

Let U be a number satisfying $0 < U \leq \phi$. Denoting by λ the least number satisfying the inequality $U < \phi_1 \cdots \phi_\lambda$, consider all the auxiliary sets, i.e. the numbers d' which run through the products of the form $p_1 \cdots p_\lambda$ and the numbers d'' which run through the products of the form $d = p_{\lambda+1} \cdots p_\beta$ (the set d'' may be the simplest set). Then, to the sets of numbers d' corresponds the number $\phi' = \phi_1 \cdots \phi_\lambda$, and to the set of numbers d'' corresponds the number $\phi'' = \phi_{\lambda+1} \cdots \phi_\beta$, where we have (for $\phi_1 \cdots \phi_{\lambda-1} \leq U$, $\phi_\lambda < N^{0,1}$)

$$U < \varphi' < UN^{0.1}, \quad \varphi'\varphi'' = \varphi.$$

If d is any number of the selected set, then the equality $d = d'd''$ is possible only in the case $(d', d'') = 1$, in which case it has

$$B = \binom{\varkappa_1 + \varkappa_2}{\varkappa_1}$$

solutions, where κ_1 is the number of factors ϕ_s of the product ϕ' which are equal to ϕ_λ, and κ_2 is the number of factors ϕ_s of the product ϕ'' which are equal to ϕ_λ.

Theorem. *We have*

$$\sum_{\substack{p \leqslant N \\ p \equiv g \pmod{a}}} \chi(p+k) \ll \frac{Nq^{\varepsilon_3}}{a} \left(\frac{q^{1/4}}{N^{1/3}} a + N^{-0,1} a^{0,5} \right).$$

Proof. Taking $\theta(x) = \chi(p + k)$, we have, by Lemma 10,

$$\sum_{\substack{p \leqslant N \\ p \equiv g \pmod{a}}} \chi(p+k) = \alpha_1 W_1 + \cdots + \alpha_{10} W_{10} + O(N^{0,9} q^{\varepsilon}).$$

Considering some particular W_s, we split the values of every d_j, $j = 1, \cdots, s$, into $< D$ sets in the way described in Lemma 10. At the same time all the values of every m_j, $j = 1, \cdots, s$, are split into $\ll \log N$ sets. One of these sets consists of a single number $m_j = 1$. For this set we assume $M_j = 1$, so that we have $m_j = M_j = 1$. Making M_j run through the values $M_j = 1, 2, 4, \cdots, 2^h$, where h is the greatest integer satisfying $2^h < N$, we construct all the other sets, assigning to the same set the values m_j satisfying $M_j < m_j \leq 2M_j$, $m_j \leq N$.

Let W_s' denote the part of the sum W_s corresponding to some system of sets of values of variables $d_1, \cdots, d_s, m_1, \cdots, m_s$. This part may be represented in the form

$$W_s' = \pm \sum_{d_1} \cdots \sum_{d_s} \sum_{m_1} \cdots \sum_{\substack{m_s \\ d_1 \ldots d_s m_1 \ldots m_s \equiv g \pmod{a} \\ d_1 \ldots d_s m_1 \ldots m_s \leqslant N}} \chi(d_1 \ldots d_s m_1 \ldots m_s + k),$$

where $d_1, \cdots, d_s, m_1, \cdots, m_s$ satisfy inequalities of the form

$$\varphi^{(1)} < d_1 \leqslant F^{(1)}, \ \ldots, \ \varphi^{(s)} < d_s \leqslant F_s, \quad F = \varphi^{1+\varepsilon_0},$$

$$M_1 < m_1 \leqslant 2M_1, \ \ldots, \ M_s < m_s \leqslant 2M_s,$$

some of which, as we have seen, may be replaced by equalities of either the form $d_j = \phi^{(j)} = 1$ or the form $m_j = M_j = 1$. Our theorem will be proved if for every s and for every sum W_s' corresponding to it we establish the estimate of the form

$$W_s' \ll \frac{Nq^{\varepsilon''}}{a}\Delta, \quad \Delta = \frac{q^{1/4}}{N^{1/3}}a + N^{-0,1}a^{0,5}. \tag{1}$$

We assume the numbers $M_1, \cdots, M_s$ to be in increasing order (since this assumption does not affect the generality of the case) and introduce the notation $\phi^{(1)} \cdots \phi^{(s)} = \phi$, $M_1 \cdots M_s = M$. We need only consider the case $\Phi M >> Nq^{\epsilon'''}\Delta/a$, since otherwise the estimate (1) is obtained trivially.

In the case $q^{1/4}N^{1/3} < M_s$ the estimate (1) follows from Lemma 1 $(Y = M_s)$.

In the case $N^{1/3}a < M_s \le q^{1/4}N^{1/3}$ the estimate (1) follows from Lemma 4 $(Y = M_s)$.

In the case $N^{1/5} < M_s \le N^{1/3}a$ the estimate (1) follows from Lemma 8 $(Y = M_s)$.

Let $M_s < N^{1/5}$. The product M may be represented in the form $M = M'M''$, where M' is the product of the numbers M_j which satisfy $M_j \le H$, and M'' is the product of the numbers M_j which satisfy $M_j > H$.

First, consider the case $N^{4/5} < M''$. Then M'' consists of not less than five factors M_j. If at least four of these factors exceed $N^{1/6}$ then the estimate (1) follows from Lemma 8, since taking $X = M_{s-3}M_{s-2}M_{s-1}M_s$ we obtain $N^{2/3} < X < N^{4/5}$. If this is not the case, then among the factors of M'' there are not less than two which do not exceed $N^{1/6}$, and the estimate (1) again follows from Lemma 8, since taking $Y = M_1M_2$ we have $N^{1/5} < Y \le N^{1/3}$.

Further, consider the case $M'' < N^{4/5}$. If we have also $N^{2/3}a^{-1} \le M''$ then the estimate (1) follows from Lemma 8, since taking $X = M''$ we shall have $N^{2/3}a^{-1} \le X \le N^{4/5}$. If $M'' < N^{2/3}a^{-1}$ but $N^{2/3}a^{-1} \le M$ then the product M' can be split into two factors $M' = M^{(3)}M^{(4)}$ so that we have

$$N^{2/3}a^{-1} \leqslant M''M^{(3)} < N^{2/3+0,1}a^{-1}.$$

The estimate (1) follows from Lemma 8, since, taking $X = M''M^{(3)}$, we obtain $N^{2/3}a^{-1} \le X < N^{4/5}$.

Finally, let $M < N^{2/3} a^{-1}$. Let δ be the least number satisfying $N^{2/3} a^{-1} < M\phi^{(1)} \cdots \phi^{(\delta)}$. Defining U by

$$N^{2/3}a^{-1} = M\varphi^{(1)}\ldots\varphi^{(\delta-1)}U \quad (\varphi^{(1)}\ldots\varphi^{(\delta-1)} = 1, \quad \text{if} \quad \delta = 1),$$

we have $0 < U < \phi^{(\delta)}$, and consequently, by Lemma 11, there exist two sets of numbers: numbers d' and numbers d'', with corresponding numbers ϕ' and ϕ'' which satisfy the condition $U < \phi' \leq UN^{0.1}$ and $\phi'\phi'' = \phi^{(\delta)}$, such that for some positive integer B we obtain all values d_δ if we select from all products $d'd''$ the ones which satisfy $(d', d'') = 1$. Assuming that

$$X = M\varphi^{(1)} \ldots \varphi^{(\delta-1)}\varphi', \qquad Y = \varphi''\varphi^{(\delta+1)} \ldots \varphi^{(s)},$$

we have

$$N^{2/3}a^{-1} < X \leqslant N^{2/3+0.1}a^{-1}, \quad XY = M\Phi,$$

where the upper bound of X is less than $N^{4/5}$. Taking $d' = x$ and $d'' = y$, and denoting by u a variable running through all products of the form $u = m_1 \cdots m_s d_1 \cdots d_{\delta-1}$ and by v a variable running through all products of the form $v = d_{\delta+1} \cdots d_s$, we reduce the sum W_s' to the form

$$W_s' = \pm \frac{1}{B} \sum_u \sum_x \sum_y \sum_v \chi(uxyv + k),$$

where the variables u, x, y, v run through values satisfying the inequalities (clearly, $X < q$, $Y < q$)

$$X < ux < cXq^{\varepsilon_0}, \qquad Y < yv < cYq^{\varepsilon_0},$$

where the summation is extended over the domain

$$uxyv \equiv g \pmod{a}, \qquad (x, y) = 1, \qquad uxyv \leqslant N.$$

The estimate (1) of the sum W_s' follows now immediately from Lemma 9.

BIBLIOGRAPHY

[1] I. M. Vinogradov, *An improvement of the estimation of sums with primes*, Izv. Akad. Nauk SSSR Ser. Mat. 7 (1943), 17–34 (Russian) MR 5, 143.

[2] ———, *New approach to the estimation of a sum of values of* $\chi(p + k)$, Izv. Akad. Nauk SSSR Ser. Mat. 16 (1952), 197–210. (Russian) MR 14, 22.

[3] ———, *The method of trigonometrical sums in the theory of numbers*, Trudy Mat. Inst. Steklov. 23 (1947); English transl., Interscience, New York, 1954. MR 10, 599; MR 15, 941.

[4] ———, *Improvement of an estimate for the sum of the values* $\chi(p+k)$, Izv. Akad. Nauk SSSR Ser. Mat. 17 (1953), 285–290. (Russian) MR 15, 855.

Translated by:
H. Alderson

FINITE GROUPS WHOSE PROPER SUBGROUPS ALL ADMIT NILPOTENT PARTITIONS

UDC 519.44

V. M. BUSARKIN AND A. I. STAROSTIN

This paper deals with nonpartitionable finite groups whose proper subgroups all admit nilpotent partitions. All such groups, with the exception of the group M_9, of order 720, are solvable.

A group is said to admit a *nilpotent* (*abelian*) partition if among its subgroups there exists a collection of nilpotent (abelian) subgroups such that every nonidentity element of the group is contained in one and only one subgroup of the collection. The structure of finite groups admitting a nilpotent partition is well known (see for example [1]–[3]). A finite group G admitting a nilpotent partition is of one of the following types (where p is a prime):

1. Nilpotent groups.

2. Frobenius groups with nilpotent complement.

3. $G = (A \times P) \leftthreetimes \{b\}$, where $P \leftthreetimes \{b\}$ is a p-subgroup with an isolated subgroup P; $b^p = e$, $A \leftthreetimes \{b\}$ is a Frobenius group.

4. The symmetric group on 4 letters.

5. $PGL\ (2, p^n)$ where $p^n > 3$, $p \neq 2$.

6. $PSL\ (2, p^n)$ where $p^n > 3$.

7. Simple Suzuki groups.

In the sequel, groups of these types are referred to, for short, as groups of type 1, 2, $\cdots$, 7. Clearly, in a group admitting a nilpotent partition, every subgroup also admits a nilpotent partition.

This paper is concerned with finite groups which do not themselves admit nilpotent partitions but whose proper subgroups all do. This class of groups is a natural generalization of the class of $\bar{A}$-groups, previously studied by these authors (see [1]), as well as of Šmidt groups [5]. By a Šmidt group we mean a nonnilpotent finite group whose proper subgroups are all nilpotent.

We shall also denote by M_9, after Zassenhaus [6], that group of order 720 which is the triply transitive permutation group on 10 letters of the smallest possible order. The group M_9 possesses only one proper normal subgroup, of order 360, isomorphic with the alternating group on 6 letters (or with the group $PSL\ (2, 9)$). A Sylow 2-subgroup of M_9 is of type (1) of the lemma below and is maximal. It should also be noted that M_9 is covered by its Sylow subgroups.

By $G(p, q, u)$ we will denote a noncommutative group of order $p^v q^u$ with normal Sylow p-subgroup whose proper subgroups are all abelian (see for example [4], p. 22); p, q and r always denote primes.

Theorem. *A finite group G satisfies the conditions:*

(a) *G does not admit a nilpotent partition;*

(b) *all proper subgroups of G admit nilpotent partitions*

if and only if it is one of the following:

1°. *Direct product of a cyclic group of order p^2 and a $G(q, p, 1)$-group.*

2°. *Direct product of a cyclic group of order r and a $G(p, q, 1)$-group, with $r \neq q$.*

3°. *Direct product of a cyclic group of order p and a Frobenius group $A \rtimes \{b\}$ where the Frobenius complement $\{b\}$ is of order p^2 and the Frobenius kernel A is an elementary abelian q-group without proper subgroups admissible with respect to $\{b\}$.*

4°. *A Šmidt group with a nontrivial center.*

5°. *$G = F \rtimes \{h\} \rtimes \{c\}$ where the order of F is p^α, $\alpha \geq 3$, $h^q = c^p = e$, and $F \rtimes \{h\}$ and $\{h\} \rtimes \{c\}$ are Frobenius groups such that, whenever F contains a proper subgroup F_1 admissible with respect to $\{h, c\}$, $F_1\{h, c\}$ is isomorphic to the symmetric group on 4 letters.*

6°. *Group M_9.*

For the proof we will need the following proposition.

Lemma. *If in a finite 2-group T there exists an involution t whose centralizer $\mathfrak{z}(t)$ is a noncyclic group of order 4, then T is of one of the following types:*

$$1)\ T = \{a, b\}, \quad a^{2^n} = b^2 = e, \quad bab^{-1} = a^{-1+2^{n-1}}, \quad n \geqslant 3;$$
$$2)\ T = \{a, b\}, \quad a^{2^n} = b^2 = e, \quad bab^{-1} = a^{-1}, \quad n \geqslant 1 \text{ (dihedral group)}.$$

Proof. We use induction on the order of T to show that T contains a cyclic subgroup of index 2. For groups of orders 4 and 8 this is obvious. Let the order of T be greater than 8. A maximal subgroup M of T containing $\mathfrak{z}(t)$ contains a cyclic subgroup $\{c\}$ of index 2 by the induction hypothesis, and $\{c\}$ is characteristic in M. Consequently, $\{c\}$ is normal in T and contains the commutator subgroup T' of T. On the other hand, since the order of $\mathfrak{z}(t)$ is 4 and the number of elements of T conjugate to t is equal to the index of $\mathfrak{z}(t)$, the index of the commutator group T' in T equals 4. Therefore $T' = \{c\}$. The factor group $T/\{c^2\}$ is noncommutative and of order 8. Therefore $c = v^2$ for some element v in T. Thus $\{v\}$ is the required subgroup of index 2 in T.

The structure of 2-groups with a cyclic subgroup of index 2 is well known (see [7], Russian p. 210) and accounts for the conclusion of the lemma.

Proof of the Theorem. Necessity. If all proper subgroups of G admit abelian partitions, than G is of one of the types 1° through 4°, as was shown in [4]. In what follows we shall therefore assume that the group G satisfies the condition:

(c) At least one proper subgroup of G does not admit abelian partitions.

A. First, let us assume that G is a nonsolvable group whose nontrivial normal subgroups are of index not greater than 2. We show that in this case G is a group of type 6°.

AI. *Any two nonprimary maximal nilpotent subgroups of G are disjoint.*

Let A and B be two nonprimary maximal nilpotent subgroups such that their intersection D is of maximal possible order. If $D \neq E$, then the normalizer $N(D)$ is a proper subgroup of G and, by the choice of A and B, $N(D)$ contains at least two nonprimary maximal nilpotent subgroups with nontrivial intersection. But in a group admitting a nilpotent partition any two nonprimary maximal nilpotent subgroups are disjoint. Thus $D = E$, and assertion AI is proved.

Remark. This proof shows that assertion AI holds in the more general case when G contains no solvable normal subgroups. This remark will be utilized later.

AII. Let us consider the case when G contains an involution t whose centralizer $\mathfrak{z}(t)$ is not nilpotent. This centralizer has a nontrival center and, being a proper subgroup of G, admits a nilpotent partition; hence $\mathfrak{z}(t)$ is a group of type 3. Let M be maximal among those subgroups of G which contain $\mathfrak{z}(t)$ and are of type 3. Then $M = (V \times T_0) \rtimes \{u\}$, $u^2 = e$.

AII. 1. *A Sylow 2-subgroup of G is a dihedral group.*

First, let us assume that $T = T_0\{u\}$ is its own normalizer in G. Then T is a Sylow 2-subgroup in G. If u lies in the center of T, then T is a abelian and coincides with the normalizer of its own center. If, on the other hand, u does not lie in the center Z of T, then Z is the center of M, and the normalizer of the center Z is M (due to the choice of M and to the fact that any solvable group that admits a nilpotent partition and has a subgroup of type 3 is itself of type 3). By the Hall-Gruen Theorem ([7], Russian p. 240) this implies that G is not 2-normal, as otherwise G would have a normal subgroup of index greater than 2. Hence G contains a subgroup T_1 conjugate to T and such that $Z \subseteq T \cap T_1$ and the center Z_1 of T_1 is distinct from Z. The normalizer of the center Z coincides with M and contains Z_1; therefore $Z_1 \subset M$. But Z is the totality of those elements in T_0 whose order does not exceed 2, which follows from the fact that the 2-group T admits partitions (see [1], p. 275). It should also be noted that T_1 is not contained in M, as M is 2-normal. If the intersection $Z_1 \cap T_0$ were nontrivial, its normalizer would contain M as a proper subgroup and be of type 3. But this contradicts the maximality of M. Thus $Z_1 \cap T_0 = E$, and the order of Z_1, and hence of Z as well, equals 2. Thus T_0 contains only one involution, i.e. T_0 is cyclic. Consequently T is a dihedral group.

Let us now assume that T is distinct from its normalizer $N(T)$. Let g be an element of $N(T)$ not contained in T. Then $M \neq M^g$ since M is its own normalizer in G, and T is its own normalizer in M. The intersection $D = M \cap M^g$ contains T. If the order of T_0 exceeded 2, then the intersection $VT_0 \cap (VT_0)^g$ would be nontrivial, which contradicts AI. Thus the order of T equals 4, and T is its own centralizer in G since the centralizer of T_0 is M and the centralizer of the subgroup $\{u\}$ in M is T. By the lemma, the Sylow 2-subgroup S of G is of type (1) or (2) of the lemma.

To prove AII. 1 it remains to show that S cannot be of type (1) of the lemma. Let us assume the opposite, i.e. that

$$S = \{a, b\}, \quad a^{2n} = b^2 = e, \quad bab^{-1} = a^{-1+2n-1}, \quad n \geqslant 3.$$

The commutator group S' of the subgroup S is equal to $\{a^2\}$ and contains the center of S, equal to $\{a^{2n-1}\}$. The preceding arguments show that the centralizer of the center of S is nilpotent as it contains S. We also note that all involutions of S that are not contained in S', are conjugate; hence the centralizer of any such involution is conjugate to M. The same, of course, is true for any Sylow 2-subgroup of G. Let S_1 denote a Sylow 2-subgroup of G distinct from S, and assume that $S \cap S_1' \neq E$. The cyclicity of S_1' implies that the intersection $S \cap S_1'$ is cyclic and contains only one involution d, which belongs to the center of S_1. The centralizer $\mathfrak{z}(d)$ is nilpotent, so that d cannot belong to $S \backslash S'$. But then d lies in S' and thus in the center of S. We arrived at a contradiction, as $\mathfrak{z}(d)$ cannot contain two distinct Sylow 2-subgroups. Thus $S \cap S_1' = E$ for any subgroup S_1 conjugate to S.

Furthermore, the normalizer $N(S)$ of the subgroup S in G is nilpotent. Indeed, a Sylow 2-subgroup in $N(S)$ is normal and contains a characteristic subgroup of order 2; therefore $N(S)$ cannot belong to any of types 2 through 7. It follows, in particular, that

$$S \cap N'(S) = S',$$

and

$$[S \cap N'(S)] \prod_{x \in G} (S \cap S'^{x}) = S'.$$

By the First Theorem of Gruen (see [7], Russian p. 238), G contains in this case a normal subgroup of index 4, contrary to assumption. This completes the proof of assertion AII.1.

AII. 2. Let s be an involution lying in the center of a Sylow 2-subgroup S of G. By AII. 1, S is a dihedral group. We may assume that the Sylow 2-subgroup $T_0\{u\}$ of M is contained in S. The centralizer $\mathfrak{z}(s)$ has a nontrivial intersection with $V \times T_0$, whence, by AI, it cannot be a nonprimary nilpotent subgroup. But then $\mathfrak{z}(s)$ contains an abelian 2-complement, and we can apply Theorem I of [8]. It implies that G is isomorphic either to a $PSL(2, p^n)$ or a $PGL(2, p^n)$ group or to the alternating group $\mathfrak{A}_7$. The first two isomorphisms are impossible, since G does not admit nilpotent partitions. Also impossible is the isomorphism $G = \mathfrak{A}_7$, as $\mathfrak{A}_7$ contains a proper subgroup which does not admit a nilpotent partition: $\{(1234)\,(56), (12)\,(56), (567)\}$.

AIII. Let the centralizer of any involution of the group G be nilpotent.

AIII. 1. *Let P be a Sylow p-subgroup of G, and let P_0 be one of its subgroups distinct from E. Then if an involution t induces a regular automorphism on P_0, it also induces a regular automorphism on P; hence P is abelian, and every subgroup of P is admissible with respect to $\{t\}$.*

To prove this assertion, let M be a maximal subgroup of G containing the normalizer of P_0. Since the normalizer of any solvable subgroup of G is itself solvable, the Sylow p-subgroup P is contained in M.

M contains the Frobenius group $P_0\{t\}$; hence M is of one of the types 2 through 4. If M is of type 4, then $P_0 = P$, and the assertion is obvious. If, on the other hand, M is of type 2 or 3, then the Sylow p-subgroup is normal in M. The subgroup $P \times \{t\}$ of M is a Frobenius group with kernel P. This implies the assertion.

AIII. 2. Assume that G contains an involution with a nonprimary centralizer. Let H be a maximal nilpotent subgroup of G containing this centralizer. It is easy to see that the centralizer of any nonidentity element of H is contained in $N(H)$; this follows from AI. The normalizer $N(H)$ cannot be a Frobenius group. Indeed, if G is not simple, then, by the hypothesis of A, it has only one proper normal subgroup N, the index of N in G is 2. Then

$N(H)$ itself has a normal subgroup of index 2, which is impossible. If, on the other hand, G is simple, then, by Theorem 3 of [9], H must be a 2-group. The normalizer $N(H)$ is not nilpotent either, since $N(H) \neq H$ by the well-known Frobenius Theorem and by AI. Consequently $N(H)$ is of type 3:

$$N(H) = (A \times P) \times \{b\}, \quad H = A \times P, \quad b^p = e, \quad p \neq 2.$$

Let u be the index of $N(H)$ in G and let v be the number of involutions in $N(H)$. Then $v \neq 1$, because $T\{b\}$ is a Frobenius group with kernel T, where T is a Sylow 2-subgroup of G contained in H. Since every involution of G is contained in some subgroup conjugate to H and since by AI all subgroups conjugate to H are pairwise disjoint, the number of involutions in G is equal to uv. This implies that $G \setminus N(H)$ contains $v(u-1)$ involutions. Among these we can find two involutions s and t belonging to the same coset of $N(H)$ in G, i.e. we can find an element g in $N(H)$ such that $s = tg$. Since $tgtg = e$ and $t^2 = e$, we have $tgt^{-1} = g^{-1}$. If $g \notin H$, then, by AIII. 1, H contains a p-subgroup F admissible with respect to t. If, however, $g \in H$, then for F we take $\{g\}$. It follows that

$$H \cap tHt^{-1} \supseteq F \neq E,$$

which contradicts AI.

AIII. 3. *Thus the centralizer of any involution of G is a 2-group.* Moreover, the hypothesis of A together with the Feit-Thompson Theorem [10] implies that G is of even order. Theorem 5 of §10 in [11], the basic result in [12], and condition (a) imply that G is isomorphic either to $PSL(3, 4)$ or M_9. Since in $PSL(3, 4)$ the normalizer of the maximal intersection of Sylow 2-subgroups does not admit a nilpotent partition, on account of condition (b) G cannot be isomorphic to $PSL(3, 4)$. Consequently G is of type 6°.

Assertion A is proved.

B. *Let G be a nonsolvable group which is not simple. Then G contains one and only one normal subgroup, and its index is* 2 (*so that assertion* A *applies to G*).

BI. Let us assume that there exist nonsolvable groups which satisfy the conditions of the theorem and contain a nontrivial solvable normal subgroup. Let H be such a group of minimal order. Let R denote a nontrivial solvable normal subgroup of H. We show that the factor group H/R is simple. All proper subgroups of H/R admit a nilpotent partition. If N/R is a proper normal subgroup of H/R, then N is solvable, as N admits a nilpotent partition and contains a nontrivial solvable normal subgroup. By the same reason, H/R

does not admit a nilpotent partition. Moreover, H/R satisfies each of the conditions (a), (b) and (c) of the theorem, due to the nonsolvability of H/R and to Lemma 3 in [4]. This, however, contradicts the minimality of the order of H.

Since the choice of R has been arbitrary, it follows that R is the only nontrivial solvable normal subgroup in H and, in particular, that R is an elementary abelian group. Let us consider two cases.

Case 1. R is contained in the center of H. Then R cannot be a Sylow p-subgroup in H, since otherwise R would be a direct factor and all proper subgroups of H/R would be nilpotent, i.e. the factor group H/R would be solvable (see [5]). Let us denote by P/R a nontrivial p-subgroup of H/R, and by K/R, its normalizer. Then K, being a subgroup of H with a nontrivial center, is of type 1° or 3°. Therefore every element of K/R whose order is prime to p belongs to the centralizer of P/R. By Theorem 14.4.7 in [7], this contradicts the simplicity of H/R.

Case 2. R coincides with its centralizer. Clearly, the order of R is greater than 4. Let Q/R denote a nontrivial q-subgroup of H/R with $q \neq p$, and let L/R denote its normalizer. The subgroup L of H admits a nilpotent partition and contains the solvable normal subgroup R, which is its own centralizer. Therefore L is of type 2°, and its Frobenius kernel is R. Hence L/R is nilpotent, and we again can apply Theorem 14.4.7 of [7], which leads to a contradiction.

Thus G contains no nontrivial solvable normal subgroups.

BII. *The group G is p-normal for all $p > 2$.*

Suppose that G is not p-normal. Then the definition of p-normality together with Lemma 14.3.1 in [7] implies that G contains Sylow p-subgroups P_1 and P_2 whose centers $Z(P_1)$ and $Z(P_2)$ are distinct and contained in $D = P_1 \cap P_2$, D being a maximal intersection of Sylow p-subgroups. Since G contains no proper solvable normal subgroup, $N(D)$ is a proper subgroup in G. Thus $N(D)$ admits a nilpotent partition, and consequently is solvable and is of type 3°. Let M be a maximal solvable subgroup of G containing $N(D)$. M is also of type 3°, and D is contained in the Fitting radical of M. Since $Z(P_1) \subset D$, by AI and the maximality of M, we have $\mathfrak{z}(Z(P_1)) = M$. Therefore $P_1 \subset M$. Similarly, $P_2 \subset M$. It follows that M is not p-normal. But this is impossible, as all groups of type 3° are p-normal.

BIII. To prove assertion B, it is sufficient to show that each proper normal subgroup of G is of index 2. Suppose that G contains a proper normal subgroup G_1 whose index exceeds 2. By BI, G_1 is not solvable. Any proper subgroup M of G containing G_1 is of type 5°. Therefore the index of G_1 in M equals 2. It follows that either $|G : G_1| = 4$ (when G_1 is not maximal in G) or $|G : G_1| = p$, where p is an odd prime. Let us consider these two cases separately.

BIII. 1. $|G : G_1| = 4$. It is easy to see that G/G_1 cannot be cyclic. Let T be a Sylow 2-subgroup of G and let M_i $(i = 1, 2, 3)$ denote maximal subgroups of G containing G_1. The groups M_i are of type 5°, as was noted above. In a group of type 5° a Sylow 2-subgroup is a dihedral group of order not less than 8. Therefore T contains three distinct maximal subgroups $T \cap M_i$ $(i = 1, 2, 3)$, each of which is a dihedral group. The intersection of these three subgroups is a Sylow 2-subgroup of G_1 and thus is itself a dihedral group. However, it is easy to see that no 2-group with such a property exists. Contradiction in the case BIII. 1 is thus obtained.

BIII. 2. $|G : G_1| = p > 2$. Since G is p-normal, the Hall-Gruen Theorem implies that the subgroup $N(Z(P))$ has a nontrivial p-factor group. Since $N(Z(P))$ admits a nilpotent partition and is not isomorphic to the symmetric group on 4 letters, the fact that it contains a nontrivial p-factor group implies the existence of a p-factor group isomorphic to a Sylow p-subgroup. It then follows from the Hall-Gruen Theorem that a Sylow p-subgroup of G is of prime order.

Consider the normalizer $N(P)$. Clearly $N(P) = P \times Q$, where $Q = N(P) \cap G_1$. Thus $N(P)$ is a nonprimary nilpotent subgroup with $Q \neq E$. Since $N(P)$ coincides with its normalizer and, by AI (or by the simplicity of P when $Q = E$), is disjoint from its conjugate subgroups, G is a Frobenius group, which is impossible.

C. Finally, let G be solvable. We show that G is of type 4° or 5°.

If all proper subgroups of G are nilpotent, then G is a Šmidt group with a nontrivial center (see [5]), since a Šmidt group without a center admits a nilpotent partition.

Let us assume that G contains at least one nonnilpotent proper subgroup. Let M be a maximal normal subgroup of G and let M be of index p. We note that M cannot be isolated in G, as otherwise G would admit partitions and thus would admit a nilpotent partition. Consequently there exists an element $g \in G \setminus M$ such that $\{g\} \cap M \neq E$.

M is of one of the types 1° through 4°. Correspondingly, we consider four separate cases.

CI. M is nilpotent. Let U be a nonnilpotent maximal subgroup of G. Then $U \cap M$ is the Fitting radical of U and its index in U is p. Therefore U cannot be of type 4°.

CI.1. Let U be a Frobenius group. Then $U \cap M$ is the Frobenius kernel of U (we denote it by F), and the Frobenius complement is cyclic of order p, i.e. $U = F\{h\}$ where $h^p = e$. Then $G = M \rtimes \{h\}$, and F is normal in G. Since G is not a Frobenius group, we can find in M an element a of prime order q which commutes with h. Thus $a \notin F$ and $G = (F \rtimes \{a\}) \rtimes \{h\}$. Here $p \neq q$, as otherwise G would be of type 3° contrary to condition (a) of the Theorem.

Let us show that the subgroup $\{a\} \times \{h\}$ is maximal in G. Assuming the opposite, let K denote a maximal subgroup of G containing $\{a, h\}$:

$$K = F_1\{a, h\},$$

where $F_1 = K \cap F$. Since the Fitting radical in K is $F_1\{a\}$, K can be of none of the types 1°, 2° or 4°. Consequently K is of type 3°. Therefore $q = p$, which is impossible, as was noted above.

From the maximality of $\{a, h\}$ in G and from the fact that $N(\{a\})$ contains $\{a, h\}$ as a proper subgroup, it follows that $\{a\}$ is normal in G. Then $G = \{a\} \times F\{h\}$ and $\{h\}$ is maximal in $F\{h\}$. Therefore all subgroups of $F\{h\}$ are commutative. But in this case G is an $\bar{A}$-group (see [4], p. 23) contrary to condition (c).

CI.2. U is of type 3°, i.e. $U = (A \times P) \rtimes \{b\}$, $b^p = e$. In this case

$$G = M \rtimes \{b\}, \quad M = M_1 \times M_p,$$

where M_p is a Sylow p-subgroup of M, $A \subseteq M_1$, $P \subseteq M_p$. Let c be a p-element of $G \setminus M$. Then $M_1\{c\}$ is not nilpotent, since otherwise the equation $M_1 \times M_p\{c\} = G$ would imply nilpotency of G. Therefore $M_1\{c\}$ is a Frobenius group. Consequently, every p-element of $G \setminus M$ induces a regular automorphism in M_1. The order of an element of G/M for which $\{g\} \cap M \neq E$, is of the form $p^\alpha n$ where $(n, p) = 1$, $\alpha \geq 1$. Since $g^p \in M$ and $g_1 = g^n$ is a p-element of $G \setminus M$, the preceding argument shows that $n = 1$. But $g^p \neq e$ and it induces the identity automorphism in M_1; thus $M_1\{g\}$ cannot be a Frobenius group. Consequently $G = M_1\{g\}$ and the Sylow p-subgroup of G is cyclic. This, however, contradicts the assumption that U is of type 3°.

Thus, in what follows we may assume that none of the maximal normal subgroups of G is nilpotent.

CII. M is of type 3°, i.e. $M = (A \times R) \rtimes \{b\}$, where $b^r = e$, $R\{b\}$ is a

Sylow r-subgroup with the isolated normal subgroup R, and $A \rtimes \{b\}$ is a Frobenius group.

CII.1. First, let $p = r$. In this case AR is of index p^2 in G, which means that the commutator subgroup G' is contained in AR. Let c be a p-element of $G \setminus M$. Then $A\{c\}$ is not nilpotent, as otherwise G would contain the nilpotent maximal normal subgroup $A \times R\{c\}$, which case was considered in CI. Therefore $A\{c\}$ is a Frobenius group. It follows that an element g of $G \setminus M$ for which $\{g\} \cap M \neq E$, is a p-element with $g^p \in M \setminus AR$. We may assume that $g^p = b$. Thus $G = AR \rtimes \{g\}$. Let A_1 be a proper characteristic subgroup of A. Then $A_1R\{g\}$ must admit a nilpotent partition, but a consideration of the types 1° through 4° shows this to be impossible. Thus A is an elementary abelian group. Let H denote a maximal subgroup of $R\{g\}$ containing $\{g\}$. The subgroup AH is not nilpotent and must admit a nilpotent partition. Therefore AH is a Frobenius group with the Frobenius complement H. Consequently $H = \{g\}$ and $R\{g\} = R \times \{g\}$ is abelian. Since all Sylow subgroups of G are abelian, any of its subgroups admitting a nilpotent partition admits an abelian partition as well. But this contradicts condition (c).

CII.2. Now let $p \neq r$. In this case $R\{b\}$ is a Sylow r-subgroup in G. It is of index p in its normalizer $N(R\{b\})$, and

$$N(R\{b\}) = R\{b\} \rtimes \{c\},$$

where c is an element of $G \setminus M$ of order p. The normalizer $N(R\{b\})$ cannot be nilpotent, as otherwise the commutativity of b and c would imply that $G' \subseteq AH$ and the subgroup $AR\{c\}$ would be a nilpotent maximal normal subgroup in G. Thus $N(R\{b\})$ is a Frobenius group.

We now show that $g^p \notin AR$, where g is an element of $G \setminus M$ such that $\{g\} \cap M \neq E$. Indeed, in the case $g^p \in AR$ the subgroup $AR\{g\}$ would be nilpotent and would contain a Sylow p-subgroup of G. Therefore the normality of R in G would imply that all p-elements of G were contained in the centralizer of R. But this is impossible, as $N(R\{b\})$ is a Frobenius group and contains p-elements.

Thus $g^p \in M \setminus AR$. We may assume $g^p = b$. Consequently the order of g is pr, and $g^r \in N(R) \cap N(\{b\}) \subseteq N(R\{b\})$. This again is impossible, since the normalizer of a Sylow r-subgroup of G contains no elements of order pr, as was shown before.

CIII. M is the symmetric group on 4 letters. Let A denote a noncyclic normal subgroup of M, of order 4. Since the centralizer $\mathfrak{z}(A)$ in G is of index 6, $\mathfrak{z}(A)$ is an abelian normal subgroup of order $4p$, and

$G = \mathfrak{z}(A) \rtimes (\{b\} \rtimes \{c\})$ where $b^3 = c^2 = e$, $cbc^{-1} = b^{-1}$. The subgroup $H = \mathfrak{z}(A)\{b\}$ is maximal and normal in G. Therefore, by CI, CII and the inequality $\mathfrak{z}(A) \neq A$, H is a Frobenius group with the kernel $\mathfrak{z}(A)$ and the complement $\{b\}$. It follows that $p \neq 2$ because 3 does not divide $8 - 1$. Therefore $\mathfrak{z}(A)$ contains a characteristic subgroup D of order p. Now $\mathfrak{z}(D)$ is normal in G, and $G \setminus \mathfrak{z}(D)$ is cyclic. Consequently $\mathfrak{z}(D)$ is maximal in G. But by CI and CII, this is impossible.

CIV. *Thus every maximal normal subgroup of G (in particular, M) is a Frobenius group.*

Let $M = F \rtimes H$, where F is the Frobenius kernel and H is the Frobenius complement. Then $N(H) = H\{c\}$, where $c \notin M$ and $c^p \in H$, and $G = F \rtimes N(H)$. Since G is not a Frobenius group, we can find in $N(H) \setminus H$ a nonidentity element which commutes with some nonidentity element of F. We may assume that this element is c and that the order of c is p.

CIV.1. $N(H)$ cannot be of type 4°, as it contains a nilpotent normal subgroup of prime index.

CIV.2. Assume that $N(H)$ is nilpotent. Then, for every maximal subgroup U of $N(H)$, the subgroup FU is normal in G and thus is a Frobenius group. We may assume that M is chosen such that its Frobenius kernel is of the greatest possible order. Then F is the Frobenius kernel of FU. Therefore elements of U induce regular automorphisms in F. It follows that the element c does not belong to any maximal subgroup of $N(H)$. This is clearly impossible.

CIV.3. Let $N(H)$ be of type 3°, i.e. $N(H) = H \rtimes \{c\}$ and $H = A \times P$, where P is a Sylow p-subgroup of H. Consider the subgroup $V = FP\{c\}$. It is easy to see that V can be of none of the types 1°, 2° or 4°. But V cannot be of type 3° either, as otherwise $\{c\}$ would be contained in the centralizer of the center of F, i.e. in $F \times \{c\}$. Thus $\{c\}$ would be normal in G and consequently also in $N(H)$, contrary to the definition of the groups of type 3°.

CIV.4. Finally, let $N(H)$ be a Frobenius group, so that c induces a regular automorphism in H.

Assume that H contains a proper characteristic subgroup H_1 and consider the subgroup $W = FH_1\{c\}$. Since FH_1 and $H_1\{c\}$ are Frobenius groups, the Fitting radical of W equals F. Therefore W can be of none of the types 1° through 3°. Thus W is the symmetric group on 4 letters. Then the order of F is 4, and the order of H_1 is 3. But since FH is a Frobenius group, the order of H must be 3 as well, contrary to the assumtion on H_1. Thus H is an elementary abelian group and is the Frobenius complement. Therefore H is a cyclic group of prime order q, and p divides $q-1$.

Assume now that F contains no proper subgroups admissible with respect to $N(H)$. Then F is an elementary abelian r-group. By the choice of the element c, the subgroup $F\{c\}$ is nilpotent. But since $\{c\}$ is not normal in G, we have $r = p$. Denote the order of F by p^α. Since $q-1 \equiv 0 \pmod{p}$ and FH is a Frobenius group, we have $\alpha \geq 2$. The equality $\alpha = 2$ is not possible as otherwise the congruences

$$q-1 \equiv 0 \pmod{p}, \quad p^2-1 \equiv 0 \pmod{q}$$

would imply that $q = 3$ and $p = 2$, and the group G would be isomorphic to the symmetric group on 4 letters. Thus in this case the group G is of type 5°.

Finally, assume that F contains a proper subgroup F_1 admissible with respect to $N(H)$. It is easy to see that F_1 is the Fitting radical of the subgroup $F_1N(H)$. Therefore F_1 cannot be of any of the types 1° through 3°, and consequently it is isomorphic to the symmetric group on 4 letters. Thus in this case G also is of type 5°.

The proof of necessity is complete.

Proof of Sufficiency. It can be readily seen that all proper subgroups of a group of each of the types 1° through 3° admit an abelian partition. It can also be easily established from the existence of a nontrivial center in groups of types 1° through 4° that these groups satisfy condition (a) of the theorem.

Consider a group G of type 5°. It is easy to see that G cannot be of any of types 1° through 4°, so that condition (a) of the theorem is satisfied by the group G. Denote by U a proper subgroup of G. If $U \subseteq F\{h\}$, then U admits a nilpotent partition. Let $U \not\subseteq F\{h\}$. We may assume that q divides the order of U and that $\{h\} \subset U$. If $\{h\}$ is its own normalizer in U, then U is a Frobenius group and thus admits a nilpotent partition. If, on the other hand, h is distinct from its normalizer in U, then $\{h, c\} \subseteq U$ and $U = (U \cap F) \rtimes \{h, c\}$ is either a Frobenius group (when $U \cap F = E$) or a group of type 4° (when $U \cap F \neq E$) by definition.

Finally, let G be of type 6°. It follows from the description of groups admitting a nilpotent partition that G satisfies condition (a) of the theorem. Let us verify condition (b). Let U be a proper subgroup of G. If U is not solvable, then, since every involution in G has a primary centralizer, Theorem 5 of §10 in [11] implies that U admits a nilpotent (even a cyclic) partition. If, on the other hand, U is solvable, then, since the order of every nonidentity element of G is a power of a prime, we have, according to [13], that U either is nilpotent or is a Frobenius group or has a normal series

$$U \supset U_1 \supset U_2 \supset U_3 = E,$$

where U_2 is the largest normal p-subgroup of U; U/U_2 is a Frobenius group with the cyclic Frobenius kernel U_1/U_2 of order $q^\alpha (p \neq q)$; and U/U_1 is a cyclic p-group. In the latter case we clearly have $p = 2$, and the structure of a Sylow 2-subgroup of G shows that U_2 is a noncyclic group of order 4, and that $q = 3$ and $\alpha = 1$. Therefore U is isomorphic to the symmetric group on 4 letters. This completes the proof of the theorem.

In conclusion, we offer two of the simplest examples of groups of type 5°.

Example 1.

$$G = \{a_1, a_2, a_3, b, c\}; \quad a_1{}^3 = a_2{}^3 = a_3{}^3 = b^{13} = c^3 = e;$$

$$a_1a_2 = a_2a_1; \quad a_1a_3 = a_3a_1; \quad a_2a_3 = a_3a_2;$$

$$ba_1b^{-1} = a_2; \quad ba_2b^{-1} = a_3; \quad ba_3b^{-1} = a_1a_2;$$

$$ca_1c^{-1} = a_2; \quad ca_2c^{-1} = a_2{}^2a_3; \quad ca_3c^{-1} = a_1a_2{}^2a_3; \quad cbc^{-1} = b^3.$$

Example 2.

$$G = \{a_1, a_2, a_3, a_4, b, c\}; \quad a_1{}^2 = a_2{}^2 = a_3{}^2 = a_4{}^2 = b^3 = c^2 = e;$$

$$a_1a_2 = a_2a_1; \quad a_1a_3 = a_3a_1; \quad a_1a_4 = a_4a_1;$$

$$a_2a_3 = a_3a_2; \quad a_2a_4 = a_4a_2; \quad a_3a_4 = a_4a_3;$$

$$ba_1b^{-1} = a_2; \quad ba_2b^{-1} = a_1a_2; \quad ba_3b^{-1} = a_4; \quad ba_4b^{-1} = a_3a_4;$$

$$ca_1c^{-1} = a_1; \quad ca_3c^{-1} = a_3; \quad ca_2c^{-1} = a_1a_2; \quad ca_4c^{-1} = a_3a_4; \quad cbc^{-1} = b^{-1}.$$

BIBLIOGRAPHY

[1] V. M. Busarkin and A. I. Starostin, *Locally finite groups with a partition*, Mat. Sb. 62 (104) (1963), 275–294. (Russian) MR **28** #130.

[2] R. Baer, *Einfache Partitionen endlicher Gruppen mit nichttrivialer Fittingscher Untergruppe*, Arch. Math. **12** (1961), 81–89. MR 25 #115.

[3] M. Suzuki, *On a finite group with a partition*, Arch. Math. **12** (1961), 241–254. MR **25** #113.

[4] V. M. Busarkin and A. I. Starostin, *Finite indecomposable groups all of whose subgroups have an Abelian partition*, Ural. Gos. Univ. Mat. Zap. **4**, tetrad' 1 (1963), 22–31. (Russian) MR **32** #1243.

[5] O. Ju. Šmidt, *Groups, all of whose subgroups are special*, Mat. Sb. **31** (1924), 366–372. (Russian)

[6] H. Zassenhaus, *Kennzeichunung endlicher linearer Gruppen als Permutationsgruppen*, Abk. Math. Sem. Univ. Hamburg **11** (1936), 17–40.

[7] M. Hall, Jr., *The theory of groups*, Macmillan, New York, 1959; Russian transl., IL, Moscow, 1962. MR 21 #1996.

[8] D. Gorenstein and J. H. Walter, *On finite groups with dihedral Sylow 2-subgroups*, Illinois J. Math. 6 (1962), 553–593. MR 26 #188.

[9] W. Feit, *On groups which contain Frobenius groups as subgroups*, Proc. Sympos. Pure Math., vol. 1, Amer. Math. Soc., Providence, R.I., 1959, pp. 22–28. MR 22 #11037.

[10] W. Feit and J. G. Thompson, *Solvability of groups of odd order*, Pacific J. Math. 13 (1963), 775–1029. MR 29 #3538.

[11] M. Suzuki, *Finite groups with nilpotent centralizers*, Trans. Amer. Math. Soc. 99 (1961), 425–470. MR 24 #A1309.

[12] ———, *On a class of doubly transitive groups*, Ann. of Math. (2) 75 (1962), 105–145. MR 25 #112.

[13] G. Higman, *Finite groups in which every element has prime power order*, J. London Math. Soc. 32 (1957), 335–342. MR 19, 633.

Translated by:
E. Sadowski

GROUPS WITH A NORMALIZER CONDITION FOR CLOSED SUBGROUPS

V. I. UŠAKOV

This work studies the connection between projective limits of nilpotent groups and topological N-groups, i.e. groups which satisfy a normalizer condition for closed subgroups, and clarifies the construction of locally compact N-groups.

In [1] V. M. Gluškov constructed a theory of locally nilpotent locally compact groups, which was a generalization, on the one hand, of the theory of commutative locally compact groups of L. S. Pontrjagin, and on the other hand, a generalization of the theory of abstract (discrete) locally nilpotent groups. Gluškov used the classical, abstract definition of nilpotency. Let us recall this definition.

Definition 1. A group G (considered abstractly) is said to be *nilpotent* in the algebraic (abstract) sense if its lower central series

$$G = G^{(0)} \supset G^{(1)} \supset \ldots \supset G^{(k)} \supset G^{(k+1)} \supset \ldots$$

attains the identity subgroup after a finite number of steps. Here $G^{(k+1)} = [G, G^{(k)}]$ is a mutual commutant of the group G and the subgroup $G^{(k)}$, $k = 0, 1, \cdots$, i.e. the subgroup generated by all possible commutators of type

$$[x, y] = x^{-1}y^{-1}xy, \quad x \in G, \quad y \in G^{(k)}.$$

The group G is said to be *locally nilpotent* in the algebraic sense, if every finite subset of G generates a nilpotent subgroup of G.

It then appeared that, for such a definition of nilpotency, which had no connection with the topology of the group, a considerable number of interesting facts from the theory of abstract locally nilpotent groups do not carry over to arbitrary locally nilpotent topological groups (not even locally compact). Here we confine ourselves to giving one example only.

V. I. Plotkin [see [5], Russian p. 408] showed that every abstract N-group, i.e. every group having the normalizer condition, is locally nilpotent. In the case of topological groups it is relevant to study groups which have the normalizer conditions for closed subgroups. Such groups will be called N-groups.

Definition 2. A topological group G is said to be an *N-group* if every one of its proper subgroups is contained as a normal divisor in some larger subgroup of G, i.e. it differs from its normalizer.

Let us recall that the normalizer of a group H which is a subgroup of a group G, is the largest subgroup of G in which H is invariant. We denote the normalizer of H by $N(H)$. If H is a closed subgroup of a topological group G, then $N(H)$ is also a closed subgroup of G. A question raised by V. S. Čarin at the Fourth All-Union Conference on General Algebra in Kiev in 1962, is, in essence, as follows: does a topological N-group satisfy the normalizer condition for all subgroups, i.e. is a topological N-group an N-group in the abstract sense? The answer to this problem proved to be a "no", and in [4] one may find an example of a compact N-group which is not even locally nilpotent.

It is possible to quote a large number of examples which justify the study of nilpotent groups from the topological point of view.

Definition 3. A topological group G is said to be *nilpotent in the topological sense* if it is a projective limit of groups which are nilpotent in the abstract sense, i.e. if in any neighborhood of its identity there is contained a member of the lower central series, for some finite index; and it is said to be *locally nilpotent in the topological sense* if every finite subset $M \subset G$ generates, in the algebraic sense, a subgroup $\{M\}$ which is nilpotent in the topological sense.

From now on, any group G which is nilpotent (locally nilpotent) in the topological sense, will be simply called nilpotent (locally nilpotent). In the discrete case, topological nilpotency becomes the usual nilpotency. Nevertheless, the class of nilpotent groups is considerably larger than the class of groups which are nilpotent in the abstract (algebraic) sense. Thus a full direct product of finite nilpotent groups with a Tihonov topology is a nilpotent group, but is not necessarily even locally nilpotent in the abstract sense.

Nevertheless, as was shown in [2], fundamental results obtained by V. M. Gluškov for locally nilpotent groups in the abstract sense* hold also for locally nilpotent locally compact groups. In this work we study the connection between nilpotent groups and N-groups and will also clarify the construction of locally compact N-groups.

Let us now make a number of preliminary remarks. If G is a nilpotent

---

* A similar generalization is due to V. P. Platonov, who also obtained a number of results on topological N-groups.

group, then any neighborhood U of its identity contains the closure of some nth member of the lower central series of G, namely the subgroup $\overline{G^{(n)}}$. This is the consequence of regularity of the space of topological groups; namely, in every neighborhood U of the identity element of G there is properly contained the closure $\overline{V}$ of some neighborhood V of the identity.

If M is a subset of G, then about the subgroup $\{\overline{M}\}$ we may say that it is topologically generated by the set M. Unlike the monograph [3], we do not confine our studies to closed subgroups of a topological group, since if we did we would have to exclude from this investigation the subgroups, important for us, which arise as the members of the lower central series. If a subgroup is closed this will be discussed separately.

It is easily seen that nilpotency is a hereditary property; in fact, all subgroups (not necessarily closed) and all quotient groups of a nilpotent group are themselves nilpotent groups. It is also clear that a quotient group of an N-group is an N-group. The problem of subgroups of N-groups is very complex, and in the general case not every subgroup of an N-group is an N-group. It is known, however, that this is the case when the N-group is discrete. A generalization of this fact is the following result.

Proposition 1. *Every open subgroup of an N-group G is an N-group.*

Proof. Clearly G is an N-group if and only if for any closed subgroup A of G there exists an increasing series of closed subgroups

$$A = A_0 \subset A_1 \subset \ldots \subset A_\alpha \subset A_{\alpha+1} \subset \ldots \subset A_\beta \subset G, \qquad (*)$$

where A_α is a normal divisor of subgroup $A_{\alpha+1}$ for all $0 \le \alpha < \beta$ (e.g. $A_{\alpha+1} = N(A_\alpha)$), and in limiting positions we have the closures of the unions of the preceding members (thus in the case of discrete groups the series (*) will be an increasing normal series, relating A with G). If A is a closed subgroup of a group H which is open in G, then A is closed in G. Hence for a given subgroup A we may construct the series (*). Let α be a first ordinal number such that the intersection of A_α and H is strictly larger than A. If the number α is limiting, then let us take an element $x \notin A$, $x \in A_\alpha \cap H$. As the subgroup H is open in G, the given element is contained in the set $H \backslash A$, which is open in G (the complement of A in H). It follows from this fact that in $H \backslash A$ there exists a $y \in A_\delta$, for some $\delta < \alpha$, and this contradicts the choice of α. Thus if α is not limiting then $\alpha - 1$ exists, and by virtue of the choice of α we have $A_{\alpha-1} \cap H = A$.

Let us take $x \in A_\alpha \cap H$, $x \notin A$. Since the subgroup $A_{\alpha-1}$ is invariant in A_α, for any $a \in A$ we have $x^{-1}ax \in A_{\alpha-1}$.

On the other hand, clearly $x^{-1}ax \in H$. Therefore we have $x^{-1}ax \in H \cap A_{\alpha-1} = A$ for all $a \in A$, which means that the element x belongs to the normalizer of A in H.

Let us construct an example demonstrating that *a subgroup of an N-group is not always an N-group* (even under the condition that the subgroup is closed).

Let $n \geq 3$ and let G_n be the nth generalized group of quaternions, i.e. the group generated by two elements a_n and b_n which are related to each other by the following relations:

$$a_n^{2^{n-1}} = 1, \quad a_n^{2^{n-2}} = b_n^2, \quad b_n^{-1}a_nb_n = a_n^{-1}.$$

The group G_n is said to be a finite nilpotent group of class $n-1$, i.e. the length of its lower central series is $n-1$.

Let us form the complete direct product $G = \Pi_{n=3}^{\infty} G_n$ of the groups G_n, $n = 3, 4, \cdots$, which is given a Tihonov topology. We thus obtain a compact totally disconnected group. Let us verify that it is an N-group.

Because in G there is always a dense subgroup A, which is a ZA-group, i.e. a group which has an increasing central series (discrete direct product of groups G_n), we can base everything on the following proposition:

Proposition 2. *If an arbitrary topological group G (not necessarily locally compact) always contains a dense subgroup A which is a ZA-group, then G is an N-group.*

Proof. Let

$$E = A_0 \subset A_1 \subset \ldots \subset A_\alpha \subset A_{\alpha+1} \subset \ldots \subset A_\gamma = A$$

be an upper central series for the subgroup A in H, where H is an arbitrary closed subgroup of G. If H coincides with its normalizer, then H would have a subgroup A_1.

Assume we have already proved that for all $\beta < \alpha$ the subgroups A_β lie in H. If α is a limiting ordinal number, then $A_\alpha = \bigcup_{\beta<\alpha} A_\beta$ is clearly contained in H. If α is not a limiting number, then there exists a subgroup $A_{\alpha-1}$ which lies in H. Let h be any element of H, x any element of A_α. Let us choose an arbitrarily small neighborhood U of the element $[x, h]$. In it there lies, necessarily, some element $[x, a]$, where $a \in A$, as $\overline{A} = G$. But $[x, a] \in A_{\alpha-1} \subseteq H$. Thus in any neighborhood of $[x, h]$ there lie elements of H. Since H is closed, the element $[x, h]$ is contained in H, whence it follows that $x^{-1}hx \in H$ for all $x \in A_\alpha$ and $h \in H$. This means that A_α belongs to the normalizer $N(H)$ of H in G; but because $N(H) = H$ we have that $A_\alpha \subseteq H$, which completes the induction. Consequently $A = A_\gamma \subseteq H$. Since H is closed and $\overline{A} = G$ we have $H = G$, which proves our proposition.

Let us consider $H = \overline{\{a, b\}}$, which is a subgroup of G and is topologically generated by elements $a = (a_3, a_4, \cdots)$ and $b = (b_3, b_4, \cdots)$. We will prove that this subgroup is an N-group. To do this, we shall show that the cyclic subgroup $\{b\}$ coincides with its normalizer in H. Let us assume that it is not so, i.e. that $\{b\} \neq N$, where N denotes the normalizer of $\{b\}$ in H. Moreover, $N \neq H$, i.e. $\{b\}$ is not a normal divisor of H. Indeed, otherwise the element b would have a finite number of conjugate elements in H, as $\{b\}$ is a group of order four. But it is known that there exists a one-to-one correspondence between conjugates of b and left cosets of H by the centralizer Z of a given element. Recall that the centralizer of an element is the set of all elements of the group which commute with this element. Consequently, since the index of Z in H is finite, there exists a natural number k such that $a^k \in Z$. But clearly no power of a commutes with b. Thus $N \neq H$, and hence in N there exists an element x of finite order which does not lie in $\{b\}$, as otherwise $\{b\}$, which is a periodic part of N, would be a characteristic part of N, and consequently invariant in $N(N) \neq N$, which would contradict the definition of the normalizer (here $N(N)$ is the normalizer of N in H; we in fact let H be an N-group, i.e. *any* proper closed subgroup, including N, is distinct from its normalizer).

Let us consider the subgroup $\{x, b\}$, which is a finite group. Moreover, because every finite-order element of G is a 2-element, i.e. its order is some power of two, we have that $\{x, b\}$ is a finite 2-group. Therefore we may assume that the quotient group $\{x, b\} / \{b\}$ has order two, i.e. $x^2 \in \{b\}$, from which it follows that the order of $\{x, b\}$ is eight.

Moreover $x^2 \neq b$, because G does not contain the square root of b. Similarly, $x^2 \neq b^3$. There remain only these cases:

1. $x^2 = 1$. In this case the element x has the form $x = (x_3, x_4, \cdots)$, where x_i is the ith component of a given element ($i \geq 3$), which is either 1 or b_i^2 (as $x_i^2 = 1$), and not all x_i are 1. No element x of this form, with the exception of b^2, lies in the subgroup $H = \overline{\{a, b\}}$. Actually, let $x \neq b^2$, $x \neq 1$. Assume otherwise, i.e. let $x \in \overline{\{a, b\}} = H$. Then in any neighborhood of this element there exists an element $y \in \{a, b\}$. As $b^{-1}ab = a^{-1}$, and $b^4 = 1$, y has the form $a^m b^l$, $0 \leq l \leq 3$.

Because $x \neq b^2$, $x \neq 1$, there exists $i \geq 3$ such that $x_i = b_i^2$, $x_{i+1} = 1$, or, the other way, $x_i = 1$, $x_{i+1} = b_{i+1}^2$. Four cases may be distinguished:

a) $x = (\cdots, b_i^2, 1, 1, \cdots)$. Let us consider a neighborhood of x defined by the ith, $(i+1)$th and $(i+2)$th places. In this neighborhood, as we have already observed, there lies the element $y = a^m b^l \in \{a, b\}$, and in the given case it follows from the equation $a_i^m b_i^l = b_i^2$ that $l = 0$ or $l = 2$. If $l = 0$, then from $a_i^m = b_i^2$ it follows that $m = t \cdot 2^{i-2}$ where $(t, 2) = 1$. But then $a_{i+1}^m = 1$, i.e. y does not belong to the exhibited neighborhood of x. If, however, $l = 2$, then $a_i^m = 1$, i.e. $m = t \cdot 2^{i-1}$. Moreover, $a_{i+1}^m b_{i+1}^2 = 1$, whence, bearing in mind that $b_{i+1}^2 = a_{i+1}^{2^{i-1}}$, we obtain

$$a_{i+1}^{2^{i-1}(t+1)} = 1. \tag{1}$$

Similarly, because $a_{i+2}^m b_{i+2}^2 = 1$ and $b_{i+2}^2 = a_{i+1}^{2^i}$, we have

$$a_{i+2}^{2^{i-1}(t+2)} = 1. \tag{2}$$

Equations (1) and (2) cannot be satisfied simultaneously, as one of the integers $t+1$, $t+2$ is odd.

b) $x = (\cdots, b_i^2, 1, b_{i+2}^2, \cdots)$. As in case a), the element $y \in \{a, b\}$ which lies in the neighborhood of the element x, defined by the ith, $(i+1)$th and $(i+2)$th places, has the form $a^m b^2$. Again, as above, we obtain $m = t \cdot 2^{i-1}$, whence it follows that

$$a_{i+1}^{2^{i-1}(t+1)} = 1, \tag{3}$$

$$a_{i+2}^{2^{i-1}t} = 1. \tag{4}$$

If t is even, equation (3) is never satisfied. If t is odd, equation (4) is never satisfied.

c) $x = (\cdots, 1, b_{i+1}^2, b_{i+2}^2, \cdots)$. Let us consider a neighborhood of x, defined by the ith, $(i+1)$th and $(i+2)$th places. If the element $y = a^m b^l$ lies in that neighborhood, then $l = 0$ or $l = 2$. If $l = 0$, i.e. $y = a^m$, then $m = 2^{i-1} t$, where $(t, 2) = 1$, because $a_{i+1}^m = b_{i+1}^2$. Now let $l = 2$. In that case $a_i^{m+2^{i-2}} = 1$, i.e.

$$m = 2^{i-2} + t2^{i-1} = 2^{i-2}(2t+1).$$

But then $a_{i+1}^m \neq 1$, i.e. $a_{i+1}^m b_{i+1}^2 \neq b_{i+1}^2$.

d) $x = (\cdots 1, b_{i+1}^2, 1, \cdots)$. Again we choose a neighborhood of the type used above, in which the element $y = a^m b^l$ must lie. If $l = 0$, i.e. $y = a^m$, then because $a_{i+1}^m = b_{i+1}^2$, $m = 2^{i-1} t$, where $(t, 2) = 1$, we obtain $a_{i+2}^m \neq 1$.

Let $l = 2$. Now we have $a_i^{m+2^{i-2}} = 1$, whence we obtain

$$m = 2^{i-2} + t2^{i-1} = 2^{i-2}(2t+1),$$

which contradicts the equation $a_{i+1}^m = 1$.

2) $x^2 = b^2$. In this case we may assume that the group $\{x, b\}$ is isomorphic to the group of quaternions. Indeed, if it is not, then $\{x, b\}$ is either a dihedral group or an abelian group decomposable into a direct sum of cyclic groups of the fourth and second orders. In both cases we are able to find in $\{x, b\}$ an element of the second order which is not in $\{b\}$. Because this element belongs to N, we now return to the case we already had considered.

Thus, $\{b, x\}$ is the quaternion group, i.e. $b^4 = 1$, $x^2 = b^2$, $xb = b^3x$. Let $x = (x_3, x_4, \cdots)$.

As $x_3 b_3 x_3^{-1} = b_3^{-1}$ and $x_3^2 = b_3^2$, we see that x_3 assumes one of the following values: a_3, a_3^{-1}, $a_3 b_3$, $a_3^{-1} b_3$. Replacing the element x by some other element of $\{x, b\}$, if necessary, we are able to reduce everything to the case where $x_3 = b_3$. Similarly, x_4 assumes one of the following values: a_4^2, a_4^{-2}, $a_4^2 b_4$, $a_4^{-2} b_4$. Let us consider a neighborhood U of x, which has been defined by its two first components x_3 and x_4. Assume that the element $y = a^m b^l \in \{a, b\}$ lies in U. As $a_3^m b_3^l = a_3$, we have that $l = 2s$, where $s = 0$ (or $= 1$), and hence $x_4 = a_4^{2\epsilon}$, $\epsilon = \pm 1$.

Consequently, we obtain two equalities: $a_3^{m+2s} = a_3$ and $a_4^{m+2s} = a_4^{2\epsilon}$, which cannot be satisfied simultaneously, because the indices of powers in the respective left parts of the equalities have equal parities, and in their right parts, unlike parities.

Thus not every subgroup of an N-group is an N-group. It is also very important to observe that the group G which we have constructed is nilpotent.

Consequently, H, a subgroup of G, is also nilpotent. Thus it has been shown that, unlike in the discrete case, here not every nilpotent group is an N-group.

The following example, communicated to the author by V. P. Platonov, shows that not every compact N-group is nilpotent, and therefore the theorem of B. I. Plotkin about local nilpotency of abstract N-groups does not carry over to arbitrary topological N-groups.

Let H be a neighborhood, and set $G = \{H, x\}$, where $x^2 = 1$ and $x^{-1}hx = h^{-1}$ for all $h \in H$. The group G contains an everywhere dense ZA-subgroup $\{A, x\}$, where A is the family of those elements which belong to H and whose orders are some powers of two.

By Proposition 2, G is an N-group. But this group will not be locally nilpotent: if $y \in H$ is an element of third order, then $[y, x] = y$.

If we invoke one additional restriction, the normalizer condition implies local nilpotency for closed subgroups, as is demonstrated by the following result.

Theorem 1. *A pure* locally compact N-group G is locally nilpotent in the abstract sense.*

Proof. Let K be the connected component of the identity of G, which is open in G and consequently, by Proposition 1, is an N-group. It follows from the results of [4] that K is nilpotent in the abstract sense.

Let us pick an arbitrary finite subset of G consisting of elements $g_1, \cdots, g_n$. We wish to establish the nilpotency of the subgroup $\{g_1, \cdots, g_n\}$. In order to do so it suffices to prove the nilpotency of the subgroup $\{K, g_1, \cdots, g_n\}$, which is an N-group, as it is open in G. Thus, let $G = \{K, g_1, \cdots, g_n\}$, be an N-group, and let it be already established that any subgroup of G generated by K and an arbitrary element $x \in G$ is nilpotent. Then, because G/K is a nilpotent group with a finite number of generators, i.e. it satisfies the maximality condition for subgroups, the element x may be included in the maximal nilpotent open subgroup N_x of G. If $N_x \neq G$ we consider the normalizer of N_x in G. Because every automorphism of N transforms N_x into a nilpotent normal divisor of N, and because the product of two nilpotent normal divisors, as was shown by B. I. Plotkin, is again a nilpotent normal divisor, we have that N_x will be invariant relative to any topological automorphisms of N, i.e. it is characteristic in N (as N_x is a *maximal* nilpotent subgroup containing x). Therefore, if $N \neq G$ we have that N_x is invariant in the normalizer N' of the group N, and $N' \neq N$, which contradicts the definition of the normalizer. Consequently $N = G$, i.e. N_x is a normal divisor of G. Thus we have proved that every element $x \in \{K, g_1, \cdots, g_n\}$ may be included in the nilpotent normal divisor N_x of this group. Thus, by the above-mentioned result of B. I. Plotkin, the group G itself proves to be nilpotent, as it is the product of its nilpotent normal divisors $N_{g_1}, \cdots, N_{g_n}$.

Thus, everything is reduced to the case when a pure N-group G has the form $G = \{K, x\}$.

---

* A topological group is said to be *pure* if it does not contain compact elements, with the exception of 1. An element is said to be compact if it generates a compact group, in the topological sense.

Our task is to establish that this group is nilpotent. We have that $\{x\} \cap K = 1$. Indeed, were it otherwise the quotient group G/K would be finite, and, because the nilpotent pure connected Lie group K has an invariant series, in G, of finite length, all of whose factors are vector groups (as members of this series we may take the closures of the members of the lower central series of K), it follows that, using Lemma 3.8 of [6] several times, we would reveal in G a nontrivial compact subgroup, which contradicts the fact that G is pure.

Let N_1 be the normalizer of $\{x\}$. This subgroup is nilpotent in the abstract sense, as it is the product of its nilpotent normal divisor $K_1 = N_1 \cap K$ and an abelian normal divisor $\{x\}$. We will prove that the subgroup K_1 is connected. In order to do that, it suffices to show that for every element $y \in K_1$ there exists a one-parameter subgroup,* properly contained in K_1, which passes through y. The fact that it is possible to have a one-parameter subgroup go through any element of a connected nilpotent Lie group has been proved by V. M. Glušškov, (see [1], §6.5). Because, on the one hand, the element y belongs to the normalizer of $\{x\}$, and because, on the other hand, the group K is invariant in G, we have $[x, y] \in K \cap \{x\} = E$.

Let n be an arbitrary natural number. In K we can find an element y_1 such that $y_1^n = y$. Let us consider the element $x^{-1}y_1$, where $x = y_2 \in K$. Then we have

$$y_2^n = x^{-1}y_1^n x = x^{-1}y\,x = y.$$

Because a nilpotent group K which has no finite-order elements is a group with unique roots (see [5], Russian p. 430), it follows from $y_1^n = y_2^n = y$ that $y_1 = y_2$, i.e. $[x, y_1] = 1$. This means that for any natural number n, the element $y_1 = y^{1/n}$ is in K_1. Consequently, the line passing through y lies in K_1, because K_1 is a closed subgroup and the exhibited elements generate an everywhere dense subgroup on the given line.

Moreover, if $N_1 \neq G$, then we consider N_2, which is the normalizer of N_1. It is again a nilpotent group, namely the product of the nilpotent normal divisors $K_2 = K \cap N_2$ and N_1. Let us establish that the group K_2 is connected, observing that $K_2 \neq K_1$, as $N_2 = K_2 \cdot \{x\} \neq N_1 = K_1\{x\}$. Here we use the following lemma, which is concerned with arbitrary connected locally compact nilpotent groups, which are not necessarily pure.

* A one-parameter subgroup is a continuous homomorphic image of an additive group of real numbers in the natural topology.

Lemma 1. *Let G be a connected locally compact nilpotent group and H a connected closed subgroup of G. Then the normalizer of H in G is connected.*

Proof. V. M. Gluškov [1] gave a description of connected nilpotent locally compact groups, from which it follows that the group G is complete in the sense of an endless derivation of the root. The subgroup H is also complete. Therefore we may use another result of Gluškov [7], which states that the normalizer of a complete subgroup of a complete nilpotent group is complete.

As a consequence of the above, we have that the group N, together with any element x which belongs to it, contains the entire one-parameter subgroup which passes through that element, and hence is connected, which proves the lemma.

Now let y be an arbitrary element of K_2 and let $y_1^n = y$, $y_1 \in K$. We wish to establish that $y_1 \in K_2$. Let us consider the normalizer N of the subgroup K_1 in K. By the lemma we know that it is connected; hence $y_1 \in N$. Let us consider the element $y_2 = x^{-1}y_1x$. This element is also contained in N. Indeed,

$$y_2^n = (x^{-1}y_1 x)^n = x^{-1}y_1^n x = x^{-1}y\, x = y\cdot[y,\, x].$$

The element $[y, x] = y^{-1}x^{-1}yx$ is contained in $K \cap N_1 = K_1$, as $y \in N_2$. This means that

$$y_2^n = y\cdot[y,\, x] \in K_2 \subseteq N,$$

whence, by the lemma, we have that $y_2 \in N$.

Let $\tilde{N}$ denote the quotient group N/K_1. Because

$$y^{-1}x^{-1}y\, x = y_1^{-n}x^{-1}y_1^n\, x = y_1^{-n}y_2^n \in K_1,$$

the elements $\tilde{y}_1 = y_1K_1$ and $\tilde{y}_2 = y_2K_1$ of this group are related by $\tilde{y}_1^{-n}\tilde{y}_2^n = \tilde{1}$ (where $\tilde{1}$ is the identity element of group $\tilde{N}$) or $\tilde{y}_1^n = \tilde{y}_2^n$. Because the group N has no elements of finite order, it is a group where extracting a root is unique; hence $\tilde{y}_1 = \tilde{y}_2$, i.e.

$$y_1^{-1}\, x^{-1}y_1\, x \in K_1.$$

This means that $y_1 \in N_2$, which we were required to prove.

Proceeding in a similar manner, we are able to construct an increasing normal series

$$1 \subset N_0 = \{x\} \subset N_1 \subset N_2 \subset \ldots \subset N_m \subset N_{m+1} \subset \ldots,$$

whose terms are nilpotent subgroups, such that $N_{m+1} = N(N_n)$. Moreover, $K_m = K \cap N_m$ is a connected group for any m, and K_m is contained in K_{m+1} as a proper subgroup. As K is finite-dimensional, we have that $K = K_s$ for some s. Then $G = N_s$, and hence G is nilpotent, which proves the theorem.

Lemma 2. *A connected component K of the identity of a locally compact N-group G is an N-group, and consequently nilpotent.*

Proof. Let H be an open subgroup of G, such that the quotient group H/K is compact. This subgroup, which is generated compactly, is Lie-projective (see [8], Theorem 8) and is an N-group by Proposition 1. Let U_α, $\alpha \in \mathfrak{M}$, be a complete system of neighborhoods of identity of group H, and let B_α be a compact normal divisor of H, which lies in U_α and which defines a Lie quotient group H/B_α. The connected component of the identity of this quotient group is open in H/B_α and therefore is nilpotent in the abstract sense (see Proposition 1 and the results of [4]).

Moreover, because B_α is compact and because H is compactly generated, the subgroup KB_α is closed in H; hence we have the topological isomorphism $KB_\alpha/B_\alpha \cong K/K \cap B_\alpha$ (see [3], §20, G), whence the group $K/(K \cap B_\alpha)$ is nilpotent in the abstract sense for any $\alpha \in \mathfrak{M}$, i.e. K is nilpotent in the abstract sense.

The following theorem, which describes the construction of locally compact N-groups, shows that some results established in [2] for locally nilpotent groups are also valid for N-groups.

Theorem 2. *Let G be a locally compact N-group, K the connected component of the identity of G, and P the set of all compact elements of G. Then the following assertions are valid.*

1) *P is a closed invariant subgroup of G, which is an N-group (periodic part of G).*

2) *The quotient group G/P does not contain compact elements, with the exception of 1, i.e. it is pure (and consequently locally nilpotent in the abstract sense).*

3) *The subgroup $P' = P \cdot K$ is open in G and defines a discrete quotient group without torsion (i.e. without finite-order elements) and contains any open subgroup which defines a torsion-free quotient group.*

4) *Every compact set B of G which consists of compact elements generates, in the topological sense, a compact subgroup.*

5) *The locally compact, totally disconnected periodic N-group $G(K = E, P = G)$ is locally nilpotent.*

Proof. First of all, let the locally compact N-group G be totally disconnected. Then it contains an open compact subgroup. Because the union of an increasing sequence of open periodic subgroups is again an open periodic subgroup, by Zorn's Lemma there exists in G a maximal open periodic subgroup P. This subgroup is invariant in G. Indeed, if the normalizer N of this subgroup ($N \neq P$) is distinct from G, then P, which is characteristic in N, is invariant in N', where N' is the normalizer of N, and because $N' \neq N$ we arrive at a contradiction to the definition of the normalizer. Why is P characteristic in N? The relevant fact to consider is that P is a periodic part of N. Indeed, if there existed a compact element $x \in N$, $x \notin P$, then, because P is open, we would have $x^n \in P$ for some n, and because P is invariant in N, we immediately would have that the subgroups $\{P, x\}$ were periodic, thus contradicting the maximality of P. Thus P is a normal divisor in G, and consequently coincides with the set of all compact elements of G. Clearly the quotient group G/P does not contain elements of finite order.

Let us now prove that conclusion 4) is satisfied for a totally disconnected locally compact N-group G. Let X be a compact subset of G whose elements are compact. Consider an open compact subgroup H of G. The group $A = \{H, X\}$ has the form

$$A = \{H, x_1, x_2, \ldots, x_n\} \quad x_i \in X, \quad i = 1, 2, \ldots, n,$$

and is periodic by what had been proved already. Moreover, it is compactly generated, and because it is open in G it is an N-group. Therefore we can construct in A a normal series

$$H = N_0 \subset N_1 \subset \ldots \subset N_\alpha \subset N_{\alpha+1} \subset \ldots \subset N_\gamma = A, \qquad (*)$$

which consists of subgroups which are open in A, where $N_{\alpha+1} = N(N_\alpha)$, and for the limiting α we have $N_\alpha = \bigcup_{\beta < \alpha} N_\beta$. Here it is clear that γ is not a limit number. Thus $\gamma = k + \delta$ where k is a natural number and δ a limiting ordinal number or 0. The quotient group $G/N_{\gamma-1}$ is a discrete periodic N-group with a finite number of generators, and therefore is finite. It was shown in [9] that if in a compactly generated locally compact group there exists a closed subgroup defining a compact quotient space, then this group is also compactly generated. By this result, $N_{\gamma-1}$ is a compactly generated N-group. Due to this fact, all terms of the series (*) contained between N_δ and $N_\gamma = A$ prove to be discrete periodic finitely generated N-groups, i.e. finite groups, and N_δ is a compactly generated group. But then δ cannot be a limit number. Consequently $N_\delta = H$ and A is a compact group. Therefore $\{\bar{X}\}$ is also a compact group.

Now let G be any locally compact N-group. Its connected component K is nilpotent, the periodic part P_1 of group K is compact, and the quotient group K/P_1 has an increasing series of finite length which consists of subgroups which are characteristic in K, hence invariant in G; moreover, all terms of this series are vector groups. If b_1 and b_2 are two arbitrary compact elements of G, then their images under the natural homomorphism of G onto the totally disconnected group G/K generates, in the topological sense, a compact subgroup A/K, by the totally disconnected case of conclusion 4) already proved by us above. Now we are able to use Lemma 3.8 of [6], by which the locally compact group G with vector normal divisor V and the compact quotient group G/V may be written as $G = V \cdot B$, where B is a compact group. Applying this lemma several times to the group $\tilde{A} = A/P_1$, where P_1 is invariant in G (since it is characteristic in K), we obtain the decomposition $\tilde{A} = \tilde{T} \cdot \tilde{B}$, where $\tilde{T} = T/P_1$ is a pure normal divisor, $\tilde{B} = B/P_1$ is a compact subgroup; and because P_1 is a compact subgroup we have that $A = T \cdot B$, where B is a compact group. Because the compact elements b_1 and b_2 lie in B, we have that $b_1^{-1} b_2 \in B \subseteq P$, i.e. P is a subgroup of G in the algebraic sense. In this case P is a subgroup of G also in the topological sense. The proof of this fact proceeds exactly as the proof for the case of locally nilpotent groups (see [2], Theorem 2).

We will now prove that P is an N-group. Let H be a closed proper subgroup of P. Denote the normalizer of H in G by $N(H)$. If $N \cap P = H$, then H, which is a periodic part of N, is a characteristic subgroup in N. But then H is invariant in the normalizer N' of N, which is distinct from N, and this contradicts the definition of the normalizer. The fact that the quotient group G/P is pure, and conclusion 3) of our theorem, are proved in the same way as the proof of the corresponding statements for locally nilpotent groups (see [2], Theorem 3.1). Conclusion 4) has already been established for the totally disconnected case, and can be proved by employing the standard argument already used by us and Lemma 3.8 of [6].

The proof of conclusion 5) follows from the fact that, by conclusion 4), every finite subset of G, together with an open compact subgroup H of G, where G is a periodic, totally disconnected, locally compact N-group, generates an open subgroup A of G. Because the subgroup A is an N-group, it is nilpotent.

BIBLIOGRAPHY

[1] V. M. Gluškov, *Locally nilpotent locally bicompact groups*, Trudy Moskov. Mat. Obšč. 4 (1955), 291–332. (Russian) MR 17, 281.

[2] V. I. Ušakov, *Topological locally nilpotent groups*, Sibirsk. Mat. Ž. 6 (1965), 581–595. (Russian) MR 31 #3538.

[3] L. S. Pontrjagin, *Continuous groups*, 2nd ed., GITTL, Moscow, 1954; English transl., Gordon & Breach, New York, 1966. MR 17, 171; MR 34 #1439.

[4] V. I. Ušakov, *Topological groups with a normalizer condition for closed subgroups*, Izv. Akad. Nauk SSSR Ser. Mat. 27 (1963), 943–948. (Russian) MR 27 #5860.

[5] A. G. Kuroš, *Theory of groups*, 2nd ed., GITTL, Moscow, 1953; English transl., Chelsea, New York, 1960. MR 15, 501; MR 22 #727.

[6] K. Iwasawa, *On some types of topological groups*, Ann. of Math. (2) 50 (1949), 507–538. MR 10, 679.

[7] V. M. Gluškov, *On normalizers of complete subgroups in a complete group*, Dokl. Akad. Nauk SSSR 71 (1950), 421–424. (Russian) MR 11, 579.

[8] ———, *The construction of locally compact groups and the fifth problem of Hilbert*, Uspehi Mat. Nauk 12 (1957), no. 2, 3–41; English transl., Amer. Math. Soc. Transl. (2) 15 (1960), 55–93. MR 21 #698; MR 22 #5690.

[9] A. M. MacBeath and S. Swierczkowski, *On the set of generators of a subgroup*, Nederl. Akad. Wetensch. Proc. Ser. A62 = Indag. Math. 21 (1959), 280–281. MR 21 #5690.

Translated by:
A. M. Scott

SOME FINITENESS CONDITIONS IN THE THEORY OF SEMIGROUPS

UDC 519.4

L. N. ŠEVRIN

In this paper the structure of semigroups with the minimality conditions for subsemigroups, semigroups of finite rank, and semigroups satisfying certain other finiteness conditions is described to within the structure of each type of group.

The present paper concerns the following finiteness conditions in the theory of semigroups: the minimality condition for subsemigroups, the maximality condition for subsemigroups in periodic semigroups, the finite rank condition, the finite width condition, and the finite length condition. We describe the structure of arbitrary semigroups with the above conditions (to within the structure of each type of group). This is the principal subject of the paper and completes our study of these conditions begun in [1-4]. The basic results of the paper indicate that many problems concerning semigroups with the above conditions are reducible to the analogous problems of group theory. For example, we pose several problems involved in ascertaining the exact relationships between the classes of semigroups under investigation. The theorems to be developed indicate that the solution of these problems reduces completely to the solution of identical problems for the appropriate classes of groups.

§1. Definitions and preliminary properties

The minimality and maximality conditions for subsemigroups are well known. A semigroup is called a *semigroup of the finite (special*) rank* r if any of its subsemigroups with a finite number of generators is generated by not more than r generators, where r is the smallest number having this property. This concept was introduced in [2] by analogy with the familiar group-theoretical concept (cf. [5]). An arbitrary semigroup of finite rank is periodic (see [3], Lemma 1), so that the intersection of the class of semigroups of finite rank with the class of groups yields exactly the class of periodic groups of finite rank.

A structure is called a *structure of the finite width* s (over intersections) if the intersection of any n of its elements, where $n > s$, is equal to the intersection of s of them, where s is the smallest number having this

* This word will henceforth be omitted, since no other ranks will be considered.

property (cf. [6]). The concept of width over unions can be defined in two ways. The width over intersections and the width over unions coincide for all structures,* so that we can speak simply of the width of a structure. If a structure does not have a finite width, we say that it has infinite width. A structure is called a *structure of the finite length* l (cf. [6]) if the lengths of all its chains are bounded in the aggregate by the number l, where l is the smallest number having this property. It is easy to prove the following lemma.

Lemma 1.1. *Every structure of finite length has a finite width not exceeding its length.*

Proof. Let the length of a structure be l. We assume that we have found elements $x_1, \cdots, x_{l+1}$ such that the intersection of any k of them, where $k \leq l$, is different from the intersection $x_1 \wedge \cdots \wedge x_{l+1}$. The elements

$$y_1 = x_1 \wedge \cdots \wedge x_{l+1},\ y_2 = x_2 \wedge \cdots \wedge x_{l+1},\ \ldots,\ y_l = x_l \wedge x_{l+1},$$
$$y_{l+1} = x_{l+1}$$

then form a chain of the length l. The element y_{l+1}, i.e. the largest element of this chain, cannot be the largest element of the structure, since if it were we would have

$$x_1 \wedge \cdots \wedge x_{l+1} = x_1 \wedge \cdots \wedge x_l,$$

which contradicts our premise. Hence the above chain can be extended by at least one element, i.e. a chain of the length $l+1$ can be constructed, and this contradicts our condition. This contradiction proves the lemma.

A semigroup if called a *semigroup of the finite width* s (cf. [4]) if the structure of its subsemigroups is of the width s. If a structure of subsemigroups is of infinite width, then the semigroup itself is also called a semigroup of infinite width. *Semigroups of finite length* are defined in the same way. Intersecting these classes of semigroups with the class of groups, we obtain the *groups of finite width* and the *groups of finite length.* These two concepts can be approached in another way; for example, it is possible to apply the term "group of finite width" to a group the structure of whose subgroups is of finite width. In fact, both definitions subsume the same class of groups. This follows from the fact that any subsemigroup in a periodic group is a subgroup, and from the following statement (cf. [4], Lemma 4.1):

---

* In the case of finite structures this statement is formulated on Russian p. 42 of [6]. For arbitrary structures it is proved in Appendix 1 of [4].

if the structure of subgroups of the group G is of finite width, then G is a periodic group.

Lemma 1.1 implies that every semigroup of finite length has a finite width not exceeding its length. The converse statement is not valid. This is evident from the simplest example of a group of the type p^∞, which is not a group of finite length, but has a width of 1.

The following proposition, which was proved essentially in [4] (cf. Theorem 3.1 in that paper) provides a definition of semigroups of finite width which is equivalent to the above definition and reflects the closeness of semigroups of finite width to semigroups of finite rank.

Lemma 1.2. *A semigroup is of the width s if and only if any of its subsemigroups with a finite number of given generators is generated by not more than s of these generators, where s is the smallest number having this property.*

This lemma implies directly that every semigroup of finite width is a semigroup of finite rank, and that its rank does not exceed its width. The converse statement is not valid. This is evident from the example of the group which is the direct product of an infinite set of cyclical groups of various prime orders. This group has the rank 1, but is clearly of infinite width. The latter example also shows that the property of having a finite rank is generally not preserved under structural isomorphisms of the semigroup. The indicated group is structurally isomorphic, for example, to the denumerable semigroup with zero in which the product of any two elements is equal to zero; the latter semigroup is of infinite rank, however. The remaining finiteness conditions considered are preserved under structural isomorphisms by virtue of their definitions.

In the course of our discussion we shall make use of the concepts of *nilelement, nilsemigroup, nilpotent semigroup, ideal extension,* and several other notions, all of which are defined in [1] (where an ideal extension is called simply an extension). We shall also adhere to the notation of [1]; specifically, by $\{M\}$ we denote a subsemigroup generated by the set M of elements of a given semigroup. As usual, we say that the semigroup Γ is *covered* by some system of its subsemigroups if Γ is the set-theoretical sum of all the subsemigroups of this system.

All the semigroups to be considered will be periodic. By K_e we denote the set of all elements of a given periodic semigroup one of whose powers is equal to the idempotent e of this semigroup. Every set K_e will be called a *torsion class*. By G_e we denote the set of all elements from K_e for which e serves as the identity element. It is well known (see [7], for example) that every periodic semigroup is the set-theoretical sum of its pairwise-nonintersecting torsion classes, which are not necessarily subsemigroups. G_e is the largest semigroup in K_e. It is also easy to see that G_e is an ideal in the subsemigroup $\{K_e\}$ and is therefore the kernel (the smallest ideal) of this semigroup.

Since the formulations of many of the statements about semigroups with the above conditions (including the formulations of the basic results of the paper) are identical, we propose to unify these formulations by introducing the following symbols:

Φ_1 –– the minimality condition for subsemigroups;

Φ_2 –– the conjunction of periodicity and the maximality condition for subsemigroups;

Φ_3 –– the finite rank condition;

Φ_4 –– the finite width condition;

Φ_5 –– the finite length condition.

It is clear that every one of conditions Φ_1–Φ_5 is hereditary for subsemigroups and homomorphic images. We also take note of the following lemmas, in which i is any fixed number from the collection 1, 2, 3, 4, 5.

Lemma A_i. *An ideal extension of a semigroup with the condition Φ_i by means of a semigroup with the condition Φ_i itself satisfies the condition Φ_i.*

Lemma B_i. *A semigroup covered by a finite number of subsemigroups satisfying the condition Φ_i itself satisfies the condition Φ_i.*

These lemmas are proved in [4] (Theorem 3.2, parts b) and c); Theorem 3.4, parts b) and c)) for $i = 1, 4$, and in [3] (Lemmas 2 and 3) for $i = 3$. For $i = 2, 5$ the proof is the same as in the case $i = 1$.

We note that for $i = 3, 4, 5$ it is possible to refine the lemmas, pointing out in Lemma A_i that the rank (width, length) of an ideal extension does not exceed the sum of the ranks (widths, lengths) of the corresponding semigroups, and in Lemma B_i that the rank (width, length) of the semigroup does not exceed the sum of the ranks (widths, lengths) of the covering subsemigroups. The indicated estimates remain exact.

To conclude this section we cite two ancillary lemmas which will be needed in the next section.

Lemma 1.3. *Let the semigroup Γ be generated by the elements $a_1, \cdots, a_n$ for which $a_i a_j = a_j$ for $i \leq j$ $(i, j = 1, \cdots, n)$. The semigroup Γ*

then consists of less than 2^n *elements.*

Proof. The elements $a_1, \cdots, a_n$ are idempotents. Hence every element from Γ which is different from them can be represented in the form $a_{i_1} \cdots \cdots a_{i_m}$, where $i_1 > \cdots > i_m$ (which implies, among other things, that all the elements from Γ are idempotents). Hence each nonempty subset of our set of generators is associated with not more than one element of the semigroup Γ. This implies the assertion of the lemma.

In the same way we can prove

Lemma 1.4. *Let the semigroup* Γ *with zero be generated by the elements* $a_1, \cdots, a_n$ *for which* $a_i a_j = 0$ *for* $i \leq j$ $(i, j = 1, \cdots, n)$. *The semigroup* Γ *then consists of not more than* 2^n *elements.*

§2. The basic lemmas

Lemma 2.1. *A semigroup generated by an infinite set of idempotents has a true subsemigroup containing an infinite set of idempotents.*

Proof. Let the semigroup Γ be generated by an infinite set of idempotents E. For an arbitrary $e \in E$ we denote by L_e the set of all $x \in \Gamma$ such that $xe = e$. It is clear that L_e is a subsemigroup, and that $L_e \neq \emptyset$, since $e \in L_e$. Let us consider the subsemigroup $\{E \backslash L_e\}$, which can be empty. It is clear that at least one of the subsemigroups L_e, $\{E \backslash L_e\}$ contains an infinite set of idempotents. Hence, if both of the indicated subsemigroups are true for some $e \in E$, we immediately obtain the required statement. Let us therefore consider the case where one of the subgroups L_e, $\{E \backslash L_e\}$ coincides with Γ for every $e \in E$. We can show that the equation $\{E \backslash L_e\} = \Gamma$ cannot hold. In fact, assuming the contrary, we obtain $e \in \{E \backslash L_e\}$, which means that the equation $e = e_1 e_2 \cdots e_m$, where $e_i \in E \backslash L_e$ $(i = 1, 2, \cdots, m)$ holds for some natural m. But then $e_1 e = e$, i.e. $e_1 \in L_e$, which is a contradiction. Thus $L_e = \Gamma$ for every $e \in E$. This means that $E = \Gamma$ is the semigroup of right-hand zeros, so that any subset of Γ is a subsemigroup. The required result follows from this.

A direct consequence of the lemma just proved is

Lemma 2.2. *A semigroup with the minimality condition for subsemigroups contains only a finite number of idempotents.*

Lemma 2.3. *A semigroup of finite rank contains only a finite number of idempotents. The number of idempotents in such a semigroup does not exceed some number which depends only on the rank.*

Proof. Let the semigroup Γ be of the rank r. The case $r = 1$ is trivial (it is clear that Γ here contains just one idempotent), so we shall assume that $r > 1$. Setting $2^r = k$, we can show that the number of idempotents in Γ does not exceed

$$(k-1)\, r^{k-2} + r^{k-3} + \ldots + r.$$

Let us assume that Γ contains $(k-1)r^{k-2} + r^{k-3} + \cdots + r + 1$ distinct idempotents, the set of which we denote by E. By hypothesis, the subsemigroup $\{E\}$ is generated by r elements. Let these elements be $x_1, \cdots, x_r$. Every element from $\{E\}$ is representable as some word in the alphabet $x_1, \cdots, x_r$. Since the set E consists of

$$(k-1)\, r^{k-2} + r^{k-3} + \ldots + r + 1$$

elements, at least one letter from $x_1, \cdots, x_r$ occurs as the first left-hand letter of above representations of not less than

$$(k-1)\, r^{k-3} + r^{k-4} + \ldots + r + 2$$

distinct elements from E. We would otherwise find that E contains not more than

$$r((k-1)r^{k-3} + r^{k-4} + \ldots + r + 1) = (k-1)r^{k-2} + r^{k-3} + \ldots + r$$

elements. Without limiting generality we can assume that the indicated letter is x_1 and denote the set of corresponding elements from E by E_1'. Since $x_1 \in \{E\}$, it follows that x_1 can be expressed as $x_1 = e_1 y_1$, where $e_1 \in E$ and $y_1 \in \{E\}$. This implies that e_1 is the left-hand identity element for all the elements from E_1'. The element e_1 may or may not belong to E_1'. In any case, we can guarantee the existence of a set $E_1 \subset E_1'$ consisting of

$$(k-1)\, r^{k-3} + r^{k-4} + \ldots + 1$$

idempotents distinct from e_1 for which e_1 serves as the left-hand identity element.

We can continue this process, considering E_1 just as we did E in the previous step, etc. After the $(k-3)$th step we have the elements $e_1, e_2, \cdots$ $\cdots, e_{k-3}$ for which $e_i e_j = e_j$ for $i \le j$ $(i, j = 1, \cdots, k-3)$, and the set E_{k-3} consisting of $(k-1)r + 1$ idempotents distinct from $e_1, e_2, \cdots, e_{k-3}$, where every element e_i serves as a left-hand identity element for any element from E_{k-3}. Let us carry out the $(k-2)$th step. The subsemigroup $\{E_{k-3}\}$ is generated by r elements. By arguments similar to those applied in the preceding paragraph, we find from this that there exists an element $e_{k-2} \in E_{k-3}$ which serves as a left-hand identity element for $k-1$ elements from E_{k-3} which are distinct from it. We denote this set of $k-1$ elements by E_{k-2}.

Let us now carry out the $(k-1)$th step. Since $k-1=2^r-1>r$ (we recall that $r>1$), we conclude as above that E_{k-2} contains at least two distinct elements e_{k-1} and e_k such that $e_{k-1}e_k=e_k$.

As a result we have the elements $e_1, \cdots, e_k$ for which $e_ie_j=e_j$ for $i \leq j$ $(i, j = 1, \cdots, k)$. Every element of the subsemigroup $H=\{e_1, \cdots, e_k\}$ is then representable in the form $e_{i_1} \cdots e_{i_m}$, where $i_1 \geq \cdots \geq i_m$. This implies that at least one of the equations $xy=y$, $yx=x$ is fulfilled for all x, $y \in H$. In fact, let

$$x = e_{i_1} \ldots e_{i_m}, \quad y = e_{j_1} \ldots e_{j_n};$$

if $i_1 \leq j_1$, then, clearly, $xy=y$; if $j_1 \leq i_1$, then $yx=x$. In particular, all the elements of the subsemigroup H are idempotents.

Let us now carry out a linear ordering of the set H in the following way. We denote by F_1 the set consisting of the element e_1 and by F_2 the set of all elements from H distinct from e_1 and representable in the form e_2e_i, where $i \leq 2$. Proceeding by induction, we denote by F_s $(s \leq k)$ the set of all elements from H not belonging to $F_1 \cup F_2 \cup \cdots \cup F_{s-1}$ and representable as $e_se_{i_1} \cdots e_{i_m}$, where $s \geq i_1 \geq \cdots$ $\cdots \geq i_m$. Some of the sets F_s may be empty. It is clear that

$$F_1 \cup F_2 \cup \ldots \cup F_k = H.$$

Let us carry out a linear ordering of the nonempty sets F_s in an arbitrary fashion; let us assume further for $x \in F_s$, $y \in F_t$, where $s \neq t$, that x precedes y if and only if $s<t$. In accordance with the resulting order, we proceed to number all the elements of H: $f_1, f_2, \cdots, f_l$. We note that $l \geq k$. From the ordering law just introduced and from what we said in the preceding paragraph we conclude that $f_if_j=f_j$ for $i \leq j$ $(i, j=1, \cdots, l)$. By hypothesis, the subsemigroup H is generated by r elements. But by virtue of Lemma 1.3 it must then consist of fewer than $2^r=k$ elements. This is contradictory, since H consists of $l \geq k$ elements. The contradiction proves the lemma.

We note that the estimate obtained in the lemma is apparently not an exact one.

Lemma 2.4. *A semigroup with the maximality condition for subsemigroups cannot contain an infinite number of idempotents.*

Proof. Let us assume that some semigroup with the maximality condition for subsemigroups contains infinitely many idempotents of E. By virtue of the maximality condition, the subsemigroup $\{E\}$ is generated by a finite number of elements, and therefore by a finite number of idempotents from E. By arguments similar to those employed at the beginning of our proof of the preceding lemma, we now find that for at least one idempotent $e_1 \in E$ there exist infinitely many elements of $E_1 \subset E$ such that $e_1 x = x$ for any $x \in E_1$. We can assume that $e_1 \notin E_1$. Continuing this process, i.e. considering the subsemigroup $\{E_1\}$, etc., we obtain an infinite sequence of elements e_1, $e_2, \cdots$, such that $e_i e_j = e_j$ for $i \le j$ $(i, j = 1, 2, \cdots)$. The subsemigroup generated by these elements is infinite. On the other hand, it must be generated by a finite number of the indicated elements, and thus cannot be infinite by virtue of Lemma 1.3. The contradiction proves the lemma.

Lemma 2.5. *A semigroup with zero generated by infinitely many nilelements has a true subsemigroup containing infinitely many nilelements.*

Proof. Let Γ be a semigroup with zero 0 generated by infinitely many nilelements N. It may be the case that $\Gamma = N$, i.e. that Γ is a nilsemigroup. The semigroup Γ then contains an infinite true subsemigroup, since (e.g. by virtue of Lemma 2.11 of [1]) it does not satisfy the minimality condition for subsemigroups.

We must therefore concentrate on the case where $\Gamma \neq N$. The set N is covered by the maximal subsemigroups which it contains. All of the latter are nilsemigroups, so that if at least one of them is infinite, we immediately have the required statement. Let us assume, therefore, that they are all finite and hence nilpotent (e.g. see Lemma 2.11 of [1]). The set of these subsemigroups is infinite because N is infinite.

Without limiting generality we can assume that the annihilator of the semigroup Γ is trivial, i.e. that it consists of zero only. In fact, in the contrary case we can construct the upper annihilator chain (see [1]) in Γ and take the union B of all its terms.* The union B is an ideal in Γ, where $B \subset N$, since B is a nilsemigroup (see [1]). But the ideal B is then finite. Hence, turning to the factor semigroup $\Gamma - B$, we obtain a semigroup satisfying the same conditions as Γ, but having a trivial annihilator.

Thus, let $[H_\alpha]$ be the set of all maximal subsemigroups contained in N; α runs over some set of indices Ω. As noted above, Ω is infinite, and all the subsemigroups of H_α are nilpotent. We denote the annihilator of the subsemigroup H for every $\alpha \in \Omega$ by A and consider the set $A = \bigcup_{\alpha\in\Omega} A_\alpha$. Two cases are possible.

1. *A is finite.* A then clearly contains at least one nonzero element belonging to the infinite set of annihilators A_α which therefore annihilates

* Carrying further the parallelism in terminology (normal divisor = ideal, center = annihilator, etc.; see [1] in this connection), we can call B the "upper hyperannihilator".

infinitely many elements from N. The subsemigroup generated by the indicated infinite set of elements is true, since the aforementioned element cannot annihilate all elements from N, since according to our assumption Γ has a trivial annihilator. Thus the required result follows immediately in this case.

Let us consider the other possible case.

2. *A is infinite*. Let us decompose A into a denumerable set of infinite classes $B_1, B_2, \cdots$,

$$A = \bigcup_{i=1}^{\infty} B_i, \quad B_i \cap B_j = \phi \quad \text{for } i \neq j.$$

If a semigroup generated by at least one of these classes is true, we immediately obtain the required statement. Let us assume, therefore, that every one of the classes generates the entire semigroup Γ. Let us take an arbitrary element $x \notin N$. It is clear that no power of this element belongs to N. We have $x \in \{B_i\}$ for all $i = 1, 2, \cdots$, i.e.

$$x = b_{11} \ldots b_{1m_1}, \ldots, x = b_{i1} \ldots b_{im_i}, \ldots,$$

where $b_{i1}, \cdots, b_{im_i}$ are elements from B_i for any i. This implies, among other things, that the elements $b_{1m_1}, b_{2m_2}, \cdots$ are all distinct. We denote the subsemigroup generated by the indicated elements by C. C is infinite. Let us complete the proof by demonstrating the truth of C. Recalling the definition of the set A, we see that the squares of all its elements are equal to zero. This implies that

$$xb_{1m_1} = 0, \ldots, xb_{im_i} = 0, \ldots$$

so that $xc = 0$ for any $c \in C$. But $x^2 \neq 0$, since $x \notin N$; hence, $x \notin C$, i.e. C is a true subsemigroup.

The lemma is proved.

A direct consequence of this lemma is

Lemma 2.6. *A semigroup with zero which satisfies the minimality condition for subsemigroups contains only a finite number of nilelements.*

Lemma 2.7. *A semigroup with zero having a finite rank contains only a finite number of nilelements. The number of nilelements in such a semigroup does not exceed some number which depends on the rank alone.*

Proof. Let the semigroup Γ be of the rank r and let it contain zero. In the case $r = 1$ it is easy to show that Γ is a cyclical nilpotent semigroup and that it consists of not more than three elements. We shall therefore assume that $r > 1$. Let us introduce the notation

$$k = 2^r, \quad m = (k-1)\, r^{k-2} + r^{k-3} + \ldots + r, \quad l = [\log_2 (2r+1)]$$

($[x]$ is the integer part of x). We can show that the number of nonzero nil-elements in Γ does not exceed $2r \cdot m^{l+1}$. As in Lemma 2.3, we note that the indicated estimate does not appear to be exact.

We shall carry out the proof in two parts.

1. We begin by showing that the number of nonzero elements whose squares equal zero does not exceed m. The proof of this statement is in many ways similar to that of Lemma 2.3, and we shall merely outline it briefly, referring the reader to the proof of the lemma in question, where the analogous points are discussed in greater detail.

Let us assume that there are $m+1$ distinct elements with the indicated properties; we denote the set of these elements by K. The subsemigroup $\{K\}$ is generated by r elements; let these elements be $x_1, \cdots, x_r$. Every element from $\{K\}$ can be represented as some word in the alphabet $x_1, \cdots$ $\cdots, x_r$. Hence, at least one of the letters $x_1, \cdots, x_r$ occurs as the first left-hand letter in the indicated representations of not fewer than

$$(k-1)\, r^{k-3} + r^{k-4} + \ldots + r + 2$$

distinct elements from K. We can assume without loss of generality that this letter is x_1. Since $x_1 = z_1 y_1$, where z_1 and y_1 are certain elements from K and $\{K\}$, respectively (y_1 may be an empty symbol; cf. [7]), and since $z_1^2 = 0$, there exists a set consisting of

$$(k-1)\, r^{k-3} + r^{k-4} + \ldots + r + 1$$

elements from z_1 distinct from K, each of which is annihilated from the left by the element z_1.

Continuing this process, we obtain k distinct nonzero elements $z_1, \cdots$ $\cdots, z_k$ for which $z_1 z_j = 0$ for $i \le j$ $(i, j = 1, \cdots, k)$. Every element of the subsemigroup $H = \{z_1, \cdots, z_k\}$ which is distinct from $z_1, \cdots, z_k$ can then be represented in the form $z_{i_1} \cdots z_{i_s}$, where $i_1 > \cdots > i_s$. By means of constructions similar to those used in the proof of Lemma 2.3, we find from this that all the elements of the subsemigroup H can be numbered as follows: $h_1, \cdots, h_n$, where $h_i h_j = 0$ for $i \le j$ $(i, j = 1, \cdots, n;\ n \ge k)$. But then, by Lemma 1.4, the subsemigroup H cannot be generated by not more than r elements, since it consists not fewer than $k = 2^r$ nonzero elements. This is the contradiction which proves the statement made at the beginning of this part of our proof.

2. Let us assume that the statement of the lemma is invalid, i.e. that there exist $2r \cdot m^{l+1} + 1$ distinct nonzero nilelements, the set of which we denote by N. If the powers $x^{r+1}, \cdots, x^{2r+1}$ of an element $x \in N$ are distinct (i.e. if $x^{2r+1} \neq 0$) then, clearly, the subsemigroup $\{x^{r+1}, \cdots, x^{2r+1}\}$ cannot be generated by fewer than $r+1$ elements. Hence $x^{2r+1} = 0$ for any $x \in N$. This and the above supposition imply that N contains at least $m^{l+1} + 1$ distinct elements of the same order. By N_1 we denote some subset of the set N consisting of $m^{l+1} + 1$ distinct elements of the same order. We can show that the set of squares of all the elements of N_1 consists of not fewer than $m^l + 1$ distinct elements. In fact, assuming the contrary, we find that the relations $x_1^2 = \cdots = x_p^2$ apply for certain elements $x_1, \cdots, x_p$ from N_1, where $p > m$. Let us consider the factor semigroup of the semigroup $\{x_1, \cdots, x_p\}$ over the ideal A generated by the element $y = x_1^2 = \cdots = x_p^2$. Since $y \in A$, the squares of the images of all the elements $x_1, \cdots, x_p$ in this factor semigroup are equal to zero. Since $p > m$, and since the ranks of the subsemigroups and homomorphic image do not exceed the rank of the initial semigroup, we arrive at a contradiction with what we showed in the first part of our proof if we show that $x_i \notin A$ for $i = 1, \cdots, p$. Thus, let us assume that $x_i \in A$ for some i. This implies that $x_i = uyv$, where $u, v \in \{x_1, \cdots, x_p\}$ (u and v may be empty symbols). The definition of the element y implies that $y = x_1^2$ lies in the center of the subsemigroup $\{x_1, \cdots, x_p\}$, so that $x_i = x_i^2 uv$. But the latter inequality means that $x_i = x_i^n (uv)^{n-1}$ for any natural n. This implies that $x_i = 0$, which is a contradiction.

Thus there exists a set N_2 consisting of $m^l + 1$ distinct elements of the same order; the elements of N_2 are squares of elements of the set N_1. Continuing this process, we arrive at the $(l+1)$th step at the set N_{l+1} consisting of $m+1$ distinct elements of the same order. The elements of N_{l+1} are 2^lth powers of elements of the set N_1. Let us show that the square of every element of N_{l+1} is equal to zero. Let $z \in N_{l+1}$ i.e. let $z = x^{2^l}$ for some $x \in N_1$. Then $z^2 = x^{2^{l+1}}$. But, recalling that $l = [\log_2 (2r+1)]$, we have

$$2^{l+1} > 2^{\log_2(2r+1)} = 2r + 1,$$

which implies that $x^{2^{l+1}} = 0$. Thus we have obtained $m+1$ distinct elements of the same order (which are therefore nonzero elements) whose squares are equal to zero. This contradicts what we proved in the first part. The contradiction completes our proof of the lemma.

Lemma 2.8. *A semigroup with zero which satisfies the maximality condition for subsemigroups contains only a finite set of nilelements.*

Proof. Let us assume that some semigroup with zero which satisfies the maximality condition for subsemigroups contains infinitely many nilelements N. By virtue of the maximality condition, the subsemigroup N is generated by a finite number of elements, and therefore by a finite number of nilelements from N. We then obtain for some $x \in N$ an infinite number of elements from N of the form $xy_1, \cdots, xy_n, \cdots$, where $y_i \in \{N\}$ $(i = 1, 2, \cdots, n, \cdots)$. Let $x^k \neq 0$ and $x^{k+1} = 0$. The element x^k then annihilates all of the indicated elements from the left. This implies that the subsemigroup H generated by the indicated elements and by x^k has a nontrivial left-hand annihilator. Let us denote the latter by A_1. It is clear that $A_1 \subset N$. Moreover, A_1 is an ideal (cf. [1], Corollary 2 of Lemma 1.4). A_1 is finite by virtue of Lemma 2.12 of [1], so that the factor-semigroup $H - A_1$ contains infinitely many nilelements. Reasoning as above, we isolate in $H - A_1$ a subsemigroup with an infinite number of nilelements which has a nontrivial left-hand annihilator. We denote its complete original under the homomorphism of H onto $H - A_1$ by A_2. We have $A_1 \subset A_2$. Continuing this process, we construct the infinite increasing sequence of subsemigroups $A_1 \subset A_2 \subset \cdots$, which is a contradiction. This contradiction proves the lemma.

§3. Theorems

Let us now formulate the principal results of our study. Since their formulations are identical, we can make use of the symbols for the finiteness conditions introduced in §1.

Theorem A_i. *In order for a semigroup Γ to satisfy the condition Φ_i it is necessary and sufficient that Γ be a periodic semigroup with a finite number of torsion classes, that the set $K_e \setminus G_e$ be finite for every class K_e, and that the group G_e satisfy the condition Φ_i.*

Proof. Sufficiency. An arbitrary class K_e is covered by the maximal subsemigroups contained in it; each of these subsemigroups contains G_e (cf. [7], Chapter 3, and [4], §8). Since the set $K_e \setminus G_e$ is not infinite, there is a finite number of these maximal subsemigroups. Every one of them is an ideal extension of G_e by means of a finite (nilpotent) semigroup and satisfies the condition Φ_i by virtue of Lemma A_i. Since the number of torsion classes in finite, Γ is covered by a finite number of subsemigroups satisfying the condition Φ_i; it follows from this by Lemma B_i that Γ satisfies the condition Φ_i.

Necessity. The periodicity of the semigroup Γ is clear (it is self-evident for $i = 1$, it is involved in the definition condition for $i = 2$, and it is proved in Lemma 1 of [3] for $i = 3$); every group G_e satisfies the condition Φ_i by virtue of the hereditary character of this condition for subsemigroups. The cases $i = 4$ and $i = 5$ need not be considered separately, since the conditions Φ_4 and Φ_5 are stronger than Φ_3. Since every torsion class contains a unique idempotent, we must prove that the set of all idempotents is finite. This is done in Lemma 2.2 for $i = 1$, in Lemma 2.4 for $i = 2$, and in Lemma 2.3 for $i = 3$. Let us consider the arbitrary class K_e. Let $K_e \neq G_e$. We can then show that the set $K_e \setminus G_e$ is finite. The image of every element from $K_e \setminus G_e$ under the homomorphism of the subsemigroup $\{K_e\}$ onto the factor semigroup $\{K_e\} - G_e$ is clearly a nilelement, and the images of distinct elements are distinct. We must therefore prove that the set of nilelements of the semigroup $\{K_e\} - G_e$ is finite. But by virtue of the hereditariness of the condition Φ_i for subsemigroups and homomorphic images, the semigroup $\{K_e\} - G_e$ satisfies the condition Φ_i. The required statement then follows from Lemma 2.6 for $i = 1$, from Lemma 2.8 for $i = 2$, and from Lemma 2.7 for $i = 3$.

The proof is now complete.

Remark 1. It follows from the results of §2 that for $i = 3, 4, 5$ we can add to the corresponding theorems the statement that the number of idempotents and the number of elements from every set $K_e \setminus G_e$ of a semigroup satisfying the condition Φ_i are bounded, respectively, by some numbers dependent only on the rank (width, length).

Remark 2. Descriptions of semigroups belonging to various classes and satisfying the conditions in question follow as corollaries from the theorems just proved. We note that Theorem A_3 subsumes the basic result of the note [3], and that Theorem A_4 subsumes Theorems 5.4 and 8.1 of [4]. One of the consequences of the theorems proved here is

Corollary 1. *Each of the conditions* Φ_1–Φ_5 *is equivalent to the finiteness of a semigroup in the class of semigroups not containing infinite subgroups.*

Theorem A_1 implies

Corollary 2. *If all the true subsemigroups in a semigroup which is not a group are finite, then the semigroup itself is finite.*

We showed in [4] that the finite width condition is equivalent to the minimality condition for subsemigroups for a whole series of classes of semigroups and groups (see [4], §5). This is true, for example, in the case of locally radical (see [8]) subsemigroups. Making use of Theorems A_1 and A_4, we infer from this that the equivalence of the above conditions is valid for semigroups locally nilpotent in the Mal′cev sense [9]. This statement subsumes Theorem 5.5 of [4], where the analogous fact was established for commutative semigroups. At the same time, we know of no specific semigroups which demonstrate the distinctness of the two classes in question. This naturally leads us to ask

S 1. *Does an arbitrary semigroup of finite width satisfy the minimality condition for subsemigroups?*

S 2. *Does an arbitrary semigroup with the minimality condition for subsemigroups have finite width?*

The latter question suggests another:

S 3. *Does an arbitrary semigroup with the minimality condition for subsemigroups have finite rank?*

We can also ask the following interrelated questions:

S 4. *Is an arbitrary semigroup of finite length a finite semigroup?*

S 5. *Is an arbitrary semigroup satisfying the minimality condition for subsemigroups and the maximality condition for subsemigroups a semigroup of finite length?*

S 6. *Is an arbitrary periodic semigroup with the maximality condition for subsemigroups a semigroup of finite length?*

S 7. *Does an arbitrary periodic semigroup with the maximality condition for subsemigroups satisfy the minimality condition for subsemigroups?*

If we replace the word "semigroup" by "group", and the word "subsemigroup" by "subgroup" everywhere in the above questions, we obtain the analogous group-theoretical questions G 1–G 7 which we need not write out here. We note that questions G 4–G 7 are related to the following question of [10]: is a group which satisfies the minimality condition for subgroups and the maximality condition for subgroups a finite group? We also note that question G 4 is Problem 43 of [6].

It is clear, for example, that a negative answer to question G 1 would imply a negative answer to question S1. Analogous relationships hold between questions G 2 and S 2, $\cdots$, G 7 and S 7. In fact, however, questions S 1–S 7 are even equivalent to the respective questions G 1–G 7 (this follows from Theorems A_1–A_5). Moreover, finding the answers to questions S 1–S 7 reduces completely to finding the answers to the respective questions G 1–G 7, since the theorems proved above imply

Corollary 3. *Let Φ_6 be the conjunction of conditions Φ_1 and Φ_2, let Φ_7 be a finiteness condition of a semigroup, and let i, j be different numbers from the collection $1, \cdots, 7$. The existence of a semigroup satisfying the condition Φ_i but not satisfying the condition Φ_j then implies that this semigroup has a subgroup with the same properties.*

BIBLIOGRAPHY

[1] L. N. Ševrin, *On the general theory of semigroups*, Mat. Sb. 53(95) (1961), 367–386. (Russian) MR 26 #3803.

[2] ———, *Nilsemigroups with certain finiteness conditions*, Mat. Sb. 55 (97) (1961), 473–480. (Russian) MR 25 #3101.

[3] ———, *Commutative semigroups of finite rank*, Uspehi Mat. Nauk 18 (1963), no. 4 (112), 201–204. (Russian) MR 27 #5845.

[4] ———, *Semigroups of finite width*, Theory of Semigroups and Appl., Saratov. University, Saratov, 1965, pp. 325–351. MR 33 #5771.

[5] A. I. Mal′cev, *Groups of finite rank*, Mat. Sb. 22(64) (1948), 351–352. (Russian) MR 9, 493.

[6] G. Birkhoff, *Lattice theory*, Amer. Math. Soc. Colloq. Publ., vol. 25, Amer. Math. Soc., Providence, R. I., 1940; rev. ed., 1948; new ed., 1967; Russian transl., IL, Moscow, 1952. MR 1, 325; MR 10, 673.

[7] E. S. Ljapin, *Semigroups*, Fizmatgiz, Moscow, 1960; English transl., Transl. Math. Monographs, vol. 3, Amer. Math. Soc., Providence, R. I., 1963; rev. ed., 1968. MR 22 #11054.

[8] B. I. Plotkin, *Generalized soluble and generalized nilpotent groups*, Uspehi Mat. Nauk 13(1958), no. 4 (82), 89–172; English transl., Amer. Math. Soc. Transl. (2) 17(1961), 29–115. MR 21 #686; MR 23 #A1713.

[9] A. I. Mal′cev, *Nilpotent semigroups*, Ivanov. Gos. Ped. Inst. Učc. Zap. Fiz.-Mat. Nauki 4 (1953), 107–111. (Russian) MR 17, 825.

[10] A. G. Kuroš and S. N. Černikov, *Solvable and nilpotent groups*, Uspehi Mat. Nauk 2 (1947), no. 3(19), 18–59; English transl., Amer. Math. Soc. Transl. (1) 1(1962), 283–338. MR 10, 677.

Translated by:
Andrew Yablonsky

SEMIGROUPS GIVEN BY IDENTITIES WITH DISTINGUISHED ELEMENTS

UDC 519.47

A. P. BIRJUKOV

Introduction

It is well known that in the theory of universal algebras, groups, rings, semigroups, lattices, etc., one of the basic objects of study are one kind or another of primitive classes of algebras (see, for example, the book [1]). In a primitive class of algebras every algebra is a homomorphic image of some free algebra of this class. Therefore, to clarify whether a primitive class of algebras has a certain property invariant under homomorphisms it is sufficient to know whether all the free algebras of the class have this property (algebras given by identities). In this paper we give an exhaustive classification of all those semigroups given by identities with a single distinguished element that are: 1) regular, 2) completely regular, 3) idempotent. Moreover, we describe certain properties of the distinguished element. In particular, these classifications can be made into algorithms to recognize known properties of semigroups given by a *finite* set of identities with distinguished elements. There is no point in raising the corresponding problem for semigroups given by identities with two or more distinguished elements on which no further restrictions are placed, because all semigroups with two generators fall into this class. And it is known that for them the majority of properties are algorithmically not recognizable (see the monograph [2]). In earlier papers [3], [4] we have given a similar classification of semigroups given by identities without distinguished elements that have the three properties listen above, and some others.

We note that the content of §1 is not new (see [1], [4], [5], [6]) and is given here only to clarify the notation. All the concepts not defined in the text can be found in the monograph [6].

§1. Initial concepts

1.1. Let $a_1, \cdots, a_m$ be a finite set of letters and let $\Xi = \{\xi_1, \xi_2, \cdots\}$ be a countable alphabet. We enumerate the elements of the alphabet $M = \Xi \cup \{a_1, \cdots, a_m\}$ in an arbitrary manner, by setting $M = \{\mu_1, \mu_2, \cdots\}$. Words in the alphabet M are denoted by $A(\mu_i)$, $B(\mu_i)$, $\cdots$, and sometimes simply by $A, B, \cdots$. The empty word is also regarded as a word in the alphabet M. We write $A \equiv B$ if the words A and B are graphically equal. We say that a word A_1 occurs in A if we can find words B and C such that $A \equiv BA_1C$. For a word $A(\mu_i)$ we denote by $\mathfrak{m}(A)$ and $\mathfrak{n}(A)$, respectively, the set of letters in M or Ξ occurring in $A(\mu_i)$; let $l_{\mu_j}(A)$ denote the number of occurrences of μ_j in A; let $l(A)$ be the length of the word. We denote the set of all possible nonempty words in the alphabet M by $\mathfrak{M}_M$. For a given $A(\mu_i)$, let ϕ be a mapping of the set $\mathfrak{n}(A)$ into $\mathfrak{M}_M$ (or, briefly, ϕ: $\mathfrak{n}(A) \to \mathfrak{M}_M$). Then the word obtained from $A(\mu_i)$ by replacing all letters of $\mathfrak{n}(A)$ occurring in it by their images is denoted by $\phi A(\mu_i)$. For words in other alphabets we shall use a similar notation.

1.2. An *identity* is a formal equality of nonempty words in the aphabet M. Here it is assumed that the identity $A(\mu_i) = B(\mu_i)$ holds in the semigroup $\mathfrak{A}$ if 1) $\mathfrak{A}$ contains elements denoted by the symbols $a_1, \cdots, a_m$; 2) under every mapping ϕ: $\mathfrak{n}(AB) \to \mathfrak{A}$ the equation $\phi A = \phi B$ holds in ($\mathfrak{A}$).

The elements $a_1, \cdots, a_m$ are called *distinguished* in the semigroup $\mathfrak{A}$ as well as in the identity $A(\mu_i) = B(\mu_i)$.

The identity $A = B$ is regarded as equal to each of the identities $\phi A = \phi B$, $\phi B = \phi A$, provided only that ϕ is a one-to-one mapping of $\mathfrak{n}(AB)$ into Ξ. Here each trivial identity, that is, one of the form $A = A$, is regarded as equal to the empty set of identities.

It is clear that the relation of equality of identities is reflexive, symmetric, and transitive. On the other hand, equal identities either hold or do not hold simultaneously in a given semigroup.

Let $\Phi = \{A_\gamma(\mu_i) = B_\gamma(\mu_i), \gamma \in \Gamma\}$ be a given collection of identities. For formal reasons it is convenient to assume that in this expression Φ

1) no trivial identities occur;

2) not more than two equal identities occur;

3) together with every identity $A_\gamma(\mu_i) = B_\gamma(\mu_i)$ there also occurs the identity equal to it $A_{\gamma'}(\mu_i) \equiv \phi B_\gamma(\mu_i) = \phi A_\gamma(\mu_i) \equiv B_{\gamma'}(\mu_i)$, where ϕ is a one-to-one mapping of $\mathfrak{n}(A_\gamma B_\gamma)$ into Ξ.

In what follows, where nothing is said to the contrary, we assume that this convention holds. We denote by $\Gamma(\Phi)$ the class of all semigroups in which the collection of identities Φ holds ($\Gamma(\Phi)$ is a primitive class of semigroups).

1.3. Let $X = \{a_1, \cdots, a_m, x_1, x_2, \cdots\}$ be an alphabet, $\Phi = \{A_\gamma(\mu_i) = B_\gamma(\mu_i), \gamma \in \Gamma\}$ a given collection of identities. A semigroup given over the generating set X by the set of all possible defining relations of the form $\phi A_\gamma(\mu_i) = \Phi B_\gamma(\mu_i)$, where ϕ: $\mathfrak{n}(A_\gamma B_\gamma) \to \mathfrak{M}_X$, is called a *semigroup given by the set of identities Φ with m distinguished elements.* This semigroup is denoted by $S(X, \Phi)$. In the primitive class of semigroups $\Gamma(\Phi)$ the semigroups $S(X, \Phi)$ exhaust to within isomorphism the set of free semigroups. The significance of the semigroups $S(X, \Phi)$ for the class $\Gamma(\Phi)$ is explained by the fact that every semigroup in $\Gamma(\Phi)$ is homomorphic image of some semigroup $S(X, \Phi)$. In the investigation of semigroups in $\Gamma(\Phi)$ it turns out convenient to use the concept of a consequence in Φ.

Every identity of the form

$$C(\mu_i) \cdot \phi A_\gamma(\mu_i) \cdot C'(\mu_i) = C(\mu_i) \cdot \phi B_\gamma(\mu_i) \cdot C'(\mu_i),$$

where $\phi\colon\ \mathfrak{n}(A_\gamma B_\gamma) \longrightarrow \mathfrak{M}_M$, $l(CC') \geq 0$, is called an *immediate consequence* in Φ. The set of all immediate consequences in Φ is denoted by Φ'.

An identity $A(\mu_i) = B(\mu_i)$ is called a *consequence* in Φ if there exists a finite chain of words

$$A(\mu_i) \equiv A_1(\mu_i),\ A_2(\mu_i),\ \cdots,\ A_n(\mu_i) \equiv B(\mu_i),$$

such that all the identities $A_j = A_{j+1} (j = 1, 2, \cdots, n-1)$ are immediate consequences in Φ. This chain of words is called a *deductive* chain leading from A to B.

In what follows it is assumed that in a deductive chain of words there are no repetitions. The set of all consequences in Φ is denoted by Φ''. Sometimes the fact that $\Psi \subset \Phi''$ is written as follows: $\Phi \longrightarrow \Psi$. If $\Phi \longrightarrow \Psi$ and $\Psi \longrightarrow \Phi$, then these sets of identities are called equivalent. This is written: $\Phi \longleftrightarrow \Psi$.

1.4. Lemma. *If in a semigroup $S(X, \Phi)$ the equation $A = B$ holds, then by replacing in this equation all letters of $X \setminus \{a_1, \cdots, a_m\}$ by letters in Ξ in one-to-one correspondence with them, we obtain an identity in Φ''*

The proof is obvious.

In what follows this lemma is constantly used, without formal reference to it. In particular, it is easy to show that for $\Phi \longrightarrow \Psi$ it is necessary and sufficient that $\Gamma(\Phi) \subset \Gamma(\Psi)$.

We recall that in this paper we discuss identities and semigroups with a *single* distinguished element a.

§2. On the distinguished element of the semigroup $S(X, \Phi)$

2.1. The properties of $S(X, \Phi)$ depend for a perfectly clear reason on the properties of the distinguished element a. Therefore in studying $S(X, \Phi)$ the first task is to investigate the properties of the distinguished element. In this section we indicate a method of computing the order of the distinguished element of the semigroup $S(X, \Phi)$; that is, the type of its monogenic subsemigroup $\mathfrak{A} = [a]$.

2.2. Let $\Phi = \{A_\gamma(\mu_i) = B_\gamma(\mu_i),\ \gamma \in \Gamma\}$. We construct a set of numbers H, including in it all numbers

$$l(A_\gamma), \text{ when } l(A_\gamma) \neq l(B_\gamma) \text{ or } \mathfrak{n}(B_\gamma) \neq \mathfrak{n}(A_\gamma); \tag{2.1}$$

$$l(A_\gamma) + l_{\xi_j}(A_\gamma), \text{ when } l_{\xi_j}(A_\gamma) \neq l_{\xi_j}(B_\gamma),\ l(A_\gamma) = l(B_\gamma),\ \mathfrak{n}(A_\gamma) = \mathfrak{n}(B_\gamma). \tag{2.2}$$

We denote by $h(\Phi)$ the smallest number in H, provided $H \neq \emptyset$, and we set $h(\Phi) = \infty$ if H is empty.

2.3. **Theorem.** *In the subsemigroup* $\mathfrak{A} = [a]$ *of* $S(X, \Phi)$ *the number of nonunit elements is* $h(\Phi) - 1$.

Proof. Let $\mathfrak{A} = [a]$ be a finite semigroup and h the least number for which there exists an $m \neq h$ such that $a^h = a^m$ in $\mathfrak{A}$. We show that $h \in H$. By analyzing a deductive chain leading from a^h to a^m we see that the equation $a^h = a^n$, with $n > h$, already holds in Φ'; that is, $a^h \equiv c \cdot \phi A_{\gamma_1} \cdot c' = c \cdot \phi B_{\gamma_1} \cdot c' \equiv a^n$. From the minimality of h it follows that $l(cc') = 0$. If $l(A_{\gamma_1}) \neq l(B_{\gamma_1})$, then from $a^{l(A_{\gamma_1})} = a^{l(B_{\gamma_1})}$, $l(A_{\gamma_1}) \leq l(\phi A_{\gamma_1}) = h$, and the minimality of h it follows that $h = l(A_{\gamma_1}) \in H$. Suppose that $l(A_{\gamma_1}) = l(B_{\gamma_1})$ and $\mathfrak{n}(A_{\gamma_1}) \neq \mathfrak{n}(B_{\gamma_1})$. Then there exists a ξ_j such that either $\xi_i \in \mathfrak{n}(B_{\gamma_1}) \backslash \mathfrak{n}(A_{\gamma_1})$ or $\xi_j \in \mathfrak{n}(A_{\gamma_1}) \backslash \mathfrak{n}(B_{\gamma_1})$. In the first case we have $a^{l(A_{\gamma_1})} = a^{l(B_{\gamma_1}) + l_{\xi_j}(B_{\gamma_1})}$ in $\mathfrak{A}$ and therefore $h = l(A_{\gamma_1}) \in H$. Similarly also in the second case $h \in H$. Suppose, then, that $l(A_{\gamma_1}) = l(B_{\gamma_1})$ and $\mathfrak{n}(A_{\gamma_1}) = \mathfrak{n}(B_{\gamma_1})$. Then from $n > h$ it follows that there exists a ξ_j such that $l(\phi\xi_j) \geq 2$. We show that $l_{\xi_j}(A_{\gamma_1}) \neq l_{\xi_j}(B_{\gamma_1})$. Suppose that this is not the case. We define $\psi\colon \mathfrak{n}(A_{\gamma_1} B_{\gamma_1}) \to \mathfrak{A}$ by setting $\psi\xi_j = a$; $\psi\xi_i = \phi\xi_i$ if $i \neq j$. Then in $\mathfrak{A}$

$$a^{l(\psi A_{\gamma_1})} \equiv \psi A_{\gamma_1} = \psi B_{\gamma_1} \equiv a^{l(\psi B_{\gamma_1})},$$

which contradicts the minimality of h. We show now that $h = l(A_{\gamma_1}) + l_{\xi_j}(A_{\gamma_1})$. For in $\mathfrak{A}$ we have $a^{l(A_{\gamma_1}) + l_{\xi_j}(A_{\gamma_1})} = a^{l(B_{\gamma_1}) + l_{\xi_j}(B_{\gamma_1})}$,

$$l(A_{\gamma_1}) + l_{\xi_j}(A_{\gamma_1}) \leq l(\phi A_{\gamma_1}) = h,\ l(A_{\gamma_1}) = l(B_{\gamma_1}),\ l_{\xi_j}(A_{\gamma_1}) \neq l_{\xi_j}(B_{\gamma_1}).$$

Therefore it follows from the minimality of h that $h = l(A_{\gamma_1}) + l_{\xi_j}(A_{\gamma_1}) \in H$; that is, $h \in H$. But from the construction of H it follows that for $h(\Phi)$ we can find an $m_1 \neq h(\Phi)$ such that $a^{h(\Phi)} = a^{m_1}$ in $\mathfrak{A}$. Consequently $h = h(\Phi)$. If, however, $\mathfrak{A}$ is an infinite semigroup, then obviously $H = \emptyset$, and the theorem is also true.

2.4. We define the numerical characteristic $d(\Phi)$ for the set of identities $\Phi = \{A_\gamma(\mu_i) = B_\gamma(\mu_i),\ \gamma \in \Gamma\}$. We denote by $d(\Phi)$ the greatest common divisor of the set of all positive numbers of the form $d(\gamma, \mu_i) = |l_{\mu_i}(A_\gamma) - l_{\mu_i}(B_\gamma)| > 0$ $(\gamma \in \Gamma,\ \mu_i \in M)$. But if this set of numbers is empty, we take $d(\Phi) = 0$.

Theorem. *The subsemigroup* $\mathfrak{A} = [a]$ *of* $S(X, \Phi)$ *is finite if and only if* $d(\Phi) \neq 0$.

The proof is obvious.

2.5. **Theorem.** *In the semigroup* $\mathfrak{A} = [a]$ *the number of regular elements is equal to* $d(\Phi)$.

Proof. Let $\mathfrak{A} = [a]$ be a finite semigroup. It is known (see [7]) that the regular elements of the monogenic semigroup $\mathfrak{A}$ form a cyclic group G and that the number of elements of G is equal to the smallest number d for which we have $a^{h(\Phi)} = a^{h(\Phi)+d}$ in $\mathfrak{A}$. We show that d is divisible by $d(\Phi)$. We choose a deductive chain leading from $a^{h(\Phi)}$ to $a^{h(\Phi)+d}$; that is,

$$a^{h(\Phi)} \equiv A_1(\mu_i),\ A_2(\mu_i), \ldots, A_s(\mu_i) \equiv a^{h(\Phi)+d},$$

where in Φ'

$$A_j(\mu_i) \equiv c_j(\mu_i) \cdot \varphi_j A_{\gamma_j}(\mu_i) \cdot c'_j(\mu_i) = c_j(\mu_i) \cdot \varphi_j B_{\gamma_j}(\mu_i) \cdot c'(\mu_i) \equiv A_{j+1}(\mu_i).$$

Then, for a suitable r,

$$d = l(A_s) - l(A_1) = \sum_{t=1}^{s-1} [l(A_{t+1}) - l(A_t)]$$

$$= \sum_{=1}^{s-1} [l(\varphi_{t1}\ B_{\gamma_{t+1}}) - l(\varphi_{t+1} A_{\gamma_{t+1}})]$$

$$= \sum_{t=1}^{s-1} \sum_{i=1}^{r} [l_{\mu_i}(B_{\gamma_{t+1}}) - l_{\mu_i}(A_{\gamma_{t+1}})]\, l(\varphi_{t+1}\mu_i).$$

Since $l_{\mu_i}(B_{\gamma_{t+1}}) - l_{\mu_i}(A_{\gamma_{t+1}})$ is always divisible by $d(\Phi)$, we see that d is divisible by $d(\Phi)$. Now we show that $d(\Phi)$ is divisible by d.

Let a^k be a generating element of G, E_G its unit element, and let $A_\gamma(\mu_i) = B_\gamma(\mu_i)$ be an identity in Φ. Then d is the group order of the element a^k in G. Let $\phi\xi_i = E_G$. Then, by multiplying the equation $\phi A_\gamma = \phi B_\gamma$ a suitable number of times by E_G and using the commutativity of G, we find that

$$(aE_G)^{l_a(A_\gamma)} = (aE_G)^{l_a(B_\gamma)}. \tag{2.3}$$

When we raise both sides of (2.3) to the power k we see that $(a^k)^{l_a(A_\gamma)} = (a^k)^{l_a(B_\gamma)}$ in G. Therefore $|l_a(A_\gamma) - l_a(B_\gamma)|$ is divisible by d. We now show that $|l_{\xi_j}(A_\gamma) - l_{\xi_j}(B_\gamma)|$ is divisible by d for every ϵ_j. Let $\psi\xi_j = a^k$ and let $\psi\xi_i = E_G$ when $i \neq j$. Then, by multiplying the identity $\phi A_\gamma(\mu_i) = \phi B_\gamma(\mu_i)$ a sufficient number of times by E_G and using the commutativity of G, we find that $(aE_G)^{l_a(A_\gamma)}(a^k)^{l_{\xi_j}(A_\gamma)} = (aE_G)^{l_a(B_\gamma)}(a^k)^{l_{\xi_j}(B_\gamma)}$. Using (2.3) we see that $(a^k)^{l_{\xi_j}(A_\gamma)} = (a^k)^{l_{\xi_j}(B_\gamma)}$ in G. Therefore $|l_{\xi_j}(A_\gamma) - l_{\xi_j}(B_\gamma)|$ is divisible by d. Consequently $d(\Phi)$ is also divisible by d, and so $d = d(\Phi)$. But if $\mathfrak{A}$ is an infinite semigroup, then the truth of the theorem follows from 2.4.

2.6. Theorems 2.3 and 2.5 have the following corollaries:

1) *If* $\Phi \longleftrightarrow \Psi$, *then* $h(\Phi) = h(\Psi)$ *and* $d(\Phi) = d(\Psi)$.

2) *The distinguished element of the semigroup* $S(X, \Phi)$ *is regular if and only if* $h(\Phi) = 1$.

3) *The distinguished element of the semigroup* $S(X, \Phi)$ *is idempotent if and only if* $h(\Phi) = d(\Phi) = 1$.

4) *The semigroup* $\mathfrak{A} = [a]$ *of* $S(X, \Phi)$ *is a semigroup with zero if and only if* $d(\Phi) = 1$.

5) *If* $d(\Phi) \neq 0$, *then* $I = a^{h(\Phi)\, d(\Phi)}$ *is an idempotent of* $\mathfrak{A}$.

2.7. We observe that if Φ is a set of identities without distinguished elements, then the type of the element $x \in X$ in $S(X, \Phi)$ is the pair of numbers $(h(\Phi), d(\Phi))$. The proof of this proposition easily follows from Theorems 2.3 and 2.5.

§3. Three lemmas

3.1. It is often necessary to solve an equation of the form $ABX = A$ in semigroups, where A and B are given elements. Generally speaking this equation does not necessarily have a solution in a semigroup.

For example, let $\mathfrak{A} = [a]$ be a monogenic semigroup of type (h, d) and

$A = a^k$, $B = a^t$ $(k, t \geq 1)$. Then the equation $ABX = A$ is solvable in $\mathfrak{A}$ for $k \geq h$ and insolvable for $k < h$. For semigroups $S(X, \Phi)$ the question reduces to that of when, for given words $A(\mu_i)$ and $B(\mu_i)$, there exists a word $C(\mu_i)$ such that the identity $ABC = A$ belongs to Φ''.

In this section we are concerned with some special solutions of the latter problem that are used in what follows.

3.2. Lemma. *Let $A_1(\mu_i)$ and $A_2(\mu_i)$ be a pair of words such that $\mathfrak{m}(A_1) = \mathfrak{m}(A_1A_2)$. Then there exists a word $C(\mu_i)$ such that the identity*

$$\xi_1 = \xi_1^2 A(\mu_i) \tag{3.1}$$

implies the identity $A_1A_2C = A_1$.

Proof. We show that by multiplying the word A_1A_2 on the right the rightmost letter of this word can be annihilated. For it follows from $\mathfrak{m}(A_1) = \mathfrak{m}(A_1A_2)$ that $A_1A_2 \equiv B_1\mu_1B_2\mu_1$. We define a mapping ϕ: $\mathfrak{n}(\xi_1 A) \to \mathfrak{M}_M$ ($\phi\xi_1 = \mu_1 B_2$; $\phi\xi_i = a$ when $i > 1$). Then by using the relation (3.1) we find that

$$A_1A_2 \cdot B_2\varphi A \equiv B_1\mu_1B_2\mu_1 \cdot B_2\varphi A \equiv B_1\varphi(\xi_1^2 A) = B_1\varphi\xi_1 \equiv B_1\mu_1B_2.$$

Applying this mapping $l(A_2)$ times in succession we arrive at the required identity. The lemma is proved.

Note that the lemma just proved can be regarded as the generalization of a lemma in [8].

3.3. Lemma. *Let $A_1(\mu_i)$, $A_2(\mu_i)$ be a pair of nonempty words such that ξ_1^2 does not occur in A_1A_2 and $\mathfrak{n}(A_1) = \mathfrak{n}(A_1A_2) = \xi_1$. Then there exists a word $C(\mu_i)$ such that the identity*

$$\xi_1 = \xi_1 a^k \xi_1 A(\mu_i), \text{ where } k \geq 1, \tag{3.2}$$

implies the identity $A_1A_2C = A_1$.

Proof. From the relation (3.2) it follows that $a = a^{d+1}$ for some $d \geq 1$. Moreover, from the conditions of the lemma it follows that either $A_1A_2 \equiv B\xi_1 a^s$, where $1 \leq s \leq d$, or $A_1A_2 \equiv B\xi_1 a^s \xi_1$, where $1 \leq s \leq d$. We show that by multiplying the word A_1A_2 on the right the rightmost letter can be annihilated. For in the first case

$$B\xi_1 a^s \cdot a^{d-s+k}\xi_1 A a^{s-1} \equiv B \cdot \xi_1 a^k \xi_1 A \cdot a^{s-1} = B\xi_1 a^{s-1}.$$

In the second case we define $\phi\xi_1 = \xi_1 a^{d+s-k}$. Then

$$B\xi_1 a^s \xi_1 \cdot a^{d+s-k} \varphi A a^k \equiv B \cdot (\xi_1 a^{d+s-k}) a^k (\xi_1 a^{d+s-k}) \varphi A a_k$$
$$\equiv B\varphi(\xi_1 a^k \xi_1 A) \cdot a^k = B\varphi\xi_1 \cdot a^k \equiv B\xi_1 a^{d+s-k} a^k = B\xi_1 a^s.$$

By applying these transformations a sufficient number of times in succession we arrive at the required word $C(\mu_i)$.

3.4. Lemma. *Let $A_1(\mu_i)$ and $A_2(\mu_i)$ be a pair of words such that $\mathfrak{n}(A_1) = \mathfrak{n}(A_1 A_2) = \xi_1$. Then there exists a word $C(\mu_i)$ such that the identity $\xi_1 = \xi_1^k a^t \xi_1 A(\mu_i)$ $(k \geq 2,\ t \geq 1)$ implies the identity $\phi A_1 \cdot \phi A_2 \cdot C = \phi A_1$, where $\phi \xi_1 = \xi_1^k$.*

Proof. From the conditions of the lemma it follows that for a suitable $d \geq 1$ we have $a = a^{d+1}$. Furthermore, we have one of the following three cases:

$$\varphi A_1 \cdot \varphi A_2 \equiv B\xi_1^2; \quad \varphi A_1 \cdot \varphi A_2 \equiv B\xi_1 a^r \xi_1 \ (1 \leqslant r \leqslant d);$$
$$\varphi A_1 \cdot \varphi A_2 \equiv B\xi_1^k a^r \quad (1 \leqslant r \leqslant d).$$

We show that by multiplying on the right the rightmost letter of the word $\phi A_1 \cdot \phi A_2$ can be annihilated. In the first and second cases this can be done according to Lemma 3.2. In the third case

$$B\xi_1^k a^r \cdot a^{d-r+t} \xi_1 A \cdot \xi_1^{k-1} a^{r-1} = B(\xi_1^k a^t \xi_1 A) \xi_1^{k-1} a^{r-1} = B\xi_1^k a^{r-1}.$$

If as a result of these transformations we obtain a word other than $\phi(A_1)$, then it can also be written in one of the three indicated forms. Therefore, by applying the transformations a sufficient number of times we arrive at the required word $C(\mu_i)$.

§4. On regularity and complete regularity of the semigroups $S(X, \Phi)$

4.1. We introduce a number of classes of sets of identities.

$\mathfrak{C}_\alpha$: we say that $\Phi \in \mathfrak{C}_\alpha$ if Φ contains the identity $\mu_1 A(\mu_i) = \mu_2 B(\mu_i)$.

$\mathfrak{C}_\alpha^*$ is its dual; that is, the class of identities symmetric relative to "left" and "right".

$\mathfrak{C}_\beta$: we say that $\Phi \in \mathfrak{C}_\beta$ if Φ contains either the identity $A(\xi_i) = B_1(\mu_i) a B_2(\mu_i)$ or $A(\xi_i) = B(\xi_i)$, where $\mathfrak{n}(A) \neq \mathfrak{n}(B)$.

$\mathfrak{C}_\gamma$: we say that $\Phi \in \mathfrak{C}_\gamma$ if Φ contains either the identity $A(\mu_i) = B(\mu_i)$, where $\mathfrak{n}(A) \neq \mathfrak{n}(B)$, or an identity $A(\mu_i) = B(\mu_i)$ such that $l_{\xi_j}(B) >$

$l_{\xi_j}(A) = 1$ for some ξ_j.

$\mathfrak{C}(\delta, m)$ $(m \geq 1)$: we say that $\Phi \in \mathfrak{C}(\delta, m)$ if Φ contains an identity $\xi_1, \cdots, \xi_t = A(\mu_i)$, where $1 \leq t \leq m$, and either $l(A) > t$ or $\mathfrak{m}(A) \neq \{\xi_1, \cdots, \xi_t\}$.

$\mathfrak{C}_\kappa$: we say that $\Phi \in \mathfrak{C}_\kappa$ if Φ contains the identity $A(\mu_i) = B(\mu_i)$, where $\mathfrak{n}(A) \neq \mathfrak{n}(B)$.

$\mathfrak{C}_\epsilon$: we say that $\Phi \in \mathfrak{C}_\epsilon$ if Φ contains the identity $A(\mu_i)\xi_1 = B(\mu_i)$, where $\xi_1 \notin \mathfrak{n}(A)$ and either $l_{\xi_1}(B) \geq 2$, or $\mathfrak{n}(A\xi_1) \neq \mathfrak{n}(B)$ or $B(\mu_i) \equiv B_1(\mu_i)\mu_1$, where $\mu_1 \neq \xi_1$.

$\mathfrak{C}_\epsilon^*$ is the dual class of sets of identities.

4.2. The significance of the classes of sets of identities introduced above is clear from the following lemma.

Lemma. *Φ implies the identities*

1) $aA(\mu_i) = \xi_1 B(\mu_i)$,

2) $\xi_1^n = B_1(\mu_i)aB_2(\mu_i)$ *for some* $n \geq 1$,

3) $a^k\xi_1 a^t = A(\mu_i)$, *where* $l_{\xi_1}(A) \geq 2$ *for certain* $k, t \geq 0$,

4) $\xi_1\xi_2 \cdots \xi_m = A(\mu_i)$, *where* $l(A) > m$,

5) $a^n = B_1(\mu_i)\xi_1 B_2(\mu_i)$ *for some* $n \geq 1$,

6) $a^n\xi_1 = A(\mu_i)$, *where* $a\xi_1 a$ *occurs in* A *for some* $n \geq 0$,

if and only if, respectively, 1) $\Phi \in \mathfrak{C}_\alpha$, 2) $\Phi \in \mathfrak{C}_\beta$, 3) $\Phi \in \mathfrak{C}_\gamma$, 4) $\Phi \in \mathfrak{C}(\delta, m)$, 5) $\Phi \in \mathfrak{C}_\epsilon$, 6) $\Phi \in \mathfrak{C}_\epsilon$.

The proof of the sufficiency of the conditions of the lemma is obvious, and the proof of their necessity is easily obtained by analyzing the corresponding deductive chains.

4.3. Lemma. *The identity* $\xi_1 = a^d\xi_1$ *for* $d \geq 1$ *and the identity* $a^k\xi_1 = a^t\xi_1 A(\mu_i)$ *for* $k, t \geq 0$ *imply the identity* $\xi_1 = \xi_1 A^d$.

Proof. By multiplying the second identity on the left by a suitable power of a and using the first identity, we see that $\xi_1 = a^s\xi_1 A$ for some $s \geq 0$. Then

$$\mu_1 A(\mu_i) = \mu_2 B(\mu_i).$$

4.4. Theorem. *The semigroup* $S(X, \Phi)$ *with* $|X| \geq 2$ ($|X|$ *is the cardinality of the set* X) *is regular if and only if* $\Phi \in \mathfrak{C}_\gamma \cap \mathfrak{C}(\delta, 1) \cap \mathfrak{C}_\epsilon \cap \mathfrak{C}_\epsilon^*$.

The proof of the necessity of the condition of the theorem is obvious.

Proof of sufficiency. Suppose that $\Phi \in \mathfrak{C}_\gamma \cap \mathfrak{C}(\delta, 1) \cap \mathfrak{C}_\epsilon \cap \mathfrak{C}_\epsilon^*$. From the fact that $\Phi \in \mathfrak{C}(\delta, 1)$ we see by Lemma 4.2 that in ϕ'' we have $\xi_1 = A(\mu_i)$, where $l(A) \geq 2$. Consequently, if the word $A(\mu_i)$ begins and ends with letters in Ξ, then $S(X, \Phi)$ is a regular semigroup. We assume, therefore, that the word $A(\mu_i)$ begins, for example, with a. Then for a suitable $d \geq 1$ it follows from this identity that

$$\xi_1 = a^d \xi_1 \quad (d \geqslant 1). \tag{4.1}$$

From $\Phi \in \mathfrak{C}_\epsilon$ we find by Lemma 4.2 that Φ'' contains $a^k \xi_1 = A_1(\mu_i) a \xi_1 a A_2(\mu_i)$ for some $k \geq 0$. Multiplying this identity on the left by $a^{(d-1)k}$ and using the identity (1) we find that in Φ''

$$\xi_1 = A_3 a \xi_1 a A_2. \tag{4.2}$$

From $\Phi \in \mathfrak{C}_\gamma$ we find by Lemma 4.2 that Φ'' contains the identity $a^t \xi_1 a^r = B(\mu_i)$, where $l_{\xi_1}(B) \geq 2$ $(r, t \geq 0)$. Furthermore, bearing in mind that $a = a^{d+1}$ we may obviously assume that $r, t \leq d$. Then, using (4.2), we find that $\xi_1 = A_3 a \xi_1 a A_2 = A_3 a^{d+1-t} \cdot a^t \xi_1 a^r \cdot a^{d+1-r} A_2 = A_3 a^{d+1-t} B a^{d+1-r} A_2 \equiv B_1$, where $l_{\xi_1}(B_1) \geq 2$. Let $B_1(\mu_i) \equiv a^s \xi_1 B_2 \xi_1 a^l$ $(s, l \geq 0)$. If $s \geq 1$, then by Lemma 4.3 the identities $\xi_1 = B_1$ and (4.1) imply the identity $\xi_1 = \xi_1 (B_2 \xi_1 a^l)^d \equiv B_3(\mu_i)$. If $t \geq 1$, then it follows from the identity $\xi_1 = B_3(\mu_i)$ and (4.1) that $\xi_1 = \xi_1 a^d$. And then, by the proposition dual to Lemma 4.3, it follows from the identities $\xi_1 = \xi_1 a^d$ and $\xi_1 = B_3(\mu_i)$ that $\xi_1 = \xi_1 B_4 \xi_1$ and, consequently, that $S(X, \Phi)$ is a regular semigroup. Dually, if the word $A(\mu_i)$ ends with the letter a, then $S(X, \Phi)$ is also a regular semigroup. This completes the proof of the theorem.

4.5. Let us clarify when a regular semigroup $S(X, \Phi)$ is completely regular. In the book [6] it is proved that a regular semigroup is completely regular if and only if the equation $A = A^2 X$ is solvable in it for every element A. In what follows this proposition and its dual are used several times.

4.6. Theorem. *The regular semigroup $S(X, \Phi)$ with $|X| \geq 2$ is completely regular if and only if Φ contains an identity*

$$A(\mu_i) \equiv \mu_{j_1} A_1 \equiv A_2 \mu_{j_2} = \mu_{j_3} B_1 \equiv B_2 \mu_{j_4} \equiv B(\mu_i), \tag{4.3}$$

for which either $\mathfrak{n}(A) \neq \mathfrak{n}(B)$ or $j_1 \neq j_3$ or $j_2 \neq j_4$, or for suitable ξ_i and ξ_j the word $\xi_i \xi_j$ occurs in B but not in A.

Proof. Necessity. Suppose that $S(X, \Phi)$ is a completely regular semigroup. Then Φ'' contains an identity $A_1(\mu_i) = A_2(\mu_i)$ such that ξ_1^2 occurs in A_2 but not in A_1. By analyzing a deductive chain leading from A_1 to A_2 we find in Φ' an identity of the form indicated above; that is, we may assume that

$$A_1(\mu_i) \equiv C \cdot \varphi A_\gamma \cdot C' = C \cdot \varphi B_\gamma \cdot C' \equiv A_2(\mu_i),$$

where ϕ: $\mathfrak{n}(A_\gamma B_\gamma) \to \mathfrak{M}_M$, $\gamma \in \Gamma$, $l(CC') \geq 0$ and ξ_1^2 occurs in A_2 but not in A_1.

We show that $A_\gamma(\mu_i) = B_\gamma(\mu_i)$ is an identity of the required form (4.3). Suppose that $\mathfrak{n}(A_\gamma) = \mathfrak{n}(B_\gamma)$ and that in the words A_γ and B_γ the leftmost and rightmost letters are equal. Then ξ_1^2 occurs in the word ϕB_γ, but not in ϕA_γ. Therefore it follows from $\mathfrak{n}(A_\gamma) = \mathfrak{n}(B_\gamma)$ that in the word B_γ there occurs a word $\xi_i \xi_j$ such that ξ_1^2 occurs in $\phi(\xi_i \xi_j)$. But since ξ_1^2 does not occur in ϕA_γ, $\xi_i \xi_j$ also does not occur in A_γ. This proves the necessity.

Sufficiency. From the fact that Φ contains the identity (4.3) we derive easily that Φ'' contains the identity

$$A(\mu_i) = B(\mu_i), \tag{4.4}$$

where ξ_1^2 occurs in B but not in A, and $\mathfrak{n}(A) = \mathfrak{n}(B) = \xi_1$. When we take the regularity of $S(X, \Phi)$ and the Remark 4.5 into account, we see that to prove the fact that $S(X, \Phi)$ is completely regular we need only examine the case when Φ'' contains the identity

$$\xi_1 = \xi_1 a^k \xi_1 A_1(\mu_i) \xi_1 a^r \xi_1, \text{ where } k, r \geq 1. \tag{4.5}$$

By Lemma 3.3 and its dual proposition there exist words $C_1(\mu_i)$ and $C_2(\mu_i)$, such that from the identity (4.5) it follows that $C_1 A C_2 = \xi_1$. Consequently $\xi_1 = C_1 B C_2 \equiv B_1(\mu_i)$, and ξ_1^2 occurs in B_1. If ξ_1^2 is the beginning of B_1, then $S(X, \Phi)$ is a completely regular semigroup. We assume therefore that $B_1 \equiv B_2 a \xi_1^2 B_3$, where ξ_1^2 does not occur in B_2. From the identity (4.5) it follows that $a = a^{d+1}$ for a suitable $d \geq 1$. We define the mapping $\phi \xi_1 = \xi_1 a^d$. Then

$$B_2 a \xi_1^2 B_3 \cdot a \equiv B_1 a = \xi_1 a = (\xi_1 a^d) a = \varphi \xi_1 \cdot a = \varphi B_1 \cdot a$$
$$\equiv \varphi(B_2 a \xi_1^2 B_3) a \equiv \varphi(B_2 a) \cdot \xi_1 \varphi(a^d \xi_1 B_3 a) = B_2 a \xi_1 \varphi(a^d \xi_1 B_3 a) \equiv B_4.$$

By Lemma 3.3 and the dual proposition the identity (4.5) implies the identities $C_3 \cdot B_2 a \xi_1 = \xi_1$ and $\xi_1 \phi(a^d \xi_1 B_3 a) \cdot C_4 = \xi_1$. And then in Φ'' we have

$$\xi_1 = \xi_1 \varphi(a^d \xi_1 B_3 a) \cdot C_4 = C_3 \cdot B_2 a \xi_1 \varphi(a^d \xi_1 B_3 a) C_4 \equiv C_3 B_4 \cdot C_4$$
$$= C_3 \cdot B_1 a C_4 \equiv C_3 B_2 a \xi_1 \cdot \xi_1 B_3 a C_4 = \xi_1 \xi_1 B_3 a C_4 \equiv \xi_1^2 B_5(\mu_i).$$

Consequently $S(X, \Phi)$ is a completely regular semigroup.

§5. On the idempotency of the semigroup $S(X, \Phi)$

5.1. **Lemma.** *Let Φ_0 be the collection of all those identities in Φ whose expression does not contain the distinguished element, and suppose that Φ does not belong to the class $\mathfrak{C}_\beta$. Then the identity $A(\xi_i) = B(\xi_i)$ follows from Φ if and only if it follows from Φ_0.*

The proof is obvious.

In this section we wish to clarify when Φ implies the identity $\xi_1 = \xi_1^2$. By the lemma we have to consider two cases: 1) in the expression of Φ the distinguished element a does not occur; 2) $\Phi \in \mathfrak{C}_\beta$. The first case was discussed in our earlier paper [4], so that Theorem 5.4 is true in the first case. Therefore it remains to consider here the case 2). Observe that a similar situation can arise in the investigation of other identities.

5.2. We define the Ξ-subword of $A(\mu_i)$ as the word $\overline{A}$ that remains when all letters a occurring in A are omitted. For example, $\overline{\xi_1 a \xi_1 a^2 \xi_2^2 \xi_1 a} \equiv \xi_1^2 \xi_2^2 \xi_1$.

Lemma. *Suppose that $\Phi = \{A_\gamma(\mu_i) = B_\gamma(\mu_i), \gamma \in \Gamma\}$, $d(\Phi) = 1$ (see §2) and that for every γ in Γ we have $\mathfrak{n}(A_\gamma), \mathfrak{n}(B_\gamma) \neq \emptyset$ and that $\phi: \Xi \to \mathfrak{M}_M$ ($\phi\xi_i = a^h \xi_i a^h$, where $h = h(\Phi) \neq 0$). If the identity $A(\xi_i)$ follows from the set of identities $\overline{\Phi} = \{\overline{A}_\gamma(\mu_i) = \overline{B}_\gamma(\mu_i), \gamma \in \Gamma\}$, then the identity $\phi A(\xi_i) = \phi B(\xi_i)$ follows from the set of identities Φ.*

Proof. We choose a deductive chain $A(\xi^i) \equiv A_1(\xi_i), A_2(\xi_i), \cdots$ $\cdots, A_{s+1}(\xi_i) \equiv B(\xi_i)$, where all the identities $A_j = A_{j+1}$ $(j = 1, \cdots, s)$ belong to $\overline{\Phi}$, so that $A_j \equiv C \cdot \psi \overline{A}_\gamma \cdot C' = C \cdot \psi \overline{B}_\gamma \cdot C' \equiv A_{j+1}$. By the condition of the lemma, according to §2, $a^h = a^{h+1}$. Using this equation we find that in Φ''

$$\varphi A_j \equiv \varphi C \cdot \varphi\psi \overline{A}_\gamma \cdot \varphi C' = \varphi C \cdot \varphi\psi A_\gamma \cdot \varphi C' = \varphi C \cdot \varphi\psi B_\gamma \cdot \varphi C'$$

$$= \varphi C \cdot \varphi\psi \overline{B}_\gamma \cdot \varphi C' \equiv \varphi(C\psi\overline{B}_\gamma \cdot C') \equiv \varphi A_{j+1}.$$

Therefore $\phi A(\xi_i) = \phi B(\xi_i)$ in Φ''.

5.3. We introduce the numerical characteristic $d_1(\Phi)$ for $\Phi = \{A_\gamma(\mu_i) = B_\gamma(\mu_i), \gamma \in \Gamma\}$. We denote by $d_1(\Phi)$ the greatest common divisor of the set of all positive numbers of the form $d(\gamma, \xi_j) = |l_{\xi_j}(A_\gamma) - l_{\xi_j}(B_\gamma)|$. If this set of numbers is empty we take $d_1(\Phi) = 0$.

Lemma. *If $\Phi \to \Psi$, then $d_1(\Psi)$ is divisible by $d_1(\Phi)$.*

The proof is obvious.

Corollary. *If Φ and Ψ are equivalent sets of identities, then $d_1(\Phi) = d_1(\Psi)$.*

5.4. Theorem. *The semigroup $S(X, \Phi)$ with $|X| \geq 2$ is a semigroup of idempotents if and only if $d_1(\Phi) = 1$ and $S(X, \Phi)$ is completely regular.*

Proof. The necessity follows from Lemma 5.3 and some obvious arguments.

Proof of sufficiency. By §2, $a = a^2$. We replace each identity $A_\gamma(\mu_i) = B_\gamma(\mu_i)$ in Φ by the identity $\overline{A}_\gamma = \overline{B}_\gamma$, provided $\overline{A}_\gamma$ and $\overline{B}_\gamma$ are nonempty words, and by the identity $\xi_1\overline{A}_\gamma = \xi_1\overline{B}_\gamma$ otherwise. So we obtain a set of identities $\Psi = \{E_\gamma(\xi_i) = F_\gamma(\xi_i), \gamma \in \Gamma\}$. Let $\phi: \Xi \to \mathfrak{M}_M$ $(\phi\xi_i = a\xi_i a)$ and $\Psi_1 = \{\phi E_\gamma(\xi_i) = \phi F_\gamma(\xi_i), \gamma \in \Gamma\}$. From the construction it follows that $S(X, \Phi) \subset \Gamma(\Psi_1)$. Furthermore, $1 = d_1(\Phi) = d(\Phi) = d_1(\Psi) = d_1(\Psi_1)$, and $1 = h(\Phi) = h(\Psi)$. Therefore, by the corresponding theorem in [4], $\xi_1 = \xi_1^2$ in Ψ''. But then, by Lemma 5.2, $a\xi_1 a = (a\xi_1 a)^2$ in Ψ_1'', and therefore $a\xi_1 a = (a\xi_1 a)^2$ in $S(X, \Phi)$.

By the Remark 5.1 it only remains to consider the case $\Phi \in \mathfrak{C}_\beta$; that is, the case when Φ'' contains the identity $A(\xi_i) = A_1(\xi_i)aA_2(\mu_i)$. Using this identity and also the fact that $S(X, \Phi)$ is completely regular, we find that in Φ''

$$\xi_1 = \xi_1^k a \xi_1 B(\mu_i), \text{ where } k \geq 2.$$

From this identity, by using $a\xi_1 a = (a\xi_1 a)^2$ we obtain in succession:

$$a\xi_1 = a \cdot \xi_1^k a \xi_1 B \equiv a\xi_1^k a \cdot \xi_1 B = a\xi_1^k a \cdot a \xi_1^k a \xi_1 B = a \xi_1^k a \cdot \xi_1;$$

$$\xi_1 = \xi_1^k a \xi_1 B \equiv \xi_1^k \cdot a \xi_1 \cdot B = \xi_1^k \cdot a \xi_1^k a \xi_1 \cdot B \equiv \xi_1^k a \cdot \xi_1^k a \xi_1 B = \xi_1^k a \xi_1;$$

$$\xi_1 a = \xi_1^k a \xi_1 \cdot a = \xi_1^k \cdot a \xi_1 a = \xi_1^k a \xi_1 a \cdot a \xi_1 a = \xi_1^k a \xi_1 \cdot a \xi_1 a = \xi_1 a \xi_1 a;$$

$$\xi_1 = \xi_1^k a \xi_1 \equiv \xi_1^{k-1} \cdot \xi_1 a \cdot \xi_1 = \xi_1^{k-1} \cdot \xi_1 a \xi_1 a \cdot \xi_1 \equiv \xi_1^k a \xi_1 \cdot a \xi_1 = \xi_1 a \xi_1;$$

$$\xi_1 = \xi_1^k a \xi_1 = \xi_1^{k-1} \cdot \xi_1 a \xi_1 = \xi_1^{k-1} \cdot \xi_1 = \xi_1^k;$$

$$\xi_1 = \xi_1 a \xi_1 = (\xi_1 a \xi_1)^k \equiv \xi_1 \cdot a \xi_1^2 a \cdot a \xi_1^2 a \ldots a \xi_1^2 a \xi_1 = \xi_i a \xi_1^2 a \cdot \xi_1$$
$$\equiv \xi_1 a \xi_1 \cdot \xi_1 a \xi_1 = \xi_1^2.$$

This completes the proof of the theorem.

5.5. We have mentioned above (see 2.6) that in case $d(\Phi) \neq 0$ the element $I = a^{h(\Phi)d(\Phi)}$ is an idempotent in $S(X, \Phi)$. Let us find out when the element I is a left unit of $S(X, \Phi)$. For example, I is a left unit of $S(X, \Phi)$ if Φ contains an identity of the form

$$\xi_1 = \mu_1 A(\mu_i), \text{where } \mu_1 \not\equiv \xi_1. \tag{5.1}$$

For formal reasons it is sometimes convenient to exclude this case from the discussion.

Theorem. *Suppose that Φ does not contain identities of the form* (5.1). *Then for $|X| \geq 2$ a certain power of a is a left unit of $S(X, \Phi)$ if and only if*

$$\Phi \in \mathfrak{C}_\alpha \cap \mathfrak{C}_\beta \cap \mathfrak{C}(\delta, 1) \cap \mathfrak{C}^*_\varepsilon.$$

The necessity is obvious.

Proof of sufficiency. From the fact that $\Phi \in \mathfrak{C}(\delta, 1)$ and that Φ does not contain identities of the form (5.1) we conclude that Φ'' contains an identity of one of the following three forms:

$$\xi_1 = \xi_1 A_1(\mu_i), \text{where } \mathfrak{m}(A_1) = \{a, \xi_1\}; \tag{5.2}$$

$$\xi_1 = \xi_1 a^d, \text{where } d \geqslant 1; \tag{5.3}$$

$$\xi_1 = \xi_1^k, \text{where } k \geqslant 2. \tag{5.4}$$

From the fact that $\Phi \in \mathfrak{C}_\beta$ and from the identity (5.4) it is easy to prove that Φ'' contains the identity (5.2). Therefore in Φ'' we have either the identity (5.2) or the identity (5.3). From $\Phi \in \mathfrak{C}^*_\epsilon$ it follows that Φ'' contains one of the following two identities

$$\xi_1 a^d = \xi_1 A_2(\mu_i), \text{where } \mathfrak{m}(A_2) = \{a, \xi_1\}; \tag{5.5}$$

$$\xi_1 a^d = a A_3(\mu_i). \tag{5.6}$$

If Φ'' contains the identities (5.3) and (5.5), then obviously also (5.2). But if Φ'' contains (5.3) and (5.6), then $I = a^d$ is obviously a left unit. We examine the only remaining possibility, when Φ'' contains (5.2). Since $\Phi \in \mathfrak{C}_\alpha$, Φ'' contains the identity

$$a B_1(\mu_i) = \xi_1 B_2(\mu_i), \tag{5.7}$$

and we may now assume that ξ_1^2 does not occur in $\xi_1 B_2$.

From the identity (5.2) we easily derive

$$\xi_1 = \xi_1^r a^s \xi_1 B_3(\mu_i), \text{where } s \geqslant 1. \tag{5.8}$$

If $r = 1$, then by 3.3 the identity (5.8) implies that $\xi_1 B_2 C_1 = \xi_1$. We then find from (5.7) that $\xi_1 = \xi_1 B_2 C_1 = a B_1 C_1$ and therefore $I = a^d$ is a left unit of $S(X, \Phi)$.

Suppose that in (5.8) $r \geq 2$ and that $\phi\xi_1 = \xi_1^r$. Then by 3.4 there exists a word $C_2(\mu_i)$ such that (5.8) implies the identity $\Phi(\xi_1 B_2) \cdot C_2 = \phi\xi_1 \equiv \xi_1^r$. Then $\xi_1 = \xi_1^r \cdot a^s\xi_1 B_3 = \phi(\xi_1 B_2) \cdot C_2 a^s \xi_1 B_3 = \phi(aB_1) \cdot C_2 a^s \xi_1 B_3 \equiv aB_4(\mu_i)$. Therefore $I = a^d$ is a left unit of the semigroup $S(X, \Phi)$.

The proof of the theorem is now complete.

From Theorem 5.5 and the proposition dual to it we easily obtain a complete answer to the question of when for $|X| \geq 2$ a power of a is a unit of $S(X, \Phi)$.

BIBLIOGRAPHY

[1] A. G. Kuroš, *Lectures in general algebra*, Fizmatgiz, Moscow, 1962; English transl., Internat. Series of Monographs in Pure and Appl. Math., vol. 70, Pergamon Press, Oxford, 1965. MR 25 #5097; MR 31 #3483.

[2] A. A. Markov, *Theory of algorithms*, Trudy Mat. Inst. Steklov 42 (1954); English transl., Israel Program for Scientific Translations, Jerusalem, 1961. MR 17, 1038; MR 24 #A2527.

[3] A. P. Birjukov, *Semigroups which are prescribed identities*, Dokl. Akad. Nauk SSSR 149 (1963), 230–232 = Soviet Math. Dokl. 4 (1963), 333–336. MR 26 #6281.

[4] ———, *Semigroups given by identities*, Učen. Zap. Novosibirsk. Gos. Ped. Inst. 18 (1963), 139–163. (Russian)

[5] Ju. I. Janov, *On systems of identities for algebras*, Problemy Kibernet. 8 (1962), 75–90. (Russian) MR 29 #5773.

[6] E. S. Ljapin, *Semigroups*, Fizmatgiz, Moscow, 1960; English transl., Transl. Math. Monographs, vol. 3, Amer. Math. Soc., Providence, R. I., 1963; rev. ed., 1968. MR 22 #11054.

[7] A. Clifford and G. Preston, *The algebraic theory of semigroups*. Vol. I, Math. Surveys, no. 7, Amer. Math. Soc. Providence, R. I., 1961. MR 24 #A2627.

[8] J. A. Green and D. Rees, *On semigroups in which* $x^r = x$, Proc. Cambridge Philos. Soc. 48 (1952), 35–40. MR 13, 720.

Translated by:
K. A. Hirsch

IDENTICAL RELATIONS ON VARIETIES OF QUASIGROUPS

UDC 519.47

A. I. MAL'CEV

The object of this note is the construction of a variety of commutative loops L defined by a finite system of identities in a single variable and such that the problem of recognizing the truth of an arbitrarily given identity in one variable on each loop of L is algorithmically insoluble. This means that in L the free loop with one generating element is nonconstructive.

In §1 we recall certain well-known definitions and facts. In §2 we construct an auxiliary variety of algebras with two unary operations having a nonrecursive free algebra. By means of this variety we construct in §4 a variety of loops and a variety of commutative loops in which free loops with one generator are nonrecursive. In §3 we prove an auxiliary lemma on the extension of partial loops.

§1. The problem of identical relations

Let $F_1, \cdots, F_s$ be an arbitrary finite system of symbols each of which is associated with a natural number, the so-called *arity* of the corresponding symbol. An algebra of signature $\sigma = \{F_1, \cdots, F_s\}$ is specified when a set A is given and to each symbol F_i a certain n_i-ary operation f_i is associated, which is defined on A with values in A, where n_i is the arity of the symbol F_i. The operation f_i is called the value of the functional symbol F_i in the algebra A. A class of algebras is an arbitrary system of algebras of a given signature.

The notion of a *term* of signature $F_1, \cdots, F_s$ in several object variables $x_1, \cdots, x_m$ is defined inductively:

a) Expressions of the form $x_1, \cdots, x_m$ are, by definition, terms;

b) If $\mathfrak{a}_1, \cdots, \mathfrak{a}_{n_1}$ are terms, then the expression $F_i(\mathfrak{a}_1, \cdots, \mathfrak{a}_{n_1})$ is also a term.

To give the value of an object variable x_j in an algebra A means to associate with the symbol x_j a certain element of A. If $\mathfrak{a}$ is a term of signature σ in the variables $x_1, \cdots, x_m$ and the values of all these variables in an algebra A of signature σ are given, then by carrying out on the values of $x_1, \cdots, x_m$ in A all those operations that are indicated in the expression of the term $\mathfrak{a}$ we obtain an element of A, which is called the *value of the term* $\mathfrak{a}$ for the given values of the variables $x_1, \cdots, x_m$.

An expression of the form $\mathfrak{a} = \mathfrak{b}$, where $\mathfrak{a}$ and $\mathfrak{b}$ are terms of a given signature σ in the variables $x_1, \cdots, x_m$, is called an *identity* (or an identical relation) of signature σ and of rank m. The identity $\mathfrak{a} = \mathfrak{b}$ is true on an algebra A if for any values of $x_1, \cdots, x_m$ in A the values of $\mathfrak{a}$ and $\mathfrak{b}$ are equal. The identity $\mathfrak{a} = \mathfrak{b}$ is true on a class K of algebras if it is true on each algebra of K (each K-algebra).

By $I_m(\mathrm{K})$ we denote the set of all identities $x_1, \cdots, x_m$ that are true on the class K. The union of all $I_m(\mathrm{K})$ is denoted by $I(\mathrm{K})$. Since an identity is a word in the alphabet $x, F_1, \cdots, F_s, =, ,, (,)$ (by x_i we code the word $xx\cdots x$), we can raise the question whether for a given class K the set $I(\mathrm{K})$ (or $I_m(\mathrm{K})$) is recursive. This question is called the problem of identical relations on the class K. It is clear that when $I_m(\mathrm{K})$ is nonrecursive for any m, then $I(\mathrm{K})$ and all $I_n(\mathrm{K})$ for $n \geq m$ are nonrecursive.

For every class K and every set of symbols $a_1, \cdots, a_m$ we can construct a special algebra, the so-called *free algebra* of rank m (or the algebra with the free generators $a_1, \cdots, a_m$) of the class K. This construction is carried out in the following manner. Let $\sigma = \{F_1, \cdots, F_s\}$ be the signature of K. We denote by $\mathfrak{A}_m$ the set of all terms of signature σ in the variables $a_1, \cdots, a_m$. We call two terms $\mathfrak{a}$ and $\mathfrak{b}$ of $\mathfrak{A}_m$ equivalent if the relation

$$\mathfrak{a}(x_1, \ldots, x_m) = \mathfrak{b}(x_1, \ldots, x_m)$$

is an identity that is true on the class K. We denote by $\mathfrak{F}_m$ the set of equivalence classes of terms and we define on $\mathfrak{F}_m$ operations $f_1, \cdots, f_s$ by

$$f_i([\mathfrak{a}_1], \ldots, [\mathfrak{a}_{n_i}]) = [F(\mathfrak{a}_1, \ldots, \mathfrak{a}_{n_i})],$$

where $[\mathfrak{a}_j]$ is the class of terms equivalent to $\mathfrak{a}_j$. The algebra $\mathfrak{F}_m$ with the basic set $\mathfrak{F}_m$ and the operations $f_1, \cdots, f_s$ is called the *free algebra* for K of rank m with the generators $a_1, \cdots, a_m$. From the construction of $\mathfrak{F}_m$ it is clear that the problem of equivalence of words in $\mathfrak{F}_m$ and the problem of identical relations of rank m on the class K come to one and the same thing.

Let $\mathfrak{S}$ be a system of identities of given signature σ and let K be any class of algebras of signature σ. By $\mathrm{K}(\mathfrak{S})$ we denote the system of all those K-algebras on which all the identities in $\mathfrak{S}$ are true. The class of algebras K_1 is called a *variety* of K-algebras if $\mathrm{K}_1 = \mathrm{K}(\mathfrak{S})$ for a suitable system of identities $\mathfrak{S}$. The class K_1 is called a *finitely-defined* variety of K-algebras if $\mathrm{K}_1 = \mathrm{K}(\mathfrak{S})$, where $\mathfrak{S}$ is a *finite* system of identities. The class K_1 is called a variety of *rank* m of K-algebras if $\mathrm{K}_1 = \mathrm{K}(\mathfrak{S})$, where $\mathfrak{S}$ is a system of identities each of which involves not more than m variables. Finally, if K is the class of all algebras of a given signature, then a variety of K-algebras is called an *absolute* variety, or simply a variety of algebras (of the given signature). Absolute finitely-defined varieties are, for example, the class of groups, of abelian groups, of rings, of associative rings, of Lie rings, of lattices, of semigroups, and so on. Algebras that are free for varieties belong to the corresponding varieties. In all the varieties listed above (groups, commutative groups, rings and so on) the free algebras are constructive; that is, they have an algorithmically solvable word problem. Apparently finitely-defined absolute varieties of algebras with an unsolvable word problem for free algebras do not occur in the literature. Below we construct finitely-defined absolute varieties of rank 1 of algebras of the simplest signatures for which the word problem in free algebras of rank 1 is insolvable.

§2. Algebras with unary operations

In a certain sense the simplest algebras are those whose signature consists only of a single unary operation $F(x)$. According to Ehrenfeucht the class of all algebras of this signature has a solvable elementary theory, and therefore the problem of identical relations for any finitely-axiomatizable class of algebras with a single unary operation is solvable. In particular, the problem of identical relations for every finitely-defined variety of algebras with a single unary operation is trivially solvable.

For algebras with two unary operations this situation changes radically.

Theorem 1. *There exists a finitely-defined variety of rank* 1 *of algebras with two unary operations for which the problem of identical relations in one variable is algorithmically insolvable.*

According to the theorem of Post-Markov (see [1]) there exists a semigroup $\mathfrak{C}$ given by a suitable system of generating elements $c_1, \cdots, c_r$ and the defining relations

$$\mathfrak{a}_i = \mathfrak{b}_i \quad (i = 1, \ldots, n), \tag{1}$$

where $\mathfrak{a}_i$ and $\mathfrak{b}_i$ are certain words in the alphabet $c_1, \cdots, c_r$ for which the problem of equivalence of words is algorithmically insolvable. Let a_1 and a_2 be new symbols. In the relation (1) we substitute for the letters c_k the corresponding words

$$c_k = a_1 a_2 a_1^{k+1} a_2^{k+1} \quad (k = 1, \ldots, r). \tag{2}$$

After this substitution the relation (1) goes over into a relation of the form

$$a_{\alpha(i,1)}\, a_{\alpha(i,2)} \ldots a_{\alpha(i,p_i)} = a_{\beta(i,1)} \ldots a_{\beta(i,q_i)}, \tag{3}$$

where $\alpha(i, j), \beta(i, j) = 1$ or 2; $i = 1, \cdots, n$. The relations (3) form a semigroup $\mathfrak{P}$ with the generators a_1 and a_2. It is clear that the formulas (2) determine an isomorphic embedding of $\mathfrak{C}$ in $\mathfrak{P}$ (see [2]). Since the problem of equivalence of words in $\mathfrak{C}$ is algorithmically insolvable, the problem of equivalence of words in $\mathfrak{P}$ is also algorithmically insolvable.

Now we consider a class of algebras with two unary basic operations A_1 and A_2. With every defining relation (3) we associate the identity

$$A_{\alpha(i,1)}(A_{\alpha(i,2)}\ldots(A_{\alpha(i,p_i)}(x))\ldots) = A_{\beta(i,1)}(\ldots(A_{\beta(i,q_i)}(x))\ldots) \tag{4}$$

in the variable x of signature A_1, A_2. Let $\mathfrak{M}$ denote the variety of those algebras of signature A_1, A_2 on which all the identities (4) are true.

Any identical relation of rank 1 in the signature A_1, A_2 has either the form

$$A_{\alpha_1}(A_{\alpha_2}\ldots(A_{\alpha_p}(x))\ldots) = A_{\beta_1}(\ldots(A_{\beta_q}(x))\ldots), \tag{5}$$

or the form

$$A_{\alpha_1}(A_{\alpha_2}\ldots(A_{\alpha_p}(x))\ldots) = x.$$

It is easy to verify that *the identity* (5) *is true on the variety* $\mathfrak{M}$ *if and only if the equation*

$$a_{\alpha_1}a_{\alpha_2}\ldots a_{\alpha_p} = a_{\beta_1}a_{\beta_2}\ldots a_{\beta_q} \tag{6}$$

is true in the semigroup $\mathfrak{P}$.

For let $\mathfrak{A}$ be an arbitrary algebra in $\mathfrak{M}$. Every unary operation defined on $\mathfrak{A}$ is a mapping of $\mathfrak{A}$ into $\mathfrak{A}$. All the mappings of $\mathfrak{A}$ into $\mathfrak{A}$ form a semigroup under the usual multiplication of mappings. In this semigroup we consider the subsemigroup $\mathfrak{P}^*$ generated by the mappings A_1 and A_2. The identities (4) indicate that in $\mathfrak{P}^*$ the elements A_1 and A_2 are connected by the relation (3). If the equation (6) is true in $\mathfrak{P}$, then it is also true in every semigroup in which the elements a_1 and a_2 are connected by the relations (3). Therefore in $\mathfrak{P}^*$ we have the true equation

$$A_{\alpha_1}A_{\alpha_2}\ldots A_{\alpha_p} = A_{\beta_1}A_{\beta_2}\ldots A_{\beta_q},$$

which is equivalent to the identity (5). Thus, (5) follows from (6).

Suppose, conversely, that (5) is true on $\mathfrak{M}$. We adjoin to $\mathfrak{P}$ an external unit e; that is, we take $e \notin \mathfrak{P}$, and we set $ee = e$ and $ex = xe = x$ for $x \in \mathfrak{P}$. We denote the new semigroup $\mathfrak{P} \cup \{e\}$ by $\mathfrak{P}_e$. On $\mathfrak{P}_e$ we define operations A_1 and A_2 by setting

$$A_1x = a_1x, \quad A_2x = a_2x. \tag{7}$$

We consider the algebra $\mathfrak{A}_e$ with the basic set $\mathfrak{P}_e$ and the operations A_1 and A_2. From (3) it follows that the identities (4) are true on $\mathfrak{A}_e$, and therefore $\mathfrak{A}_e$ belongs to $\mathfrak{M}$. By assumption the identity (5) is true on $\mathfrak{M}$ and consequently also on $\mathfrak{A}_e$. When we set $x = e$ in it and use (7), we obtain (6), as required.

Thus, on the variety $\mathfrak{M}$, which is defined by a finite number of identities of rank 1, the problem of identical relations of rank 1 is algorithmically insolvable and Theorem 1 is proved. However, the variety $\mathfrak{M}$ has a further property which we shall need later.

Lemma. *Let $\mathfrak{M}$ be the variety of algebras we have constructed above with the basic operations A_1 and A_2 satisfying the identities* (4). *Denote by $\mathfrak{M}_0$ the class of infinite algebras in $\mathfrak{M}$ in which there exists an element* 0 *such that for any elements x and y*

α) $A_2^{i+1}x = x \leftrightarrow x = 0,$

β) $A_1A_2^ix = x \leftrightarrow x = 0,$

γ) $A_1x = A_2y \leftrightarrow x = y = 0 \qquad (i = 0, 1, \ldots),$

δ) $A_2x = A_2y \leftrightarrow x = y,$

ε) $A_1A_2^{i+1}x = A_1x \leftrightarrow x = 0.$

Then every identity of the form (5) *that is true on all $\mathfrak{M}_0$-algebras is true on all $\mathfrak{M}$-algebras, and therefore the problem of identical relations on the class $\mathfrak{M}_0$ is algorithmically insolvable.*

Suppose that the identity (5) is true on every algebra of the class $\mathfrak{M}_0$. We consider the algebra $\mathfrak{A}_e$ constructed in the process of proving Theorem 1. We adjoin to it a new element 0 and set $A_10 = A_20 = 0$. We denote the extended algebra by $\mathfrak{A}_0$. For $x = 0$ the relations (4) are trivially true. Therefore the identities (4) are true on $\mathfrak{A}_0$, and $\mathfrak{A}_0 \in \mathfrak{M}$. We show that $\mathfrak{A}_0$ has the properties α)–ε).

The elements of $\mathfrak{A}_0$ are 0, e and the elements of the semigroup $\mathfrak{P}$, which can be represented by nonempty words in the alphabet a_1, a_2. Here two words are equivalent, that is, represent one and the same element of $\mathfrak{P}$, if and only if one of them can be carried into the other by the elementary transformations corresponding to the defining relation (3). Let us call a word in the alphabet a_1, a_2 *regular* if it splits (graphically) into the compositum of words of the form $a_1a_2a_1^{k+1}a_2^{k+1}$ ($k > 0$). The pecularity of the relations (3) consists in the fact that their left-hand and right-hand sides are regular words. Hence it follows, in particular, that a regular word can be equivalent only to another regular word. Moreover, every letter of any word can occur in not more than one subword of the form $a_1a_2a_1^{k+1}a_2^{k+1}$. Therefore every word in the alphabet a_1, a_2 splits uniquely into the compositum of its maximal regular pieces, which are joined by irregular segments not containing subwords of the form $a_1a_2a_1^{k+1}a_2^{k+1}$. In the process of elementary transformations only the maximal regular pieces are changed and the irregular segments remain unchanged.

Let us verify property γ). Let $A_1x = A_2y$ and $x \neq 0$ or $y \neq 0$. Then $x \neq 0$ and $y \neq 0$ and therefore $a_1x = a_2y$. Under elementary transformation the initial letter of a word does not change. Therefore the equivalence $a_1x = a_2y$ is impossible, and $x = y = 0$.

Let $A_2x = A_2y$. If $x = 0$ or $y = 0$, then $x = y = 0$. But if $x \neq 0$ and $y \neq 0$, then we have $a_2x = a_2y$ in $\mathfrak{P}_e$. But the letter a_2 standing at the beginning of a word cannot be affected by elementary transformations. Therefore $x = y$, and property δ is true. Finally, if $A_1x = A_1A_2x$, $x \neq 0$, then $a_1x = a_1a_2x$ in $\mathfrak{P}_e$. The left-hand and right-hand sides of the defining relations (3) are words of even length. Hence the difference of the lengths of equivalent words is an even number. The difference of the lengths of the words a_1x and a_1a_2x is 1 and therefore the equivalence $a_1x = a_1a_2x$ is impossible in $\mathfrak{P}_e$. The remaining properties are checked similarly.

Thus, the algebra $\mathfrak{A}_0$ belongs to the class $\mathfrak{M}_0$ and hence the identity (5) is true on $\mathfrak{A}_0$. When we set $x = e$ in it, we obtain the equation (6), from which, as we have shown above, it follows that the identity (5) is true on every algebra of the variety $\mathfrak{M}$.

§3. Partial quasigroups

An algebra Q with a single binary operation $\circ$ is called a *quasigroup* if for any a, x, y in Q we have the cancellation rules

$$a \circ x = a \circ y \to x = y, \tag{8}$$

$$x \circ a = y \circ a \to x = y \tag{9}$$

and if the equations

$$a \circ x = b, \quad y \circ a = b$$

are solvable in Q for arbitrary a, b, in Q.

An element 0 of a quasigroup Q is called a zero in Q if

$$x \circ 0 = 0 \circ x = x \quad (x \in Q). \tag{10}$$

A set Q containing a fixed element 0 and equipped with a *partial* binary operation $\circ$ is called a *partial* quasigroup with 0 relative to the operation $\circ$ if the products $x \circ 0$ and $0 \circ x$ are defined in Q for all x and are equal to x, and if the cancellation rules (8) and (9) hold in Q, provided that all the products occurring in them are defined in Q.

A quasigroup with the operation $\circ$ is called *commutative* if the identity $x \circ y = y \circ x$ holds on it. A partial quasigroup is called *commutative* if $y \circ x$ is defined whenever $x \circ y$ is defined, and $x \circ y = y \circ x$.

Theorem 2. *Let Q_0 be a countable commutative partial quasigroup with 0 satisfying the conditions:*

1) *for every $a \neq 0$ the set of those x for which $a \circ x$ or $x \circ a$ is defined is finite;*

2) *for every $r \in Q_0$ there exist infinitely many $p \in Q_0$ such that the equation $p \circ x = r$ is not solvable in Q_0.*

Then the operation $\circ$ can be extended to an everywhere defined operation on Q_0 such that Q_0 becomes, relative to the new operation, a commutative quasigroup with 0.

Without loss of generality we may assume that the elements of Q_0 are nonnegative integers and that the number 0 is the zero element of Q_0. We arrange all pairs of nonnegative integers in the form of a sequence, for example in the following way:

$$(0,0),\ (0,1),\ (1,0),\ (0,2),\ (1,1),\ \ldots. \tag{11}$$

Let M_0 be the set of those pairs (x, y) for which the product $x \circ y$ is defined in the partial semigroup Q_0. Next we extend the set M_0 in several steps by supplementing the definition of the operation $\circ$ suitably and seeing to it that the set Q_0 with the extended operation remains a partial commutative quasigroup satisfying the conditions 1) and 2).

Suppose, then, that after the nth step we have obtained a set of pairs M_n for which the product $\circ$ is defined, and that the extended partial commutative quasigroup Q_n with the domain of definition M_n satisfies the conditions 1) and 2). In the sequence (11) we take the first pair (a, b) that does not occur in M_n. From the commutativity of Q_n and the properties of the zero element it follows that $(b, a) \notin M_n$, $a \neq 0$, $b \neq 0$. According to 1), the set of those x in Q_n for which $a \circ x$ or $x \circ a$ is defined is finite. Therefore we can find a number c different from all values assumed by the products $a \circ x$ and $x \circ b$ in Q_n. Setting $a \circ b = b \circ a = c$, we extend the operation $\circ$ and obtain a partial commutative quasigroup Q_n^* with the domain of definition $M_n^* = M_n \cup \{(a, b), (b, a)\}$. The conditions 1) and 2) clearly hold in Q_n^*.

Now we look for the first pair (p, r) in the sequence (11) for which the equation $p \circ x = r$ is insolvable in Q_n^*. By virtue of 1) and 2) there exists an element q such that the equation $q \circ x = r$ is insolvable and $p \circ q$ is not defined in Q_n^*. We extend the operation $\circ$ by setting $p \circ q = q \circ p = r$. Let Q_{n+1} denote the new partial quasigroup with the domain of definition $M_{n+1} = M_n^* \cup \{(p, q), (q, p)\}$. The conditions 1) and 2) obviously hold in Q_{n+1}.

So we have obtained a sequence of partial quasigroups $Q_0, Q_1, Q_2, \cdots$, defined on the set of nonnegative integers, which extend the operation $\circ$ step by step. The limit algebra Q is the required commutative quasigroup.

We observe that Theorem 2 is also true for arbitrary (noncommutative) quasigroups. It can easily be generalized to the noncountable case.

§4. Varieties of quasigroups

In §2 we have considered algebras with two unary operations. Apart from these, in a certain sense the simplest are the algebras with a single binary operation; that is, groupoids. The situation for groupoids turns out to be the same as for algebras with unary operations.

Theorem 3. *There exists a finitely-defined variety of rank* 1 *of groupoids* $\mathfrak{G}$ *such that the problem of identical relations of rank* 1 *is algorithmically insolvable for every class of groupoids contained in* $\mathfrak{G}$ *and containing the set* $\mathfrak{G}_0$ *of all infinite commutative quasigroups with zero belonging to* $\mathfrak{G}$.

We consider the system of identities (4) that occurs in the lemma of §2. In every identity (4) we replace the expressions of the form $A_2(z)$, $A_1(z)$ successively by terms of signature $\circ$ by means of the formulas

$$A_2(z) = z \circ z, \tag{12}$$

$$A_1(z) = (z \circ z) \circ z. \tag{13}$$

For example, the expression $A_2(A_1(x))$ goes over under these transformations into the term

$$((x \circ x) \circ x) \circ ((x \circ x) \circ x). \tag{14}$$

As a result we obtain from the identities (4) identities of the form

$$G_i(x) = H_i(x), \tag{15}$$

where G_i and H_i are terms in the variable x of signature $\circ$. We denote by $\mathfrak{G}$ the variety of groupoids satisfying the identities (15), and by $\mathfrak{G}_0$ the class of all commutative quasigroups with zero contained in $\mathfrak{G}$.

We choose an arbitrary identity of the form (5). Transforming it in the manner indicated we obtain an identity of the form

$$F(x) = G(x). \tag{16}$$

To prove Theorem 4 it is sufficient to show that 1) the validity of (16) on the class $\mathfrak{G}_0$ implies the validity of the identity (5) on the class $\mathfrak{M}_0$ of algebras with the operations A_1 and A_2; and 2) the validity of (5) on the class $\mathfrak{M}_0$ implies the validity of the identity (16) on the variety $\mathfrak{G}$.

We begin with the second assertion. Suppose that the identity (5) is true on the class $\mathfrak{M}_0$ (in the lemma of §2). According to this lemma it is then also true on the variety $\mathfrak{M}$ defined by the identities (4). We choose an arbitrary groupoid G in the variety $\mathfrak{G}$. Defining on G new operations A_1 and A_2 by the formulas (12) and (13) we turn G into an algebra of the signature A_1, A_2. Since the identities (15) are true on G, the identities (4) are true on the algebra G; that is, G belongs to the variety $\mathfrak{M}$, and therefore the identity (5) is true on G, which indicates, by (12) and (13), that the identity (16) is true on the groupoid G.

The first assertion is somewhat more complicated to prove. Suppose that the identity (16) is true on the class $\mathfrak{G}_0$, and let $\mathfrak{A}$ be an arbitrary algebra of the class $\mathfrak{M}_0$ satisfying the conditions α)–ϵ) of the lemma in §2. In particular, $\mathfrak{A}$ contains an element 0 satisfying the conditions listed. On the set of elements of $\mathfrak{A}$ we define a partial operation $\circ$ by means of (12) and (13) and the conventions

$$z \circ (z \circ z) = A_1 z, \tag{17}$$

$$0 \circ z = z \circ 0 = z. \tag{18}$$

From the properties α)–ϵ) it follows that (12), (13), (17) and (18) do not contradict each other. For (12) determines the product for the diagonal pairs (z, z) and for these pairs it is in accordance with (18), because $A_2 0 = 0$. The pair $(z \circ z, z)$ can only be diagonal when $z \circ z = z$, that is, $A_2 z = z$. By property α) it follows that $z = 0$, and by β) that $A_1 0 = 0$. Finally, the pair $(z \circ z, z)$ for $z \neq 0$ cannot be the pair $(0, z)$ by the same property α). Thus, by means of (12), (13), (17) and (18) a certain *partial groupoid* $\mathfrak{Q}_0$ is defined on the set of elements of $\mathfrak{A}$. This partial groupoid is commutative and contains a zero element 0. Now we verify the cancellation laws. Suppose that for certain a, x, y we have $a \circ x = a \circ y$ in Q_0. If $a = 0$, then $x = y$. Therefore we assume further that $a \neq 0$. The product $a \circ x$ is defined in Q_0 only in the four cases

$$x = 0, \ x = a, \ x = a \circ a, \ a = x \circ x.$$

Similarly, the product $a \circ y$ is defined only in the cases

$$y = 0, \ y = a, \ y = a \circ a, \ a = y \circ y. \tag{19}$$

We have to discuss all possible combinations of these cases. If $x = 0$, then in the last three cases of (19) we obtain, respectively,

$$a = a \circ a = A_2 a, \ a = A_1 a, \ a = A_1 y = A_2 y,$$

that is, by properties α)–γ), $y = a = 0$. Similarly we obtain in all the remaining cases that $x = y$. Thus Q_0 is a partial commutative quasigroup with zero 0.

It is easy to verify that Q_0 satisfies the conditions 1) and 2) of Theorem 2. For suppose that $a \in Q_0$, $a \neq 0$. Then the product $a \circ y$ is defined only in the cases (19). But the equation $y \circ y = A_2 y = a$, by δ), has not more than one solution. Therefore there exist not more than four values of y for which $a \circ y$ is defined in Q_0, and condition 1) is true for Q_0.

On the other hand, let $r \in Q_0$ be an arbitrary element. If $r = 0$, then the equation $p \circ x = r$ is impossible in Q_0 for any $p \neq 0$. Let $r \neq 0$. We consider the elements $A_2 r, A_2^2 r, \cdots$. By α) all these elements are distinct. The equation $(A_2^{i+1} r) \circ y = r$ can have a solution y only in the cases (19). But in these cases we obtain, respectively,

$$A_2^{i+1} r = r, A_2^{i+2} r = r, A_2 y = A_2^{i+1} r \rightarrow y = A_2^i r, A_1 A_2^i r = r,$$

from which we see by α) and β) that $r = 0$.

Thus the partial commutative quasigroup Q_0 satisfies the conditions of Theorem 2, and therefore its definition can be supplemented to make it a commutative quasigroup Q with a zero element 0. The identity (4) holds on $\mathfrak{A}$. Therefore, by the relations (12) and (13) the identity (15) holds on Q; that is, $Q \in \mathfrak{G}_0$, and so the identity (16) is true on Q. By virtue of the same relations (12) and (13) this means that the identity (5) holds on $\mathfrak{A}$, as required.

A *loop* (primitive loop) is an algebra with an operation of multiplication and operations of division $/, \backslash$ connected by the identities

$$(xy)/y = y\backslash(yx) = y(y\backslash x) = (x/y)y = x, \quad x\backslash x = y/y.$$

A loop satisfying the identity $xy = yx$ is called commutative.

Every loop is a quasigroup with zero element under the operation of multiplication. Conversely, on every quasigroup with zero element operations of division can be defined so as to make it into a loop. Therefore an immediate consequence of Theorem 4 is

Corollary. *There exists a finitely-defined variety of rank* 1 *of loops* L *such that the problem of the truth of identities of rank* 1 *is algorithmically insolvable on every class of loops contained in* L *and containing all the commutative loops of* L.

In particular, the problem of the truth of identities of rank 1 is algorithmically insolvable on the variety of all commutative loops contained in L.

Apart from the question of the algorithmic solvability of the problem of identical relations on every finitely-defined variety of groups it would be interesting to solve the question of the existence of a finitely-defined variety of loops in which the finite loops form a class with an algorithmically insolvable problem of identical relations.

BIBLIOGRAPHY

[1] A. A. Markov, *The theory of algorithms*, Trudy Mat. Inst. Steklov. 42 (1954); English transl., Israel Program for Scientific Translations, Jerusalem, 1961. MR 17, 1038; MR 24 #2527.

[2] Marshall Hall, Jr., *The word problem for semigroups with two generators*, J. Symbolic Logic 14 (1949), 115–118. MR 11, 1.

Translated by:
K. A. Hirsch

ON A THEOREM ON INFINITE DIMENSIONALITY OF AN ASSOCIATIVE ALGEBRA

UDC 519.4

È. B. VINBERG

In the case when the defining relations of an associative algebra are not homogeneous, a weaker sufficient condition for infinite dimensionality is found than in the recent article by E. S. Golod and I. R. Šafarevič.

In article [1] E. S. Golod and I. R. Šafarevič proved that if an associative algebra A (over an arbitrary field) with g generators is defined by homogeneous relations of degree ≥ 2, in which the number of relations of degree n does not exceed r_n and the formal series

$$\left(1 - gt + \sum_{n=2}^{\infty} r_n t^n\right)^{-1} \tag{1}$$

has nonnegative coefficients, then the algebra A is infinite dimensional. The results of article [1], and also E. S. Golod's article [2], are based on this proposition.

The purpose of the present article is to prove an analogous assertion for the case when the defining relations of the algebra A are not necessarily homogeneous. We shall call the least degree of the terms entering into a relation the *degree of the relation.*

Theorem 1. *Let an associative algebra A with g generators be defined by relations of degree ≥ 2, where the number of relations of degree n does not exceed r_n. If the formal series*

$$(1 - t)\left(1 - gt + \sum_{n=2}^{\infty} r_n t^n\right)^{-1} \tag{2}$$

has nonnegative coefficients, then the algebra A is infinite dimensional.

Proof. Let L be the free associative algebra with g generators, i.e. the algebra of noncommuting polynomials in g variables. Let L_n be the subspace of L generated by the homogeneous polynomials of degree n. It is obvious that $\dim L_n = g^n$. We set

$$L^{(n)} = \sum_{k=n}^{\infty} L_k.$$

The defining relations of the algebra A can be considered as elements of the algebra L. We let R designate their linear span. The algebra A is isomorphic to the quotient algebra L/I, where $I = LRL$ is the two-sided ideal generated by R.

We set $I^{(n)} = I \cap L^{(n)}$ and choose the subspaces $B_n \subset L_n$ $(n = 0, 1, 2, \cdots)$ in a way so that

$$L^{(n)} = I^{(n)} \dotplus B_n \pmod{L^{(n+1)}}.$$

Let $B = \sum_{n=0}^{\infty} B_n$. Then for any n

$$L = I \dotplus B \pmod{L^{(n)}}. \tag{3}$$

(If the defining relations are homogeneous we have exact equality.) We prove that $\dim B = \infty$, from which it follows that the algebra A is infinite dimensional. We note for future reference that

$$b_n = \dim B_n = g^n - \dim I^{(n)}/I^{(n+1)}. \tag{4}$$

The proof of the theorem is based on the following representation of the ideal I:

$$I = IL_1 + BR \pmod{L^{(n)}} \tag{5}$$

for any n. (If the defining relations are homogeneous we have exact equality.) This representation is obtained from noting that

$$I = LRL = (LRL)\, L_1 + LR = IL_1 + LR$$

and, on the other hand, that for any n

$$LR = IR + BR \subset ILL_1 + BR = IL_1 + BR \pmod{L^{(n)}}.$$

We consider the case when the defining relations of the algebra A are homogeneous. Then

$$R = \sum_{n=2}^{\infty} R_n, \quad I = \sum_{n=2}^{\infty} I_n,$$

where

$$R_n = R \cap L_n, \quad I_n = I \cap L_n,$$

and it follows from (5) that

$$I_n = I_{n-1} L_1 + \sum_{k=2}^{n} B_{n-k} R_k.$$

Hence

$$\dim I_n \leqslant (\dim I_{n-1})\, g + \sum_{k=2}^{n} b_{n-k}\, r_k.$$

Since $b_n = g^n - \dim I_n$ (see (4)), we have

$$b_n \geqslant g^n - (\dim I_{n-1})\, g - \sum_{k=2}^{n} b_{n-k} r_k = b_{n-1} g - \sum_{k=2}^{n} b_{n-k} r_k.$$

Thus, in the case of homogeneous relations

$$\delta_n = b_n - b_{n-1} g + \sum_{k=2}^{n} b_{n-k}\, r_k \geqslant 0. \qquad (6)$$

Using generating functions, this fact can be written as follows:

$$\left(\sum_{n=0}^{\infty} b_n t^n\right)\left(1 - gt + \sum_{n=2}^{\infty} r_n t^n\right) \geqslant 1,$$

where the inequality is interpreted coefficient-wise. If series (1) has nonnegative coefficients, then, multiplying both sides of the inequality by it, we obtain

$$\sum_{n=0}^{\infty} b_n t^n \geqslant \left(1 - gt + \sum_{n=2}^{\infty} r_n t^n\right)^{-1} \geqslant 1. \qquad (7)$$

As was shown in [1], series (1) cannot be a finite series having nonnegative coefficients. It follows from (7) that

$$\dim B = \sum_{n=0}^{\infty} b_n = \infty. \qquad (8)$$

Now suppose the defining relations of the algebra A are not necessarily homogeneous. We set $R^{(n)} = R \cap L^{(n)}$. It follows from the condition of the theorem that

$$\dim R/R^{(n+1)} \leqslant \sum_{k=2}^{n} r_k. \qquad (9)$$

Further, we find from (5) that

$$I/I^{(n+1)} = (I/I^{(n)})\, L_1 + \sum_{k=2}^{n} B_{n-k}\, (R/R^{(k+1)}). \qquad (10)$$

(The reader can easily make sense of the products in the right side.) Consequently,

$$\dim I/I^{(n+1)} \leqslant (\dim I/I^{(n)})\, g + \sum_{k=2}^{n} b_{n-k} \sum_{l=2}^{k} r_l.$$

Since by (4) we have

$$\sum_{k=1}^{n} b_k = \sum_{k=1}^{n} g^k - \dim I/I^{(n+1)},$$

it follows that

$$\sum_{k=1}^{n} b_k \geqslant \sum_{k=1}^{n} g^k - (\dim I/I^{(n)})\, g - \sum_{k=2}^{n} b_{n-k} \sum_{l=2}^{k} r_l$$

$$= \Big(\sum_{k=1}^{n} b_{k-1}\Big) g - \sum_{k=2}^{n} \Big(\sum_{l=2}^{k} b_{k-l} r_l\Big).$$

Using the notation (6), we can write this inequality in the form

$$\sum_{k=1}^{n} \delta_k \geqslant 0, \tag{11}$$

or, using generating functions, in the form

$$(1-t)^{-1} \sum_{n=1}^{\infty} \delta_n t^n \geqslant 0.$$

Since $\delta_0 = 1$, we finally obtain

$$\sum_{n=0}^{\infty} b_n t^n \cdot \frac{1 - gt + \sum_{n=2}^{\infty} r_n t^n}{1-t} \geqslant \frac{1}{1-t}. \tag{12}$$

If series (2) has nonnegative coefficients, multiplying both sides of inequality (12) by it leads to inequality (7) and then to (8). (We note that if series (2) has nonnegative coefficients, then series (1) has this property all the more.) The theorem is proved.

Keeping the earlier notation, we consider the closure $\bar{I}$ of the ideal I in the natural topology of the space L. (A polynomial $f \in L$ belongs to $\bar{I}$ if and only if for any n there can be found a polynomial $f_n \in I$ such that $f - f_n \in L^{(n)}$.) It is easily seen that

$$L = \overline{J \dotplus B}. \tag{13}$$

Thus, we have actually proved more than the infinite dimensionality of the algebra A. We proved that even the quotient algebra L/J is infinite dimensional; but if the algebra A is infinite dimensional from infinite dimensionality of J/I, our theorem does not detect this.

Let $\overline{L}$ be the algebra of noncommuting formal series in g generators, and let $\overline{L}^{(n)}$ be the subspace of $\overline{L}$ generated by series not containing terms of degree $< n$. Replacing L by $\overline{L}$ and $L^{(n)}$ by $\overline{L}^{(n)}$ in the proof of Theorem 1, and considering the preceding remark, we obtain the following stronger theorem.

Theorem 2. *Let the associative algebra A be represented in the form $\overline{L}/J$, where J is the ideal of the algebra $\overline{L}$, topologically generated by the subspace $R \subset \overline{L}^{(2)}$. Set $R^{(n)} = R \cap \overline{L}^{(n)}$. If* $\dim R^{(n)}/R^{(n+1)} \leq r_n$ *and series* (2) *has nonnegative coefficients, then the algebra A is infinite dimensional.*

We shall need the following simple lemma in what follows.

Lemma. *If the nonnegative integers r_n, s_n $(n = 2, 3, \cdots)$ are such that $\sum_{k=2}^{n} r_k \leq \sum_{k=2}^{n} s_k$ for any n and the series*

$$(1-t)(1-gt+\sum_{n=2}^{\infty} s_n t^n)^{-1}$$

has nonnegative coefficients, then series (2) *also has nonnegative coefficients.*

Proof. It follows from the condition of the lemma that

$$(1-t)^{-1}\sum_{n=2}^{\infty}(s_n - r_n)\,t^n \geqslant 0.$$

Therefore

$$(1-t)\left(1-gt+\sum_{n=2}^{\infty} r_n t^n\right)^{-1}$$

$$=(1-t)\left(1-gt+\sum_{n=2}^{\infty} s_n t^n\right)^{-1}\left(1-\frac{\sum_{n=2}^{\infty}(s_n-r_n)\,t^n}{1-gt+\sum_{n=2}^{\infty} s_n t^n}\right)^{-1} =$$

$$= \sum_{m=0}^{\infty} \left(\frac{1-t}{1-gt+\sum\limits_{n=2}^{\infty} s_n t^n} \right)^{m+1} \left(\frac{\sum\limits_{n=2}^{\infty} (s_n - r_n) t^n}{1-t} \right)^m \geqslant 0.$$

Corollary. *If under the conditions of Theorem* 1 *or* 2 *we have*

$$r = \sum_{n=2}^{\infty} r_n \leqslant \frac{1}{4} g^2,$$

then the algebra A is infinite dimensional.

Proof. We set $s_2 = \frac{1}{4} g^2$, $s_3 = s_4 = \cdots = 0$. It is obvious that the first condition of the lemma is then fulfilled. Therefore it suffices to verify that for $g \geq 2$

$$\frac{1-t}{1-gt+\frac{1}{4}g^2t^2} = \frac{1-t}{\left(1-\frac{1}{2}gt\right)^2} \geqslant 0.$$

We have

$$\frac{1-t}{\left(1-\frac{1}{2}gt\right)^2} = \frac{1}{1-\frac{1}{2}gt} + \frac{\left(\frac{1}{2}g-1\right)t}{\left(1-\frac{1}{2}gt\right)^2} \geqslant 0,$$

and the corollary is proved.

(In article [1], by replacing nonhomogeneous relations by their homogeneous components, it is shown that the algebra A is infinite dimensional if $r \leq \frac{1}{4}(g-1)^2$.

BIBLIOGRAPHY

[1] E. S. Golod and I. R. Šafarevič, *On class field towers*, Izv. Akad. Nauk SSSR Ser. Mat. 28 (1964), 261–272; English transl., Amer. Math. Soc. Transl. (2) 48 (1965), 91–102. MR 28 #5056.

[2] E. S. Golod, *On nil-algebras and finitely approximable p-groups*, Izv. Akad. Nauk SSSR Ser. Mat. 28 (1964), 273–276; English transl., Amer. Math. Soc. Transl. (2) 48 (1965), 103–106. MR 28 #5082.

Translated by:
N. Koblitz

INVARIANT MEASURES ON BOOLEAN ALGEBRAS

UDC 512.9

D. A. VLADIMIROV

Let X be a Boolean algebra. Under what conditions on this algebra can we define a countably additive essentially positive measure? If we consider the elements of the algebra as events, then our problem has the formulation: when does there exist on the set of events X a countably additive probability equal to zero only for an impossible event? The goal of this work is to establish some conditions for the existence of such a measure; we will call them conditions for normability.* A similar problem has been considered by many authors (see [6], [7]).

Our approach to the problem of normability will be connected with the study of the set of automorphisms of the algebra X. We will find out which properties of this collection imply the existence of the measure. Namely, such an approach is the basis of the so-called "classical determination of probability", if the set of events X is finite. For example, for throwing a homogeneous playing die we have the algebra X consisting of 2^6 elements and the probability is generated by the group of automorphisms of this algebra, isomorphic to the group of rotations of the cube: the probability is completely determined by the requirement that "congruent" events be equiprobable. This situation, as it turns out, is typical for other than finite algebras. We will see that under certain conditions a group of automorphisms of the given algebra X generates a measure on X that is invariant relative to the automorphisms of the initial group, and that every measure arises in such a manner.

In the axiomatic construction of the theory of probability the measure on the algebra of events is supposed given a priori. However, instead of postulating the existence of the measure, we can assume the existence of a group of automorphisms with sufficiently simple properties.

Establishing conditions under which a group of automorphisms of a given Boolean algebra generates a measure, we incidentally obtain proofs of certain theorems of ergodic theory relating to the existence of invariant measures (theorems of Hopf, Hahn-Kakutani, and others).

---

*Some results of this report were given without proofs in the author's notes [1], [2].

§1

In this report, a Boolean algebra, as usual, is a distributive lattice, containing a largest element 1 and a smallest element 0, where for each element x there is a unique complement Cx, such that $x \wedge Cx = 0$, $x \vee Cx = 1$. As is standard, we will denote by $x \vee y$ the supremum of x and y, and by $x \wedge y$ their infimum. For an arbitrary set $\{x_\alpha\}$ we use the symbols $\bigvee_\alpha x_\alpha$ and $\bigwedge_\alpha x_\alpha$. If $x \wedge y = 0$, we say that x and y are disjoint. The supremum of a set of pairwise disjoint elements is called their sum, and we write $+$ and Σ. The formula $z = x - y$ means that $x \geq y$ and $x = y + z$. Finally, by definition, $|x - y| = (x \wedge Cy) \vee (y \wedge Cx)$; this is the symmetric difference of x and y. (The choice of notation is justified by the fact that under embedding the algebra into the overlying K-space the symmetric difference is converted into the modulus of the usual difference.) As a rule, we will consider complete algebras, i.e. those in which any set has extrema, supremum and infimum. The only topology on a Boolean algebra that we will need is the order topology or (o)-topology. It is defined as follows. For any net $\{x_\alpha\}$ of elements of a complete Boolean algebra we set

$$\overline{\lim_\alpha}\, x_\alpha = \inf_\alpha \sup_{\beta \geqslant \alpha} x_\beta, \quad \underline{\lim_\alpha}\, x_\alpha = \sup_\alpha \inf_{\beta \geqslant \alpha} x_\beta.$$

If $\overline{\lim}_\alpha x_\alpha = \underline{\lim}_\alpha x_\alpha = y$ we say that the net $\{x_\alpha\}$ (o)-converges to the element y. We write $x_\alpha \xrightarrow{(o)} y$ or $(o)\text{-}\lim_\alpha x_\alpha = y$. A set F is called closed if it contains the limits of all (o)-convergent nets of its elements. With this we have defined the (o)-topology. For the notation of the convergence in this topology (we call it (t)-convergence) we use the symbols lim and $\longrightarrow$.

A measure on a Boolean algebra will always be a real function μ such that:

1) $\mu(x) > 0$ if $x \neq 0$;
2) if $x \wedge y = 0$, then $\mu(x + y) = \mu(x) + \mu(y)$;
3) if $x_n \geq x_{n+1}$ $(n = 1, 2, \cdots)$ and $x_n \longrightarrow 0$, then $\mu(x_n) \longrightarrow 0$;
4) $\mu(1) = 1$.

A complete algebra X is called *normable* if we can define on it a measure with properties 1)–4).

The (o)-topology on a normable Boolean algebra coincides with the topology of the metric space which is gotten if we define the distance between elements x and y of the given algebra by the formula $\rho(x, y) = \mu(|x - y|)$. (t)-convergence is convergence in measure.

Among the properties which are possessed by a normable Boolean algebra, note the following two:

(c). Any set of pairwise disjoint elements is at most countable.

(d). If $x_n^m \geq x_{n+1}^m$ $(n, m = 1, 2, \cdots)$ and $x_n^m \longrightarrow 0$ as $n \longrightarrow \infty$ $(m = 1, 2, \cdots)$, then there exists a "diagonal" sequence $x_{n_m}^m$ such that $(o)\text{-}\lim_m x_{n_m}^m = 0$.

A complete algebra satisfying conditions (c) and (d) is called regular. The notion of regularity was introduced by L. V. Kantorovič in 1936, for linear semiordered spaces ([8]; see also [3], [4]). The fact that a normable algebra is always regular is the basis of many facts in the theory of functions of a real variable, such as the theorems of N. N. Luzin, D. F. Egorov and others. There is no known example of a regular but not normable Boolean algebra (there are examples of nonnormable algebras possessing property (c); see for example [9]). However, an attempt to prove the existence of a measure on any regular algebra goes into great set-theoretical difficulties, as was shown in [6]:* this hypothesis is not weaker than the well-known hypothesis of Suslin.

Algebras satisfying condition (c) are called *algebras of countable type*. In such algebras every set contains an at most countable subset with the same supremum. (This is equivalent to (c).)

In an algebra of countable type a sequence $\{x_n\}$ $(n = 1, 2, \cdots)$ (t)-converges to y if and only if for any $n_1 < n_2 < \cdots$ there exist $k_1 < k_2 < \cdots$ such that $(o)\text{-}\lim_i x_{n_{k_i}} = y$.

All the necesssry facts on Boolean algebras can be found in the books [3], [4], [9] and [10]. We end this introductory section with two lemmas of a general character.

We agree to write $\overline{\lim}_n \operatorname{abs} x_n = y$ if $\overline{\lim}_k x_{n_k} = y$ for any subsequence $\{x_{n_k}\}$, $n_1 < n_2 < \cdots$.

Lemma 1. *Let X be a complete Boolean algebra of countable type. The sequence $\{x_n\}$ does not converge to zero in the order topology if and only if there exists a subsequence $\{x_{n_k}\}$, $n_1 < n_2 < \cdots$, with* $\overline{\lim}_k \operatorname{abs} x_{n_k} > 0$.

For the proof let us order the set of all strictly increasing sequences of natural numbers $\tau = \{n_k^\tau\}$ by the rule $\tau' \succ \tau''$ if there exists an $n_0 = n_0(\tau', \tau'')$ such that $\{n_k^{\tau'}\}_{k > n_0} < \{n_k^{\tau''}\}$ and at the same time $\overline{\lim}_k x_{n_k^{\tau'}} < \overline{\lim}_k x_{n_k^{\tau''}}$. Consider any ordered set $T = \{\tau\}$. Let $y = \inf_{\tau \in T} \overline{\lim}\, x_{n_k^\tau}$. In view of the countability of the type, there exist $\tau_1, \tau_2, \cdots$ such that $\tau_m \in T$ $(m = 1, 2, \cdots)$ and $y = \inf_m \lim_k x_{n_k^{\tau_m}}$.

We introduce the notation

---

* The terminology in [6] is somewhat different from ours.

$$\tau_k^* = \max_{s\leqslant k} \tau_s, \quad n_m = n_m^{\tau_m^*}, \quad \tau_0 = \{n_m\} \ (m = 1, 2, \dots).$$

It is easy to check that either $\tau_0 \succ \tau$ for any $\tau \in T$, or already among the elements of T there is a largest one in the sense of our order. By Zorn's Lemma, in the class of all sequences τ there is a maximal element $\tau^* = \{n_k^*\}$, $n_1 < n_2 < \cdots$. This means that $\overline{\lim}_k \, x_{n_k^*} = \overline{\lim}_s \, x_{n_{k_s}^*}$ for any $k_1 < k_2 < \cdots$. All that was said above is valid for any sequence $\{x_n\}$. Suppose now that $x_n \not\to 0$. This means that there exists a partial sequence $\{x_{n_k}\}$ from which we can extract no subsequence (o)-converging to zero. Applying the earlier arguments to it, we obtain the desired subsequence $\{x_{n_{k_i}}\}$ such that $\overline{\lim}_i$ abs $x_{n_{k_i}} > 0$. This proves the necessity; the sufficiency is clear.

Lemma 2. *Let X be a regular Boolean algebra, $\{x_{nm}\}$ $(m, n = 1, 2, \cdots)$ a double sequence of its elements satisfying the conditions*

1) $x_{nm} \xrightarrow[n\to\infty]{(o)} x_n \ (n = 1, 2, \cdots)$,

2) $\overline{\lim}_n$ abs $x_n = y$.

Then there exists a sequence $\{m_n\}$, $m_1 < m_2 < \cdots$ such that

$$\overline{\lim_n} \operatorname{abs} x_{nm_n} = y.$$

For the proof we introduce the notation $z_{nm} = |x_{nm} - x_n|$ and $u_{nm} = \bigvee_{s\geq m} z_{ns}$. From 1) it follows that $u_{nm} \searrow 0$ as $m \to \infty$, $n = 1, 2, \cdots$. Since the algebra satisfies condition (d) in the definition of regularity, there exists a sequence $\{u_{nm_n}\}$ (o)-converging to zero. Take any sequence $\{n_n\}$, $n_1 < n_2 < \cdots$. For each k we have $x_{n_k m_{n_k}} \leq x_{n_k} \vee u_{n_k m_{n_k}}$. Hence for all s we have

$$\bigvee_{k\geqslant s} x_{n_k m_{n_k}} \leqslant \bigvee_{k\geqslant s} x_{n_k} \vee \bigvee_{k\geqslant s} u_{n_k m_{n_k}}$$

and finally

$$\overline{\lim_k} \, x_{n_k m_{n_k}} = (o)\text{-}\lim_s \bigvee_{k\geqslant s} x_{n_k m_{n_k}} \leqslant (o)\text{-}\lim_s \bigvee_{k\geqslant s} x_{n_k} \vee 0 = \overline{\lim_k} \, x_{n_k} = y.$$

(We have used the "(o)-continuity" of the operation $\vee$, which holds in any algebra.)

Exchanging in this argument $\{x_{n_k m_{n_k}}\}$ and $\{x_{n_k}\}$, we get $\overline{\lim} \, x_{n_k m_{n_k}} \geq y$. Thus $\overline{\lim}_k \, x_{n_k m_{n_k}} = y$ for any sequence $\{n_k\}$ $(n_1 < n_2 < \cdots)$. And this means that $\overline{\lim}_n$ abs $x_{nm_n} = y$. The lemma is proved.

§ 2

An *automorphism* of a Boolean algebra X is a one-to-one mapping A of X onto itself which preserves the order (this means that $x < y$ and $Ax < Ay$ are equivalent). It is clear that every automorphism is continuous relative to the (o)-topology.

Let $\mathfrak{A}$ be some group of automorphisms of a complete Boolean algebra X. Let us list a series of important conditions relative to the properties of this group.

(A_1'). For any $x > 0$, $\bigvee_{A \in \mathfrak{A}} Ax = 1$.

(A_1''). For any x and $y \neq 0$, there is an $A \in \mathfrak{A}$ such that $Ax \wedge y > 0$.

(A_2). For any $x \neq 0$ there are a finite number of automorphisms $A_1, A_2, \cdots, A_n \in \mathfrak{A}$ such that $\bigvee_{i=1}^n A_i x = 1$.

(A_1'''). If $Ax = x$ for all $A \in \mathfrak{A}$, then either $x = 0$ or $x = 1$ (the "ergodicity" of the group).

These conditions (call them conditions of type (A)) say that $\mathfrak{A}$ is sufficiently rich. It is clear that (A_1') is equivalent to (A_1'') and that (A_2) is stronger than the preceding ones. It is easy to check also that (A_1''') is equivalent to (A_1') and (A_1'').

Let us mention briefly also one property equivalent to (A_2). It is well known that X is isomorphic to the algebra of all open-closed sets of a certain bicompact Q (see [3]). With each automorphism of X there is quite naturally connected some homeomorphism of the bicompact, and with the group $\mathfrak{A}$, a group of homeomorphisms $\mathfrak{A}'$. It is easy to show that (A_2) is equivalent to the following: for any $q \in Q$ the set $\{Cq\}$, $C \in \mathfrak{A}'$, is everywhere dense in Q.

Note one obvious but important fact. Let μ be a finitely additive function on X, nonnegative for all x. Further, let μ be invariant relative to all automorphisms of the group $\mathfrak{A}$ satisfying (A_2). (This means that $\mu(Ax) = \mu(x)$ for all x and A.) Then μ either is identically zero or is essentially positive, i.e. $\mu(x) = 0$ implies $x = 0$.

Before formulating new conditions, we introduce some terminology and concepts important later. Elements x and y are called *congruent* (if need be, $\mathfrak{A}$-congruent) if $y = Ax$ for some $A \in \mathfrak{A}$. We call x and y *equicomposite* ($\mathfrak{A}$-equicomposite) if they can be written in the form

$$x = \sum_{\alpha} x_\alpha, \quad y = \sum y_\alpha, \tag{$*$}$$

where x_α and y_α are congruent for every α in the set of indices (which can have any power). If in $(*)$ both sums are finite, we say that x and y are *finitely equicomposite*. Congruency is denoted by $\approx$ or $\approx_{\mathfrak{A}}$, equicomposability by $\sim$ or $\sim_{\mathfrak{A}}$, finite equicomposability by $\simeq$ and $\simeq_{\mathfrak{A}}$. These relations are reflexive, symmetric and transitive. Let us show only the transitivity of $\sim$. Let $x \sim y$ and $y \sim z$. This means that $x = \sum_\alpha x_\alpha$ and $y = \sum_\alpha y_\alpha$, where $y_\alpha = A_\alpha x_\alpha$. At the same time $y = \sum_\beta y_\beta'$ and $z = \sum_\beta z_\beta$, where $y_\beta' = A_\beta' z_\beta$. Set

$$y_{\alpha\beta} = y_\alpha \wedge y'_\beta,\ z_{\alpha\beta} = A'^{-1}_\beta y_{\alpha\beta},\ x_{\alpha\beta} = A^{-1}_\alpha y_{\alpha\beta}.$$

It is clear that

$$x = \sum_{\alpha,\beta} x_{\alpha\beta},\ z = \sum_{\alpha,\beta} z_{\alpha\beta} \text{ and } z_{\alpha\beta} = A'^{-1}_\beta A_\alpha x_{\alpha\beta},\ \text{i.e., } x \sim z.$$

We continue now with our listing of basic conditions.

(B_1). $x_n \to 0$ always implies that $A_n x_n \to 0$, for any $A_n \in \mathfrak{A}$.

(B_2). $x_n \approx x_{n+1}$ $(n = 1, 2, \cdots)$ and $x_n \wedge x_m = 0$ $(n \neq m)$ implies that $x_1 = x_2 = \cdots = 0$.

(B_3). If $x_n = \Sigma_\alpha x_{n\alpha}$, $x_n \to 0$, then always $\bigvee_\alpha A_{n\alpha} x_{n\alpha} \to 0$ for any $A_{n\alpha} \in \mathfrak{A}$. (Here the index α runs over any set depending on n.)

(B_4). The relations $x < y$ and $x \sim y$ are simultaneously impossible.

These conditions (call them conditions of type (B)) relate to the kind of equipotent continuity of the automorphisms in the group $\mathfrak{A}$ (with the peculiarity (B_1)). While conditions of type (A) hold for extensions of the group, conditions of type (B), conversely, hold for restrictions of the group; they say that there are "not too many" automorphisms in $\mathfrak{A}$.

It is obvious that (B_3) is stronger than (B_1), while (B_1), in turn, is stronger than (B_2). If in the formulation of (B_3) we change the sequence $\{x_n\}$ to an arbitrary net $\{x_\gamma\}$ then we get a still stronger condition which we denote (B_3^+). It is easy to show that if the algebra X is regular, then (B_3) is equivalent to (B_3^+); however in this work we will not use this remark.

Lemma 3. *(B_3) implies (B_4).*

Proof. Suppose (B_4) is not satisfied. Then there exist x and y such that $x < y$ and

$$x = \sum_\alpha x_\alpha,\ y = \sum_\alpha y_\alpha,\ y_\alpha = A_\alpha x_\alpha,\ A_\alpha \in \mathfrak{A}.$$

Set

$$y'_\alpha = y_\alpha \wedge x,\ x'_\alpha = A^{-1}_\alpha y'_\alpha,\ y''_\alpha = y_\alpha \wedge (y - x),$$

$$x''_\alpha = A^{-1}_\alpha y''_\alpha,\ x' = \sum_\alpha x'_\alpha,\ x'' = \sum_\alpha x''_\alpha = x - x'.$$

(It is clear that x'_α and x''_α are pairwise disjoint, and the use of Σ is justified.) Note that $x = \Sigma_\alpha y'_\alpha = \Sigma_\alpha A_\alpha x'_\alpha \sim x'$. By the transitivity of $\sim$ we have $y \sim x'$. Further, $y - x = \Sigma_\alpha y''_\alpha = \Sigma A_\alpha x''_\alpha \sim x''$. Now set $y - x = x_1$, $x'' = x_2$ and repeat our argument, taking as x the element x'. Continuing this, we define inductively a sequence of elements $\{x_n\}$ with the properties

$$\text{a) } x_n \wedge x_m = 0\ (n \neq m), \quad \text{b) } x_{n+1} \sim \sum_{k=1}^{n} x_k.$$

It is clear that $x_n \to 0$. From (B_3) it follows that $\Sigma_{k=1}^{n} x_k \to 0$. But then $x_1 = 0$ and $x = y$. The lemma is proved.

Remark 1. From the proof of the lemma it is clear that (B_4) follows from a proposition weaker than (B_3): there do not exist sequences of nonzero pairwise disjoint and pairwise equipotent elements. We call this proposition condition (B_3^-).

Remark 2. Let X be regular. Then (B_3^-) (and hence also (B_4)) follows from the yet weaker condition:

(B_3^{--}). $x_n \simeq x_{n+1}$ $(n = 1, 2, \cdots)$ and $x_n \wedge x_m = 0$ $(n \neq m)$, imply $x_1 = x_2 = \cdots = 0$.

In fact let (B_3^-) not hold. This means there exists a sequence of pairwise disjoint elements $\{x_n\}$, $n = 1, 2, \cdots$, such that $x_n \neq 0$ $(n = 1, 2, \cdots)$, and

$$x_n = \sum_{k=1}^{\infty} x_{nk},\ x_1 = \sum_{k=1}^{\infty} y_{nk},\ y_{nk} = A_{nk}x_{nk},\ A_{nk} \in \mathfrak{A}.$$

Since $\Sigma_{k=1}^{s} y_{nk} \nearrow x_1$ as $s \to \infty$ for each $n = 1, 2, \cdots$, then, by the regularity, there is a sequence $\{s_n\}$ such that $x_1 = (o)\text{-}\lim_n \Sigma_{k=1}^{s_n} y_{nk}$. There exists an element $y > 0$ such that $\Sigma_{k=1}^{s_n} y_{nk} \geq y$ for $n \geq n_0$. Set

$$\widetilde{x}_n = \sum_{k=1}^{s_n} A_{nk}^{-1}(y \wedge y_{nk}),\ n = n_0,\ n_0+1,\ \ldots .$$

We see that the pairwise disjoint elements $\widetilde{x}_n$ are finitely equicomposite with y, and hence with each other. This is incompatible with (B_3^{--}). And thus, with regularity, (B_3^{--}) implies (B_4).

Lemma 4. *If X is regular, then (B_4) (and hence also (B_3^{--})) implies (B_3).*

Proof. Suppose the lemma is false. Then there is a sequence $x_n \to 0$ such that $x_n = \Sigma_{k=1}^{\infty} x_{nk}$ and yet $\bigvee_{k=1}^{\infty} A_{nk} x_{nk} \not\to 0$ for some choice of automorphisms $A_{nk} \in \mathfrak{A}$. Lemma 1 gives us the right to assume that $\overline{\lim}_n$ abs $\bigvee_{k=1}^{\infty} A_{nk} x_{nk} > 0$. We can also assume than that $x_n \xrightarrow{(o)} 0$, since otherwise we simply thin out the sequence without changing $\overline{\lim}$ abs. By Lemma 2 there exists a sequence $y_n = \bigvee_{k=1}^{k_n} A_{nk} x_{nk}$ such that $\overline{\lim}_n$ abs $y_n > 0$. Set $u_{nm} = y_n - \bigvee_{i \geq m} \bigvee_{k=1}^{k_n} A_{nk}(x_{nk} \wedge x_i)$. Since $x_i \xrightarrow{(o)} 0$, we also have $\bigvee_{k=1}^{k_n} A_{nk} x_i \xrightarrow{(o)} 0$. But then, since $x_{nk} \wedge x_i \leq x_i$, we have $u_{nm} \xrightarrow{(o)} y_n$ $(n = 1, 2, \cdots)$. Again applying Lemma 2, form a sequence $m_1 < m_2 < \cdots$ so that $\underline{\lim}_n$ abs $u_{nm_n} > 0$. Introduce the notation $\widetilde{x}_n = x_n \wedge C \bigvee_{i \geq m_n} x_i$. Construct a sequence of natural numbers $\{n_j\}$ by the rule $n_1 = 1$, $n_{j+1} = m_{n_j}$ $(j = 1, 2, \cdots)$. It is clear that the $\widetilde{x}_{n_j}$ are pairwise disjoint. Also, for each $j = 1, 2, \cdots$ we have

$$\widetilde{x}_{n_j} = \sum_{k=1}^{k_{n_j}} \widetilde{x}_{jk}, \text{where } \widetilde{x}_{jk} = x_{n_jk} - x_{n_jk} \wedge \bigvee_{i \geq m_{n_j}} x_i.$$

It is easy to show that $u_{n_jm_{n_j}} \leq \bigvee_{k=1}^{k_{n_j}} A_{n_jk} \widetilde{x}_{jk}$. Hence

$$y = \overline{\lim_j} \operatorname{abs} \bigvee_{k=1}^{k_{n_j}} A_{n_jk} \widetilde{x}_{jk} \geq \overline{\lim_j} \operatorname{abs} u_{n_jm_{n_j}} > 0.$$

Decompose the natural series of numbers into a countable number of nonintersecting infinite sequences $\{j_{si}\}_{s=1}^{\infty}$ $(i = 1, 2, \cdots)$. By the basic property of $\overline{\lim}$ abs, for each i we have

$$\bigvee_{s=1}^{\infty} \bigvee_k \overline{A}_{j_{si}k} \widetilde{x}_{j_{si}k} \geq \overline{\lim_s} \operatorname{abs} \bigvee_k \overline{A}_{j_{si}k} \widetilde{x}_{j_{si}k} = y,$$

where for convenience we have introduced the notation $\overline{A}_{jk} = A_{n_jk}$. Fixing an arbitrary index i, arrange all the $\overline{A}_{j_{si}k} \widetilde{x}_{j_{si}k}$ in a simple sequence $\{y^{(m)}\}$. Corresponding to this, the $\overline{A}_{j_{si}k}$ are also arranged in a sequence $\{A^{(m)}\}$. Further, we set

$$\widetilde{y}^{(1)} = y^{(1)}, \ \ldots \ \widetilde{y}^{(s)} = y^{(s)} \wedge C \bigvee_{q<s} y^{(q)}.$$

It is clear that all the $\widetilde{y}^{(s)}$ are pairwise disjoint and their sup for each i is y. Setting $x_i^{(s)} = A^{(s)-1} \widetilde{y}^{(s)}$, we see that all the $x_i^{(s)}$ are pairwise disjoint, as are the $\overline{x}_i = \Sigma_s x_i^{(s)}$. Also, $x_i \sim y$ for any i. Hence all the $\overline{x}_i$ are pairwise equicomposite. Finally set $\overline{x}_1 + \overline{x}_2 + \cdots = z$, $\overline{x}_2 + \overline{x}_4 + \overline{x}_6 + \cdots = z_1$. Clearly $z \sim z_1$. This is incompatible with (B_4). The lemma is proved.

Remark 3. From the proof of the lemma it is seen that with regularity (B_2) implies $\overline{\lim}_n A_n x > 0$, for any $x > 0$ and any $A_n \in \mathfrak{A}$.

Remark 4. Also we see from the proof that if $A_n^{-1} x \to 0$, $A_n \in \mathfrak{A}$, $x \neq 0$, then there exists an $x' \in X$, $0 < x' \leq x$, such that the $A_n^{-1} x'$ are pairwise disjoint.

In the proof of Lemma 4 we have used the regularity of X essentially. Now let us go back to an arbitrary algebra. We agree first to write (A_1) for any of the equivalent conditions (A_1'), (A_1''), (A_1'''). The following lemma will play a basic role later.

Lemma 5. *Let X be a complete Boolean algebra, $\mathfrak{A}$ a group of automorphisms of it satisfying* (A_1) *and* (B_4). *There exists a group $\mathfrak{A}^*$ of automorphisms of the algebra containing $\mathfrak{A}$ and such that:*

a) *$x \sim_{\mathfrak{A}} y$, $x \sim_{\mathfrak{A}^*} y$, and $x \approx_{\mathfrak{A}^*} y$ are equivalent;*

b) *the group $\mathfrak{A}^*$ possesses properties* (A_1) *and* (B_4).

Proof. Let $x \sim_{\mathfrak{A}} y$. This means that

$$x = \sum_\eta x_\eta,\ y = \sum_\eta y_\eta,\ y_\eta = A_\eta x_\eta,\ A_\eta \in \mathfrak{A}.$$

Set $\bar{x} = Cx$, $\bar{y} = Cy$. By (A_1), there is a $B_1 \in \mathfrak{A}$ with $\bar{y} \geq B_1\bar{x} \wedge \bar{y} = \bar{y}_1 > 0$. Set $\bar{x}_1 = B_1^{-1}\bar{y}_1$. It is clear that $\bar{x}_1 \leq \bar{x}$. Suppose that we have constructed $\bar{x}_1, \cdots, \bar{x}_\alpha, \bar{y}_1, \cdots, \bar{y}_\alpha$, with $\bar{x}_\alpha \wedge \bar{x}_{\alpha'} = 0$ and $\bar{y}_\alpha \wedge \bar{y}_{\alpha'} = 0$ for $\alpha \neq \alpha'$, $\Sigma_{\alpha<\alpha_0} \bar{x}_\alpha < \bar{x}$, $\Sigma_{\alpha<\alpha_0} \bar{y}_\alpha < \bar{y}$, and $\bar{x}_\alpha = B_\alpha^{-1}\bar{y}_\alpha$, $B_\alpha \in \mathfrak{A}$. We can apply to the elements $C(\Sigma_\eta x_\eta + \Sigma_{\alpha<\alpha_0}\bar{x}_\alpha)$ and $C(\Sigma_\eta y_\eta + \Sigma_{\alpha<\alpha_0}\bar{y}_\alpha)$ arguments similar to those applied to $\bar{x}$ and $\bar{y}$. This will give us elements $\bar{x}_{\alpha_0}$ and $\bar{y}_{\alpha_0}$ and an automorphism $B_{\alpha_0} \in \mathfrak{A}$ such that $\bar{x}_{\alpha_0} \leq C(\Sigma_\eta x_\eta + \Sigma_{\alpha<\alpha_0}\bar{x}_\alpha)$ and $\bar{y}_{\alpha_0} \leq C(\Sigma_\eta y_\eta + \Sigma_{\alpha<\alpha_0}\bar{y}_\alpha)$. This process stops when either $\Sigma_{\alpha<\alpha_0}\bar{x}_\alpha = \bar{x}$ or $\Sigma_{\alpha<\alpha_0}\bar{y}_\alpha = \bar{y}$. Let us show that in fact these equalities can only hold together. In fact, if, for example, $\Sigma_{\alpha<\alpha_0}\bar{x}_\alpha = \bar{x}$, then, arranging all the x_1, $x_2, \cdots; \bar{x}_1, \bar{x}_2, \cdots$ into one transfinite sequence $\{z_\gamma\}$ and the operators A_1, $A_2, \cdots, B_1, B_2, \cdots$ into a sequence $\{C_\gamma\}$, we see that $\Sigma_\gamma z_\gamma = 1$. Then, by (B_4), $\Sigma_\gamma C_\gamma z_\gamma = 1$, and clearly $\Sigma_{\alpha<\alpha_0}\bar{y}_\alpha = \bar{y}$. Similarly the second equality implies the first.

Construct an automorphism U of X by the rule

$$Uz = \sum_\gamma C_\gamma (z \wedge z_\gamma). \tag{1}$$

The correctness of this definition is almost obvious. It is easy to check that the terms in the sum are pairwise disjoint and that (1) in fact determines an automorphism. We clearly have $y = Ux$. The set of all automorphisms having a similar structure will be denoted by $\mathfrak{A}^*$. More precisely, an operator T is in $\mathfrak{A}^*$ if and only if there exist:

a) a family of pairwise disjoint elements $\{u_\xi\}$, whose sum is the unit of the algebra;

b) a family of automorphisms $\{A_\xi\} \subset \mathfrak{A}$ with the property $A_\xi u_\xi \wedge A_{\xi'} u_{\xi'} = 0$, $\xi \neq \xi'$, such that $Tu = \Sigma_\xi A_\xi(u \wedge u_\xi)$ for all $u \in X$.

All such operators are automorphisms, as is easy to see. The inverse automorphism is defined by the formula $u = T^{-1}v = \Sigma_\xi A_\xi^{-1}(v \wedge v_\xi)$, where $v_\xi = A_\xi u_\xi$. It is also in $\mathfrak{A}^*$. It is clear that $\mathfrak{A} \subset \mathfrak{A}^*$.

Let $A, B \in \mathfrak{A}^*$, $AB = C$, $Ay = \Sigma_\xi A_\xi(y \wedge y_\xi)$ and $Bx = \Sigma_\eta B_\eta(x \wedge x_\eta)$. Let us show that $C \in \mathfrak{A}^*$. It is easy to check that $Cx = \Sigma_{\xi,\eta} A_\xi B_\eta(z_{\xi\eta} \wedge x)$, where $z_{\xi\eta} = x_\eta \wedge B_\eta^{-1} y_\xi$. Here $\Sigma_{\xi,\eta} z_{\xi\eta} = 1$, and the $A_\xi B_\eta z_{\xi\eta}$ are pairwise disjoint.

Thus $C \in \mathfrak{A}^*$. We see that $\mathfrak{A}^*$ is a group of automorphisms. The equivalence of $x \sim_{\mathfrak{A}} y$, $x \sim_{\mathfrak{A}^*} y$, and $x \approx_{\mathfrak{A}^*} y$ follows immediately from the definition of $\mathfrak{A}^*$.

Just as obvious is the fact that $\mathfrak{A}^*$ possesses properties (A_1) and (B_4). The lemma is proved.

Let us establish one important property of our group $\mathfrak{A}^*$.

Lemma 6. *For any two elements $x, y \in X$ we always have exactly one of the following three relations:*

1) $Ax < y$ *for some* $A \in \mathfrak{A}^*$;

2) $Ax > y$ *for some* $A \in \mathfrak{A}^*$;

3) $Ax = y$ *for some* $A \in \mathfrak{A}^*$.

Proof. By (A_1) there is an $A_1 \in \mathfrak{A}^*$ such that $A_1 x \wedge y = y_1 > 0$.* Set $x_1 = A_1^{-1} y_1$. It is clear that $x_1 \leq x$, $y_1 \leq y$ and $x_1 \approx y_1$. Suppose that $x_1, x_2, \cdots, x_\alpha, \cdots, y_1, y_2, \cdots, y_\alpha, \cdots; A_1, A_2, \cdots, \alpha < \alpha_0$, are already constructed such that

$$x_\alpha \wedge x_{\alpha'} = 0, \quad y_\alpha \wedge y_{\alpha'} = 0, \qquad \alpha \neq \alpha',$$

$$x_\alpha = A_\alpha^{-1} y_\alpha, \quad \sum_\alpha x_\alpha \leqslant x, \quad \sum_\alpha y_\alpha \leqslant y.$$

If $\bar{x}_{\alpha_0} = x - \Sigma_{\alpha<\alpha_0} x_\alpha > 0$ and $\bar{y}_{\alpha_0} = y - \Sigma_{\alpha<\alpha_0} y_\alpha > 0$, then by ($A_1$) we find an A_{α_0} such that $y_{\alpha_0} = A_{\alpha_0} \bar{x}_{\alpha_0} \wedge \bar{y}_{\alpha_0} > 0$; and then we set $x_{\alpha_0} = A_{\alpha_0}^{-1} y_{\alpha_0}$. This transfinite process stops when either $\bar{x}_{\alpha_0} = 0$, $\bar{y}_{\alpha_0} \neq 0$, or $\bar{x}_{\alpha_0} \neq 0$, $\bar{y}_{\alpha_0} = 0$, or $\bar{x}_{\alpha_0} = \bar{y}_{\alpha_0} = 0$. And this corresponds respectively to the cases 1), 2), and 3). For example, $x \sim \Sigma_{\alpha<\alpha_0} y_\alpha < y$ for $\bar{y}_{\alpha_0} \neq 0$, or what is the same, $Ax < y$, $A \in \mathfrak{A}^*$. Note that the automorphisms A_α can be taken from $\mathfrak{A}$.

In view of (B_4) and the transitivity of $\sim$, we can encounter only one of the three cases 1), 2), 3).

We will write $x \succ_{\mathfrak{A}^*} y$ if for some $A \in \mathfrak{A}^*$ we have $x \geq Ay$. It is clear that with (A_1) and (B_4) we always have either $x \approx_{\mathfrak{A}^*} y$, or $x \succ_{\mathfrak{A}^*} y$ or $x \prec_{\mathfrak{A}^*} y$. We will express this property of the group by saying that the group compares any two elements (in this case, of the group $\mathfrak{A}^*$).

Lemma 7. (A_1) *and* (B_4) *together imply* (B_3).

The proof of this lemma will be given in the following section.

Let us formulate yet another property. First let us agree to call positive additive functionals on X *quasi-measures*. More precisely, a quasi-measure is a real function μ with the properties:

1) $0 \leq \mu(x) < +\infty$ for all $x \in X$;

---

* We consider only the interesting case when $x, y \neq 0$.

2) if $x \wedge y = 0$, then $\mu(x + y) = \mu(x) + \mu(y)$.

μ is invariant relative to the group $\mathfrak{A}$ if $\mu(Ax) = \mu(x)$ for all $x \in X$, $A \in \mathfrak{A}$.

(B_5). For any $x > 0$ there is a quasi-measure μ invariant relative to $\mathfrak{A}$ such that $\mu(x) > 0$.

(B_5) can be formulated differently by saying that the group $\mathfrak{A}$ admits a sufficient number of invariant quasi-measures.

Lemma 8. (B_5) *implies* (B_3^{--}) (*and, in a regular algebra, all conditions of the group* (B)).

Proof. If (B_3^{--}) does not hold then there is a sequence $\{x_n\}$, $x_1 \simeq x_2 \simeq x_3 \simeq, \cdots, x_n \wedge x_m = 0$, $n \neq m$, $x_n \neq 0$. Denoting any of these elements by x, we use the quasi-measure μ whose existence is assured by (B_5). It is clear that $\mu(x_1) = \mu(x_2) = \cdots = \mu(x) > 0$. At the same time the series $\Sigma_n \mu(x_n)$ must converge, since $\Sigma_{n=1}^m \mu(x_n) \leq \mu(1)$ for all m. This contradiction proves the lemma.

Let us mention here a well-known theorem: in any Boolean algebra, for any $x > 0$ there is a quasi-measure μ with $\mu(x) > 0$.

§3

Definition. A complete Boolean algebra is called *completely homegeneous* if there is a group of automorphisms of it possessing properties (A_1) and (B_3).

Noting Lemma 7, we can here change (B_3) to (B_4). We will see later that a completely homogeneous algebra is always regular. If we suppose regularity beforehand, then in the definition (B_3) can be changed to any of the properties (B_4), (B_3^-), (B_3^{--}), (B_5), (B_3^+). The complete homogeneity of X means there is a group of automorphisms $\mathfrak{A}$ on X, which is on the one hand sufficiently rich (ergodic), but also possesses the property of strengthened equipotent continuity. Namely, there must exist a basis of neighborhoods of zero in the (o)-topology each of which together with the element x contains all elements of the form $\mathbf{V}_\alpha A_\alpha x_\alpha$, $\Sigma\, x_\alpha = x$, $A_\alpha \in \mathfrak{A}$ (and, as is easy to see, this is (B_3^+); in the regular case we can restrict to finite sums). In this section we give a complete listing of all completely homogeneous algebras.

As usual, an element $x \neq 0$ is called *discrete* (or an "atom"), if it cannot be written in the form $x = x_1 + x_2$, where $x_1, x_2 \neq 0$. We will say that the set $E \subset X$ is *minorant*, if for any $x > 0$ there is a $y \in E$ with $y \leq x$. An algebra in which the set of all atoms is minorant is called *discrete*. If there are no discrete elements in X then we call X *continuous* (or nonatomic).

It is easy to see that a completely homogeneous algebra, thanks to (A_1), can be either discrete or continuous. It is also easy to see that a discrete algebra is completely homogeneous if and only if it is finite.

Theorem 1. *A completely homogeneous algebra is always normable. Here, if $\mathfrak{A}$ is a group of automorphisms of the algebra satisfying* (A_1) *and* (B_3), *then there exists one measure, and only one, invariant relative to* $\mathfrak{A}$.

Proof. We will rely on (A_1) and (B_4). ((B_4) holds by Lemma 3.) The basic Lemmas 5 and 6 give us the right to assume that $x \approx y$ and $x \sim y$ are equivalent and that $\mathfrak{A}$ compares elements. The latter, as we agreed, means that for any $x, y \in X$ we always have exactly one of the relations $x \succ y$, $x \prec y$, or $x \approx y$. Finally, we will assume that X is continuous (otherwise, as was said, the algebra is finite and the theorem is trivial).

First let us make an obvious remark. If some set $E \subset X$ is minorant, then any nonzero element x can be written in the form $x = \Sigma_\alpha x_\alpha$, where all x_α are in E. (This is easily established by a transfinite induction.)

We will say that an element $x \in X$ is *divisible* if it can be written in the form of a sum of two congruent terms: $x = x_1 + x_2$, $x_1 \approx x_2$.

Lemma 9. *If $x = \Sigma_\alpha x_\alpha$, where all the x_α are divisible, then x is also divisible.*

Proof. We have $x_\alpha = x_\alpha' + x_\alpha''$ and $x_\alpha = A_\alpha x_\alpha''$. Set $x_1 = \Sigma_\alpha x_\alpha'$ and $x_2 = \Sigma_\alpha x_\alpha''$. It is clear that $x = x_1 + x_2$ and $x_1 \approx x_2$.

Lemma 10. *Every element $x \in X$ is divisible.*

Proof. By Lemma 9, it suffices to show that the divisible elements form a minorant set in X.* Take any $x \neq 0$. By the continuity, we can write it in the form of a sum of two nonzero terms: $x = y_1 + y_2$. We can assume that $y_1 \succ y_2$. This means that $y_1 = y' + y''$, $y'' \approx y_2$. The element $y = y'' + y_2$ is divisible, and $y \leq x$. The lemma is proved.

By Lemma 10 there is a sequence $\{x_n\}$, $n = 0, 1, \cdots$, with the properties

1) $x_0 = 1$,

2) $x_n - x_{n+1} \approx x_{n+1}$.

We will use the notation $x_n - x_{n+1} = x_{n+1}'$.

Lemma 11. *There is a sequence $\{\bar{x}_n\}$ $(n = 1, 2, \cdots)$, the terms of which are pairwise disjoint and $\bar{x}_n \approx x_n$, for all n.*

Indeed, we can take $\bar{x}_n = A_1 A_2 \cdots A_{n-1} x_n$, $n \geq 2$, $\bar{x}_1 = x_1$ where A_i is the automorphism such that $A_i x_i = x_i'$.

---

* The divisibility of zero is obvious.

Lemma 12. *Let* $x = \Sigma_{k=1}^{p} y_k$, $y_k = B_k x_{n_k}$ $(k = 1, 2, \cdots, p)$, $B_k \in \mathfrak{A}$, $\sigma = \Sigma_{k=1}^{p} 1/2^{n_k}$, $m \geq \max_{k=1,\cdots,p} n_k$. *Then there exists the representation* $x = x^{(1)} + x^{(2)} + \cdots + x^{(s)}$, $x^{(i)} \approx x_m$ $(i = 1, 2, \cdots, s)$, *where* $s = 2^m \sigma$.

In fact, it is easy to check that as $x^{(i)}$ we can take all the elements of the form

$$z = B_k A_{i_1} A_{i_2} \ldots A_{i_q} x_m,$$

$$n_k < i_1 < i_2 < \ldots < i_q \leqslant m, \quad 0 \leqslant q \leqslant m - n_k, \quad k = 1, 2, \ldots, p,$$

where A_i is the same as above.

Lemma 13. *Let* $x = \Sigma_{k=1}^{r} u_k$, $y = \Sigma_{k=1}^{s} v_k$, $u_k \approx x_{n_k'}$ $(k = 1, 2, \cdots, r)$,

$$v_k \approx x_{n_k''} \ (k = 1, 2, \ldots, s), \quad \sigma' = \sum_{k=1}^{r} \frac{1}{2^{n_k'}}, \quad \sigma'' = \sum_{k=1}^{s} \frac{1}{2^{n_k''}}, \quad \sigma' \leqslant \sigma''.$$

Then $x \prec y$. *Here if* $\sigma' < \sigma''$, *then* x *and* y *cannot be taken congruent.*

For the proof, set m equal to the greatest of $n_1', \cdots, n_s''$. By the preceding lemma we have

$$x = x^{(1)} + \ldots + x^{(\mu)}, \quad y = y^{(1)} + \ldots + y^{(\nu)},$$

$$x^{(1)} \approx x^{(2)} \approx \ldots \approx y^{(1)} \approx \ldots \approx y^{(\nu)}.$$

Since $\mu = 2^m \sigma'$ and $\nu = 2^m \sigma''$ we have $\mu \leq \nu$. It is clear that $x \prec y$ and that for $\sigma' < \sigma''$ the elements x and y are not congruent.

Corollary. *If* $x \approx y$, $x = \Sigma_{k=1}^{\mu} u_k$ *and* $y_k = \Sigma_{k=1}^{\nu} v_k$, *where* $u_k \approx x_{n_k'}$ *and* $v_k \approx x_{n_k''}$, *then* $\Sigma_{k=1}^{\mu} 1/2^{n_k'} = \Sigma_{k=1}^{\nu} 1/2^{n_k''}$.

Lemma 14. *Let*

$$x = \sum_{k=1}^{p} u_k, \quad y = \sum_{k=1}^{q} v_k, \quad u_k \approx x_{n_k'}, \quad v_k \approx x_{n_k''}, \quad x > y,$$

$$\sigma' = \sum_{k=1}^{p} \frac{1}{2^{n_k'}}, \quad \sigma'' = \sum_{k=1}^{q} \frac{1}{2^{n_k''}}.$$

There is a representation of $w = x - y$ *in the form of a finite sum,* $w = \Sigma_{k=1}^{r} w_k$, *where* $w_k \approx x_{n_k}$. *Here* $\Sigma_{k=1}^{r} 1/2^{n_k} = \sigma' - \sigma''$.

Proof. By Lemma 12 we have

$$x = \sum_{k=1}^{\mu} z_k, \quad y = \sum_{k=1}^{\nu} z_k', \quad x_m \approx z_1 \approx z_2 \approx \ldots \approx z_1' \approx z_2' \approx \ldots,$$

where $m = \max(n_1', \cdots, n_p', n_1'', \cdots, n_q'')$. From (B_4) we have $\mu > \nu$. Set $\bar{x} = z_1 + \cdots + z_\nu$ and $\bar{\bar{x}} = z_{\nu+1} + \cdots + z_\mu$. We see that $\bar{x} \approx y$. But then $\bar{\bar{x}} \approx w$.*

* In general, if $a = u + v$, $b = u' + v'$, $a \approx b$, $u \approx u'$, then $v \approx v'$. Indeed, if, for example, $v \approx v'' < v'$, then $b \approx a = u + v \approx u' + v'' < b$, which is impossible in view of (B_4). Just as impossible is $v \approx v'' > v'$. But then $v \approx v'$, since the group compares elements.

The second assertion of the lemma follows from the Corollary to Lemma 13.

Lemma 15. *Let* x *have two representations in the form of a finite or infinite sum* $x = \Sigma_k u_k$, $x = \Sigma_k v_k$, $u_k \approx x_{n'_k}$, $v_k \sim x_{n''_k}$. *Then* $\Sigma_k 1/2^{n'_k} = \Sigma_k 1/2^{n''_k}$.

Proof. Suppose the lemma is false. Then we can suppose that $\sigma' < \sigma''$. Take m so that $2^{-m} < \sigma'' - \sigma'$. If the sum $\Sigma_k v_k$ contains a finite number of terms we can suppose that its last on v_{k_0} is such that $n''_{k_0} < m$. (Otherwise, we can write it as $v_{k_0} = \Sigma \bar{v}_i$, where $\bar{v}_1 \approx \bar{v}_2 \approx \cdots \approx \bar{v}_{2^{m-n''_{k_0}}} \approx \bar{x}_m$.) We can now assume that for some p_0

$$\sigma' < \sum_{k=1}^{p_0} \frac{1}{2^{n''_k}}, \quad \bar{x}_1 = \sum_{k=1}^{p_0} v_k < x.$$

By Lemma 13, there is $\bar{u}_1 \approx u_1$, such that $\bar{u}_1 < \bar{x}_1$. Set $\bar{x}_2 = \bar{x}_1 - \bar{u}_1$. By Lemma 14, we have $\bar{x}_2 = \Sigma_{k=1}^s w_k$, $w_k \approx \bar{x}_{n_k}$, where

$$\sum_{k=1}^{s} \frac{1}{2^{n_k}} = \sum_{k=1}^{p_0} \frac{1}{2^{n''_k}} - \frac{1}{2^{n'_1}} > 0.$$

There exists an element $\bar{u}_2 \approx u_2$, $\bar{u}_2 < \bar{x}_2$. Now set $\bar{x}_3 = \bar{x}_2 - \bar{u}_2$. Repeating this argument we form a sequence (finite or infinite) $\bar{u}_1, \bar{u}_2, \cdots$, such that $\Sigma_k \bar{u}_k \leq \bar{x}_1$, $\bar{u}_k \approx u_k$, $k = 1, 2, \cdots$. Then $\Sigma \bar{u}_k \approx \Sigma u_k = x$ and $\Sigma \bar{u}_k \leq \bar{x}_1 < x$, which contradicts (B_4). The lemma is proved.

Lemma 16. *The set* $\{Ax_n\}$, $A \in \mathfrak{A}$, $n = 1, 2, \cdots$, *is minorant.*

Proof. If the lemma is false, then, since the group compares elements, there is a sequence $\{A_n\}$, $n = 1, 2, \cdots$, such that $x = \bigwedge_n A_n \bar{x}_n > 0$. Set $x_n = A_n^{-1} x$; we see that all the $\bar{\bar{x}}_n$ are pairwise disjoint and that $\bar{\bar{x}}_1 \approx \bar{\bar{x}}_2 \approx \cdots \approx x > 0$. But then $\Sigma_{n=1}^{\infty} \bar{\bar{x}}_n \approx \Sigma_{n=1}^{\infty} \bar{\bar{x}}_{2n}$, which contradicts (B_4).

Now the proof of Theorem 1 does not require strengthenings. Each element $x > 0$ is representable, by Lemma 16, as a sum $x = \Sigma_\alpha u_\alpha$, where each u_α is congruent to one of the x_n. This sum contains at most a countable number of terms, since otherwise there would be an infinite sequence of pairwise disjoint and congruent terms, which is impossible (see, for example, the proof of Lemma 16). Thus we always have

$$x = \sum_k u_k, \; u_k \approx x_{n_k}. \tag{**}$$

The sum $\Sigma_k 1/2^{n_k}$ does not (by Lemma 15) depend on the way of writing x in the form (**). Denote its value by $\mu(x)$. It is clear that:

a) $\mu(x) > 0$,

b) $\mu(\Sigma_k x_k) = \Sigma_k \mu(x_k)$ (this follows from Lemma 15),

c) $\mu(1) = 1$ (also by Lemma 15), and

d) $\mu(x) \leq \mu(1) < +\infty$.

Setting also $\mu(0) = 0$, we see that we have constructed a measure [property 3) in the definition of measure is equivalent, as we know, to b)]. The invariance and uniqueness of μ are obvious. The theorem is proved.

Remark. Together with the theorem we have also proved Lemma 7. Indeed, we have for the proof relied on (A_1) and (B_4). But now we see that if $x_n \to 0$, $x_n = \Sigma_\alpha x_{n\alpha}$ and $y_n = \mathsf{V}_\alpha A_{n\alpha} x_{n\alpha}$,* then $\mu y_n \to 0$, since $\mu y_n \leq \Sigma\, \mu A_{n\alpha}\, x_{n\alpha} = \Sigma_\alpha\, \mu x_{n\alpha} = \mu x_n$, i.e. $y_n \to 0$. Thus the group $\mathfrak{A}$ also satisfies (B_3).

Consider now a series of examples of completely homogeneous algebras.

1. All finite algebras, we recall, are completely homogeneous.

2. The algebra R_I of all Lebesgue measurable sets of the interval $I = [0, 1]$ with the usual identification is completely homogeneous. The role of $\mathfrak{A}$ can be played by the group of all automorphisms preserving measure. We can take an even smaller group: the group of all powers of any ergodic measure-preserving automorphism, the group of all motions (i.e. automorphisms generated by mappings of the interval by the formulas $A_\alpha t = t + \alpha$, $\alpha + t \leq 1$, $A_\alpha t = t + \alpha - 1$, $\alpha + t > 1$) and so forth.

3. Let $\Gamma = \{\gamma\}$ be any set of power τ. Consider the family $\{I_\gamma\}$ of τ copies of $I = [0, 1]$ and form their Cartesian product $\Pi_\gamma I_\gamma$. In this Cartesian product it is well known (see [16]) how to introduce a measure, the product of the Lebesgue measures on the intervals. After identifying equivalent sets, there arises the normed algebra R_Γ which is called the product of the algebras R_{I_γ}. Without difficulty we can prove the ergodicity of the group of all measure-preserving automorphisms, and hence also the complete homogeneity of R_Γ.

It turns out that algebras of type R_Γ exhaust the entire class of completely homogeneous algebras, so that we have obtained their axiomatic description, not assuming a priori the existence of a measure.

Theorem 2. *A complete Boolean algebra is isomorphic to one of the algebras R_Γ if and only if it is completely homogeneous and infinite.*

We can easily prove this theorem with Theorem 1 and a fundamental fact due to D. Maharam (see [5]). Let us introduce some known definitions. For any x_0 set $Z_{x_0} = \{x;\ x \leq x_0\}$. Such sets are called *components* or *principal ideals*. We say that X is the *union* or *direct sum* of the components Z_{x_α} if

---

* Since the algebra is normable these sums contain at most a countable number of terms.

$1 = \Sigma_\alpha x_\alpha$. With the natural ordering, each $Z_{x\alpha}$ is a Boolean algebra, where the unit is x_α. Let us associate with each $x \neq 0$ a cardinal number $\tau(x)$ equal to the smallest power of a set dense in Z_x (relative to the (o)-topology). Following Maharam, we call the algebra *homogeneous* if $\tau(x) = \tau(1)$ for any $x \neq 0$. (In [5] this definition is given for algebras with a measure.)

Lemma 17. *An infinite completely homogeneous algebra is homogeneous.*

We omit the almost obvious proof of this lemma. It is based on the elementary properties of cardinal numbers and Lemma 6. Lemma 17 is not invertible, as is shown by simple examples.

D. Maharam showed that an infinite homogeneous normable algebra is isomorphic to one of the algebras R_Γ where the power of Γ is equal to $\tau(1)$. Now Theorem 2 is clear, since by Theorem 1 a completely homogeneous algebra is always normable.

In the same work, Maharam showed that every normable algebra is isomorphic to the union of homogeneous algebras. From this and Theorem 1 follows

Theorem 3. *For a complete Boolean algebra to be normable it is necessary and sufficient that it be the union of at most a countable set of completely homogeneous algebras.*

Every separable normable homogeneous algebra is continuous, and hence isomorphic to R_I (see [4]). From this and the preceding follows

Theorem 4. *In order that a complete Boolean algebra be isomorphic to R_I it is necesssry and sufficient that it be infinite, completely homogeneous and separable in the (o)-topology.*

In conclusion of this section let us give yet another characterization of completely homogeneous algebras, using the terminology of the theory of semiordered spaces. We will not introduce here the basic definitions of this theory, referring the reader to the monographs [3], [4].

We denote by S_X the extended K-space for which the complete Boolean algebra X is a basis.

Lemma 18. *Let X be a complete Boolean algebra of countable type, and $\mathfrak{A}$ a group of its automorphisms. Let $\mathfrak{A}$ satisfy the condition*

(B_6). *For any $A_k \in \mathfrak{A}$, $k = 1, 2, \cdots$, the inequality $\Sigma_k x_k \leq 1$ always implies that $\Sigma_k A_k x_k < +\infty$ in S_X.*

Then (B_4) *holds.*

Proof. By Remark 1, it suffices to show that (B_3^-) holds. The denial of (B_3^-) would mean that there are nonzero pairwise disjoint x_k, $x_k = \Sigma_i x_{ki}$

$(k = 1, 2, \cdots)$, $x_1 = \Sigma_i A_{ki} x_{ki}$, $A_{ki} \in \mathfrak{A}$ $(k, i = 1, 2, \cdots)$. Then we would have $\Sigma_{k,i} x_{ki} \leq 1$, but at the same time $\Sigma_{ki} A_{ki} x_{ki} = +\infty$, which contradicts (B_6).

An assertion similar to Lemma 18 is valid also in any complete algebra, only in (B_6) we need to change the countable sums to arbitrary ones.

Thus (B_6) and (A_1) together ensure the complete homogeneity of the algebra, and if it is infinite they also ensure its being isomorphic to one of the R_Γ. Here the sums $\Sigma_k A_k x_k$ will be restricted to the corresponding space of summable functions, and not only on S_X, as assumed in (B_6). In general, in these terms it is easy to give also an abstract characterization of L-spaces (KB-spaces with an additive norm).

§ 4

In this section we vary the character of the requirements on the family of automormphisms $\mathfrak{A}$. Namely, we will assume that $\mathfrak{A}$ is a commutative or at least a solvable group (see [11], [12], [13]).

The following is a well-known result.

Theorem (A. A. Markov). *Let Q be a set, E the Boolean algebra of all its subsets, and $T = \{t\}$ some collection of pairwise permutable mappings of Q into itself. Then there is a quasi-measure μ on E, invariant relative to all $t \in T$ in the sense that $\mu(te) = \mu(e)$ and $\mu(Q) = 1$.*

For a proof of this theorem, see [3]. The theorem remains valid if we assume that T is a solvable group of mappings (see, in this connection, [13]). First let us use this theorem, taking as Q a group $\mathfrak{A}$ of automorphisms of the Boolean algebra X and as T the group of right translations: $A \to AP$. Assuming that $\mathfrak{A}$ is commutative or solvable, let us denote the integral of the measure, whose existence is ensured by the theorem of A. A. Markov, by ${}_{\mathfrak{A}}\int f(A)dA$. It exists for any bounded real function and possesses the usual properties of a positive linear functional. For any quasi-measure p, the function $m(e) = {}_{\mathfrak{A}}\int p(Ae)dA$ is, as we can easily see, an invariant quasi-measure on X, where $m(1) = p(1)$. We will always assume that $m(1) = 1$. Let us attempt to find conditions under which m is a measure, i.e. possesses the additional properties of countable additivity and essential positivity. The function m is certainly essentially positive, if the group $\mathfrak{A}$ satisfies (A_2) (see the remark on page 247). And thus, we have

Theorem 5. *Let X be a complete Boolean algebra, and let $\mathfrak{A}$ be a commutative or solvable group of its automorphisms, satisfying (A_2). Then there*

is a quasi-measure m invariant relative to $\mathfrak{A}$ *and such that* $m(x) > 0$ *for* $x \neq 0$.*

A. G. Pinsker showed that if the algebra X is regular and there is an essentially positive quasi-measure on X, then there is a measure (see [4], page 428); this result was essentially repeated by J. Kelley [7]. An addition to the theorem of A. G. Pinsker is

Theorem 6. *Let X be a regular Boolean algebra,* $\mathfrak{A}$ *a commutative or solvable group of its automorphisms. Further, let there exist a quasi-measure m invariant relative to* $\mathfrak{A}$ *such that* $m(x) > 0$ *for* $x \neq 0$. *Then there is a measure on X invariant relative to* $\mathfrak{A}$.

Proof. By Pinsker's theorem there is a measure ϕ on X. Set $\mu(e) = {}_{\mathfrak{A}}\int \phi(Ae)dA$, $e \in X$, and let us see that μ is a measure. From the hypothesis of the theorem it follows that $\mathfrak{A}$ possesses property (B_5), and hence (by Lemma 8) all the properties of type (B), in particular, (B_1). For any $x \neq 0$ we have $\inf_A \phi(Ax) > 0$. Indeed, if not, then there is a sequence $\{A_n\}$ such that $\phi(A_n x) \to 0$, i.e. $A_n x \to 0$, whence, by (B_1), $x = 0$. Hence always for $x \neq 0$ we have $\mu(x) \geq \inf_A \phi(Ax) > 0$ and μ is essentially positive. Further, if $x_n \to 0$, then $\sup_A \phi(Ax_n) \to 0$ and $0 \leq \mu(x_n) = \sup_A \phi(Ax_n)$, whence $\mu(x_n) \to 0$. Thus μ is continuous. The invariance of μ is obvious, and the theorem is proved.

The following theorem is near in character to the one just proved.

Theorem 7. *Let X be a normable Boolean algebra,* $\mathfrak{A}$ *a commutative group of its automorphisms,* $\mathfrak{A}_1$ *a semigroup of automorphisms contained in* $\mathfrak{A}$ *and such that for any* $A \in \mathfrak{A}$ *either A or* A^{-1} *is in* $\mathfrak{A}_1$. *Finally suppose that from the relations* $x_n \wedge x_m = 0$, $n \neq m$, $A_n x_n = x$, $A_n^{-1} \in \mathfrak{A}_1$, $n = 1, 2, \cdots$, *it always follows that* $x = 0$ *(a weakened form of* (B_2)*). Then there is on X a measure invariant relative to* $\mathfrak{A}$.

Proof. Again we use the theorem of A. A. Markov, taking as Q the semigroup $\mathfrak{A}_1$, as T the semigroup of right translations corresponding to the elements of $\mathfrak{A}_1$. Now set $\mu(x) = {}_{\mathfrak{A}_1}\int \phi(Ax)dA$, where ϕ is any of the measures on X. As above, we have a quasi-measure invariant relative to $\mathfrak{A}_1$, and thus invariant relative to $\mathfrak{A}$. By the last hypothesis of the theorem and Remark 4 (§3), for $x \neq 0$ we always have $\inf_{A \in \mathfrak{A}_1} \phi(Ax) > 0$, i.e. $\mu(x) > 0$. Hence μ is essentially positive. Now use Theorem 5.

---

* Theorem 5, as is clear from the proof, remains valid for any group on which there is a Banach mean. Conditions for the existence of a Banach mean are given in [12].

In case $\mathfrak{A}$ consists of the powers of a single automorphism A, and $\mathfrak{A}_1 = \{A^n\}$, $n = 0, -1, -2, \cdots$, Theorem 7 becomes the theorem of Kahn-Kakutani on finite invariant measure (see [14]), formulated in the standard "algebraic" language of this report. This theorem belongs to a number of recent results solving the rather popular problem in ergodic theory of invariant measures. One of the first was the result of E. Hopf [15]. In our terms the theorem of Hopf is formulated as follows: if X is a normable algebra and $\mathfrak{A}$ is the group of powers of a single automorphism A satisfying (B_4), then there is an invariant measure on X. Of course, this theorem is also contained in Theorem 7. Theorem 1 shows that the condition of ergodicity (A_1) allows us to drop the requirement of the normability of the algebra, as well as the assumptions relative to the algebraic structure of the group. We do not know whether the existence of an invariant measure follows in the case of a normable algebra from condition (B_4) (or equivalently, by Lemma 4, (B_3)). However it is easy to show that the answer is affirmative, if, not changing the form of (B_4), we give it a somewhat different interpretation, putting this condition not only on the algebra, but also on some K-space containing this algebra. In fact, it is well known that for any normable algebra X there is a KB-space with an additive norm for positive members, such that X is a basis. (If X is realized as an algebra of measurable sets, then the space of all summable functions will be such a space.) Let us now agree to denote this space by L_X. We have

Theorem 8. *Let X be a complete normable Boolean algebra, $\mathfrak{A}$ a group of its automorphisms, such that if $x = \Sigma_k x_k \in L_X$, $x_k \in X$, $y = \Sigma A_k x_k$, $A_k \in \mathfrak{A}$ $(k = 1, 2, \cdots)$, $y \in L_X$ then $x < y$ is impossible. Then there is an invariant measure on X.*

We only outline the proof of this theorem. Set

$$\mu(x) = \inf_{\substack{\Sigma x_k = x \\ A_k \in \mathfrak{A}}} \sum_{k=1}^{\infty} \varphi(A_k x_k),$$

where ϕ is a measure on X. This is the desired measure. Let us check its essential positivity. If $\mu(x) = 0$ for some $x \neq 0$, then there exists a sequence of families $\{A_{nk}\}_{n=1}^{\infty}$ $(k = 1, 2, \cdots)$ and a decomposition of x into members $\{x_{nk}\}$ such that $\Sigma_n \phi(A_{nk} x_{nk}) < 1/k^2$, $k = 1, 2, \cdots$. Then we have $\|\Sigma A_{nk} x_{nk}\| < 1/k^2$,* and the series $\Sigma_k \Sigma_n A_{nk} x_{nk}$ converges in the K-space L_X. Hence

$$\sum_{k=1}^{\infty} \sum_n A_{nk} x_{nk} > \sum_{k=1}^{\infty} \sum_n A_{n2k} x_{n2k}.$$

---

* $\|x\| = \varphi(x)$ for $x \in X$.

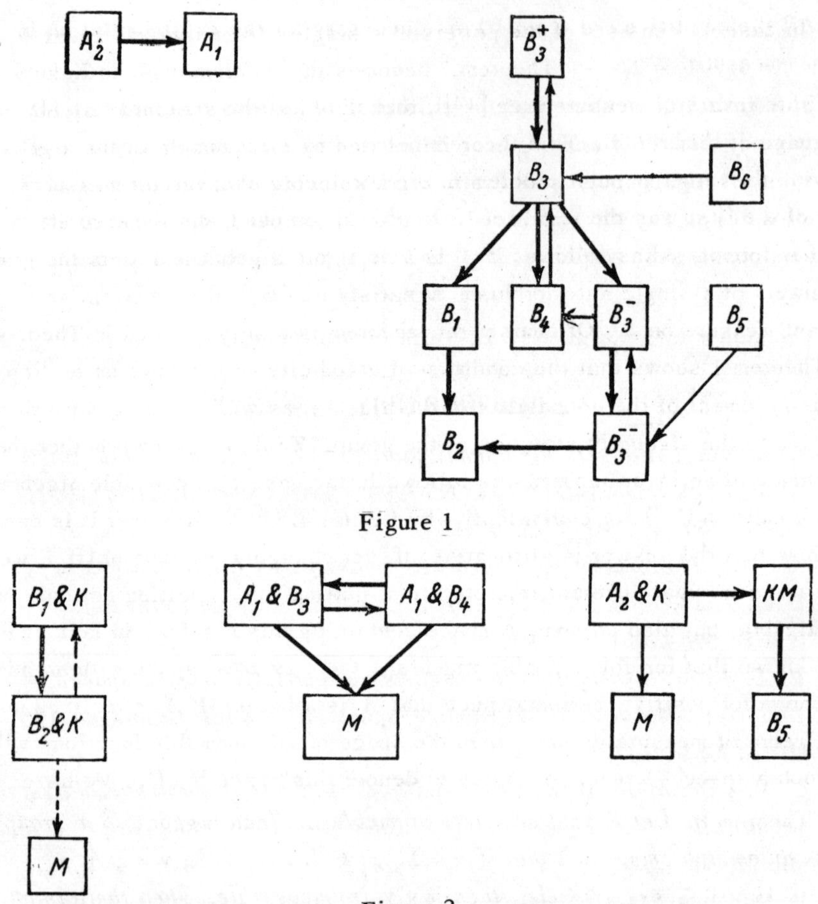

Figure 1

Figure 2

It is easy to show that this is incompatible with the hypothesis of the theorem. The finiteness of μ follows from the obvious inequality $\mu(x) \leq \phi(x)$; the verification of the countable additivity has a standard character. Also obvious is the invariance of the constructed measure.

On Figures 1 and 2 we give schemes illustrating the results of §§2 and 3, and partially §4. On them we use our previous notation. In addition,

M means: "there is an invariant measure",

KM means: "there is a invariant essentially positive quasi-measure",

K means: "the group $\mathfrak{A}$ is commutative or solvable",

$\boldsymbol{\longrightarrow}$ means: "implies in any complete algebra",

$\longrightarrow$ means: "implies in any regular algebra",

$\dashrightarrow$ means: "implies in any normable algebra",

& is the sign of logical conjunction.

All the conditions of type (B) are necessary for the existence of an invariant measure.

The natural conjecture that the existence of an invariant measure follows already from (A_1) and (B_1) can be refuted by the example of the algebra X_o of all regular open sets (sets which coincide with the interior of their closure) on any circumference. As can be checked, the group of all rotations possesses properties (A_1) and (B_1), but X_o is not a normable algebra.

The author is grateful to B. Z. Vulih for fruitful talks.

BIBLIOGRAPHY

[1] D. A. Vladimirov, *On the normalizability of a Boolean algebra*, Dokl. Akad. Nauk SSSR 146 (1962), 987–990 = Soviet Math. Dokl. 3 (1962), 1407–1409. MR 27 #69.

[2] ———, *The existence of invariant measures on Boolean algebras*, Dokl. Akad. Nauk SSSR 157 (1964), 764–766 = Soviet Math. Dokl. 5 (1964), 998–1001. MR 30 #226.

[3] B. Z. Vulih, *Introduction to the theory of partially ordered spaces*, Fizmatgiz, Moscow, 1961; English transl., Noordhoff, Groningen, 1967. MR 24 #A3494; MR 37 #121.

[4] L. V. Kantorovič, B. Z. Vulih and A. G. Pinsker, *Functional analysis in partially ordered spaces*, GITTL, Moscow, 1950. (Russian) MR 12, 340.

[5] D. Maharam, *On homogeneous measure algebras*, Proc. Nat. Acad. Sci. U.S.A. 28 (1942), 108–111. MR 4, 12.

[6] ———, *An algebraic characterization of measure algebras*, Ann. of Math. (2) 48 (1947), 154–167. MR 8, 321.

[7] J. L. Kelley, *Measures on Boolean algebras*, Pacific J. Math. 9 (1959), 1165–1177. MR 21 #7286.

[8] L. V. Kantorovič, *On the properties of linear semiordered spaces*, C. R. Acad. Sci. Paris 202 (1936), 813–816.

[9] G. Birkhoff, *Lattice theory*, Amer. Math. Soc. Colloq. Publ., vol. 25, Amer. Math. Soc., Providence, R. I., 1940; rev. ed., 1948; new ed., 1967; Russian transl., IL, Moscow, 1952. MR 1, 325; MR 10, 673.

[10] R. Sikorski, *Boolean algebras*, Academic Press, New York and Springer-Verlag, Berlin, 1964. MR 31 #2178.

[11] A. A. Markov, *On the existence of an integral invariant*, Dokl. Akad. Nauk SSSR 17 (1937), 455–458. (Russian)

[12] G. M. Adel'son-Vel'skiĭ and Ju. A. Šreĭder, *The Banach mean on groups*, Uspehi Mat. Nauk 12 (1957), no. 6 (78), 131–136. (Russian) MR 20 #1238.

[13] N. Bourbaki, *Espaces vectoriels topologiques sur un corps valué. Espaces vectoriels topologiques*, ch. 1, Fasc. XVIII, XIX, Elements de Mathematiques, Livre V, Actualités Sci. Indust. nos. 1229, 1230, Hermann, Paris, 1955; Russian transl., IL, Moscow, 1959. MR 17, 1109.

[14] P. Halmos, *Recent progress in ergodic theory*, Bull. Amer. Math. Soc. 67 (1961), 70–80; Russian transl., Matematika 6 (1962), no. 3, 17–27. MR 23 #A290.

[15] E. Hopf, *Theory of measure and invariant integrals*, Trans. Amer. Math. Soc. 34 (1932), 373–393.

[16] N. Dunford and J. T. Schwartz, *Linear operators. I: General theory*, Pure and Appl. Math., vol. 7, Interscience, New York and London, 1958; Russian transl., IL, Moscow, 1962. MR 22 #8302.

Translated by:
K. Hamdani